Contact and Long-Term Behavior of Current-Carrying Connections in Electrical Power Engineering

Stephan Schlegel · Michael Gatzsche ·
Christian Hildmann · Toni Israel

Contact and Long-Term Behavior of Current-Carrying Connections in Electrical Power Engineering

Theory and Practice on Behavior, Calculation Approaches, as well as Design and Dimensioning Criteria

Springer

Stephan Schlegel
Dresden, Germany

Michael Gatzsche
Zürich, Switzerland

Christian Hildmann
Dresden, Germany

Toni Israel
Dresden, Germany

ISBN 978-3-662-69643-9 ISBN 978-3-662-69644-6 (eBook)
https://doi.org/10.1007/978-3-662-69644-6

Translation from the German language edition: "Kontakt- und Langzeitverhalten stromführender Verbindungen in der Elektroenergietechnik" by Stephan Schlegel et al., © Der/die Herausgeber bzw. der/die Autor(en), exklusiv lizenziert an Springer-Verlag GmbH, DE, ein Teil von Springer Nature 2022. Published by Springer Berlin Heidelberg. All Rights Reserved.

This book is a translation of the original German edition "Kontakt- und Langzeitverhalten stromführender Verbindungen in der Elektroenergietechnik" by Stephan Schlegel, published by Springer-Verlag GmbH, DE in 2022. The translation was done with the help of an artificial intelligence machine translation tool. A subsequent human revision was done primarily in terms of content, so that the book will read stylistically differently from a conventional translation. Springer Nature works continuously to further the development of tools for the production of books and on the related technologies to support the authors.

This Springer imprint is published by the registered company Springer-Verlag GmbH, DE, part of Springer Nature.
The registered company address is: Heidelberger Platz 3, 14197 Berlin, Germany

Preface

The book is intended for students, engineers, and scientists who deal with current-carrying contacts and connections in the context of education and training, design, or research. The summarized results and findings are based on over 40 years of research at the Chair of High Voltage and High Current Technology at the Dresden University of Technology. The authors see themselves as part of a research group that has carried out unique work on the topic of contact and long-term behavior of stationary connections during this time. The starting point for this was the first experiments in the 1970s on screw connections with conductors made of aluminum by Mr. PD Dr.-Ing. habil. Helmut Löbl and the book Medium Voltage Technology by Mr. Prof. Dr.-Ing. habil. Helmut Böhme, which contains a first summary of the findings at that time. Building on this, numerous of publications and qualification works have emerged (Fig. 1).

This book outlines the basics of contact theory and presents possibilities for analytical and numerical modeling of current-carrying connections. Furthermore, approaches are presented that enable the estimation of complex relationships in engineering terms. The boundary conditions and limits are named, making open scientific questions clear. Selected relationships are illustrated and made comprehensible through calculation examples. The focus is always on the relationship between mechanical, electrical, and thermal quantities. A variety of standard works from the literature, which deal particularly with mechanical design, are referenced.

The second part addresses the long-term behavior of current-carrying connections, which was a major research focus at the chair and represents a unique feature in the national and international context of the scientific field. Experiments lasting over 20 years make it possible to understand aging effects almost in real-time.

Building on this, the design and testing of current-carrying connections are examined in more detail. Individual design steps for the creation of screwed, plug-in, and compression connections are presented in a structured manner. The aim is to link the content with the already existing guidelines and specifications. In addition, standardized test procedures are evaluated and supplemented with further developed approaches, with the goal of obtaining generally valid, transferable results or specific, adapted procedures.

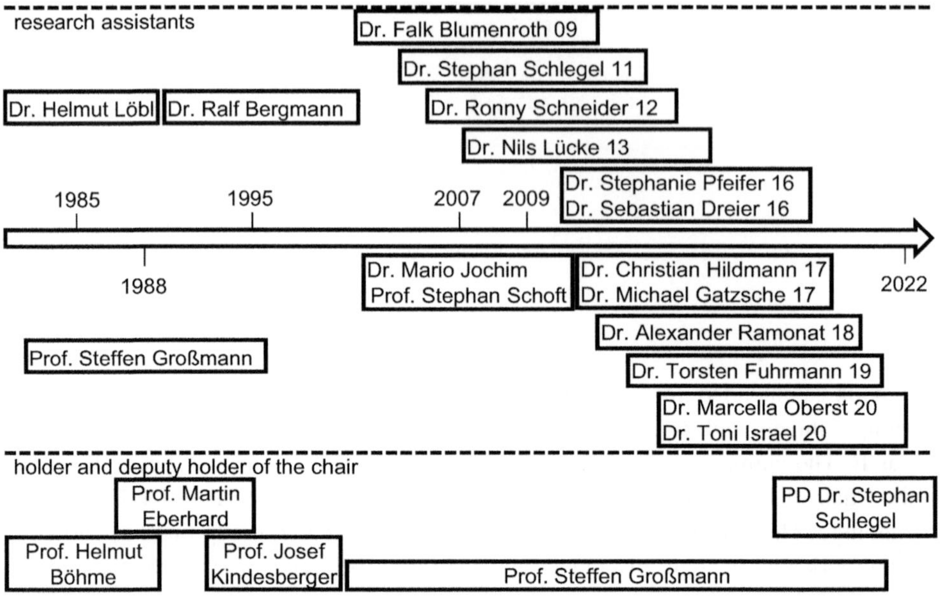

Fig. 1 Evaluated dissertations and habilitations by researchers at the chair (behind the name date of publication)

The author team thanks all researchers who contributed to this book through their work in the form of study, diploma, doctoral, and habilitation theses. Without the interest and passion to work on this interdisciplinary topic and the willingness to add individual pieces of the puzzle, it would not have been possible to compile this summary.

We also thank all industry partners and project sponsors who, through their trust in this work and the associated financial support, have made continuous research in this field possible.

Stephan Schlegel

Michael Gatzsche
Christian Hildmann
Toni Israel

Symbols, Names and Units

Symbols

Symbol	Unit[a]	Meaning
a	m	distance between two current-carrying conductors
a_r	m	groove distance
A	m^2	cross-sectional area
A^*		proportionality factor
A_3	m^2	cross-sectional area of a regular hexagon
A_B	m^2	area of the borehole
A_{Ers}	m^2	equivalent cross-sectional area between the contact partners
A_{qm}	m^2	quasi-metallic contact area
A_H	m^2	cross-sectional area of a crimping sleeve
A_L	m^2	conductor cross-section
$A_{L\,0}$	m^2	conductor cross-section before crimping
$A_{L\,V}$	m^2	conductor cross-section after crimping
A_K	m^2	cross-sectional area of the core of a conductor
A_p	m^2	supporting surface of the screw head
A_S	m^2	apparent contact area
A_t	m^2	mechanically load bearing contact area
A_w	m^2	metallic contact area
A_{w0}	m^2	metallic contact area in the initial state
b	J/mol	factor to be determined experimentally (ageing process)
b_0	m	remaining height of a sphere contact after wear
B	T	magnetic field vector
B‘		material-specific parameter to be determined experimentally
c	mol/m^3	concentration

Symbol	Unit[a]	Meaning
c_p	J/(kgK)	specific heat capacity
C		proportionality factor
C'		specific material constant
C_B		parameter to be determined experimentally
d		parameter to be determined experimentally (Ageing Absolute"?>thermodynamic temperature
T_1	K	thermodynamic of the contact parts
T_E	K	thermodynamic at the constriction
T_{max}	K	thermodynamic maxmimum contact temperature
T_S	K	thermodynamic melting temperature
U_K	V	contact voltage
$U_{K,r}$	V	rated contact voltage
U_V	V	voltage drop across the connection
v_k		constant of the speed of a chemical reaction
V		crimp ratio
V_r		wear volume
Z^*q		effective load of atomic cores
$ü$		oversize
α	1/K	linear thermal expansion coefficient
α_{el}	1/m	inverse exponential constant of the electrical current distribution
α_{20}	1/K	temperature coefficient of electrical resistance
ε_S		strain rate in the secondary section of creep
ϑ	°C	temperature
ϑ_0	°C	ambient temperature
ϑ_{ED}	°C	steady-state softening temperature
ϑ_{EK}	°C	short-time softening temperature (short circuit)
ϑ_{max}	°C	limiting upper temperature
ϑ_S	°C	melting temperature
κ	m/(Ωmm^2) = 10^6 S/m	electrical conductivity
κ_{20}	m/(Ωmm^2)	electrical Cconductivity at a temperature of 20 °C
λ	W/(mK)	thermal conductivity
μ_G		coefficient of friction between thread and screw
μ_{GL}		coefficient of kinetic friction
μ_H		coefficient of static friction
μ_K		coefficient of friction between screw head and supporting surface

Symbol	Unit[a]	Meaning
μ_r		relative permeability
ν		Poisson's Ratio
ρ	$(\Omega mm^2)/m$ $= 10^{-6}\ \Omega/m$	specific electrical resistance
ρ_F	$(\Omega mm^2)/m$	specific electrical film resistivity
ρ_{IMP}	$(\Omega mm^2)/m$	specific electrical resistivity of the intermetallic phase
δ	kg/m^3	density
ρ_0	$(\Omega mm^2)/m$	specific electrical resistance due to free-moving electrons
ρ_T	$(\Omega mm^2)/m$	portion of the specific electrical resistance due to temperature-dependent crystallographic oscillation
ρ_T	$(\Omega mm^2)/m$	portion of the specific electrical resistance due to crystallographic defects
σ_H		specific skin resistance
σ_m	N/mm^2	average mechanical tension between contact partners
φ	V	electrical potential
φ_u	1	degree of deformation
Φ	eV	height of the rectangular potential barrier

[a] Unless otherwise specified, values in SI units should be used for calculations

Constants

Symbol	Unit	Meaning
L	$(V/K)^2$	Lorenz number for pure metals $(2.2 \ldots 2.6) \cdot 10^{-8}$
R	J/(Kmol)	universal gas constant (8.314 J/(Kmol))
k_B	J/K	Boltzmann constant ($1.38064852 \cdot 10^{-23}$ J/K)
μ_0	N/A^2	magnetic field constant ($1.25663706212 \cdot 10^{-6}\ N/A^2$)

Abbreviations

Symbol	Meaning
RoHS	restriction of hazardous substances
hBN	hexagonal boron nitride
HTLS	high temperature low sag
IACS	international annealed Copper standard
MTF	mean time to failure

Contents

List of Figures

List of Tables

1 Introduction

Currently, the electrical power supply system, consisting of generators, transmission and distribution networks, and consumers, is undergoing a fundamental change. A new structure of few central and many decentralized electrical power generation plants, combined with an increasing volatility of feed-in power due to the use of renewable energy sources, is facing an increasing demand for electrical energy, for example due to the electrification of mobility. Electrical energy is increasingly being transported over greater distances between producers and consumers, and existing operating resources are therefore significantly more heavily loaded. The requirements for availability, safety, resilience, and cost of modern electrical installations have increased considerably. Examples of this are switching and distribution systems, but also new devices such as charging stations for or the power on-board network in electric vehicles. In addition to the safe operation of the electrical power supply network through stable switching states, no failures in the transmission and distribution network or between individual plant components should occur.

This can only be achieved through long-term stable and, if possible, maintenance-free or low-maintenance operation of each individual component. Sensitive assemblies, such as transformers, power or disconnect switches, should be serviced at regular intervals. Maintenance of all stationary current-carrying connections is not possible due to the very high number and their accessibility (Fig. 1.1). A large number of the current-carrying connections are therefore assembled and must function maintenance-free over the required lifetime of the plants of several decades. To ensure this under the new requirements, the correct conductor and coating materials must be chosen, a robust design and reliable construction ensured, the correct assembly carried out and a physically determined limit temperature for the desired lifetime taking into account the environmental conditions must be known. Current-carrying connections can be classified according to their properties as follows (Fig. 1.2).

S. Schlegel et al., *Contact and Long-Term Behavior of Current-Carrying Connections in Electrical Power Engineering*, https://doi.org/10.1007/978-3-662-69644-6_1

Fig. 1.1 Current-carrying connections in electrical energy systems using the example of an open-air switchgear

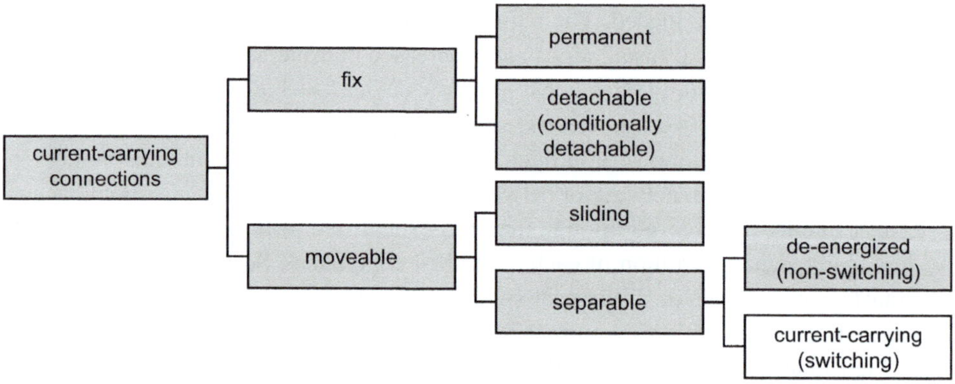

Fig. 1.2 Classification of current-carrying connections according to their properties

While previous technical literature has mainly dealt with the switching connections and the effect of the arc, this book considers the behavior of the significantly larger number of current-carrying connections that do not have to carry an arc (Fig. 1.2—gray). It describes the theory of contact physics, the most important conductor and coating materials, the influencing factors on the electrical contact behavior such as the relationship between mechanical and electrical parameters, the physics of aging and the effect

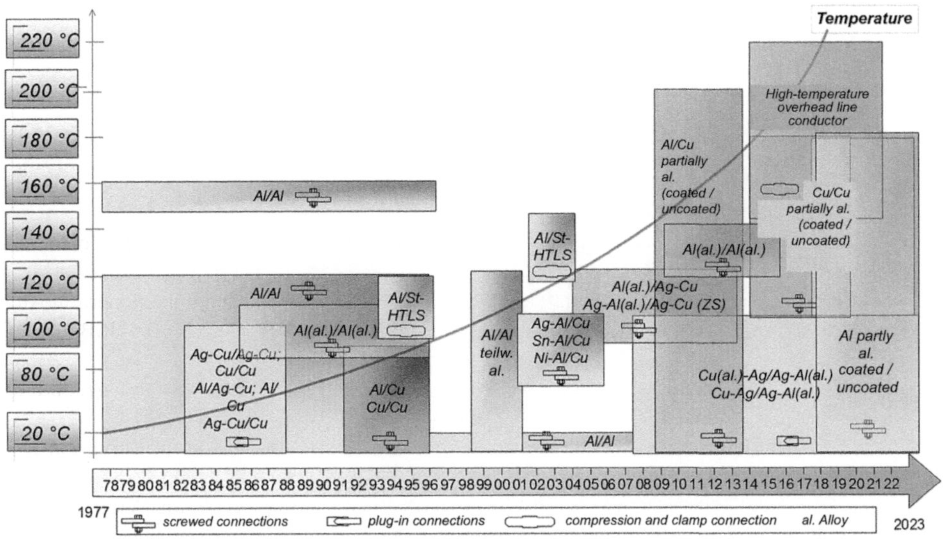

Fig. 1.3 Overview of the materials and types of connections studied at the Chair of High Voltage and High Current Engeneering at the Dresden University of Technology

on long-term behavior under different application and environmental conditions, as well as the testing procedures and the design criteria for construction. For this purpose, basic investigations on screwed, compression and plug-in connections are presented and generalizable findings are shown, which can also be transferred to other types of current-carrying connections. The aim of this book is to make scientifically secured statements about the design, assembly, and long-term behavior of current-carrying connections. Based on this, approaches to the numerical modeling of the electrical-mechanical-thermal behavior are described as well as evaluation criteria and testing procedures for qualification.

The findings of this book are based at their core on results that have been developed in over 40 years of research in this field at the Chair of High Voltage and High Current Technology at Dresden University of Technology (Fig. 1.3).

2 Types, Function and Requirements for Current-Carrying Connections

An electrically conductive connection has both an electrical and a mechanical function. Electrically, a current-carrying contact must be established between the conductors. Mechanically, the conductors should either be firmly connected (e.g., compression connection) or movably joined (e.g., sliding connection). When connecting the conductors, a reproducible and low contact resistance should be achieved, which does not increase impermissibly during the operating time [1, 2]. This ensures the galvanic coupling between the conductors via electrical contacts and a current can be carried permanently. In medium and high voltage technology, the design of the connection must also be such that no impermissible partial discharges occur. The required lifetime of current-carrying connections in electrical power engineering is usually several decades. Stable contact and long-term behavior are therefore prerequisites for the safe and reliable operation of all components and systems.

The electrical contact in current-carrying connections is fundamentally established and maintained with three types of closure (Fig. 2.1).

While in form-fit connections the cohesion of two interlocking contact partners is created by the force to be transmitted itself, in force-fit connections an externally applied force ensures the cohesion of the contact partners. An atomic bond is created by the atomic or molecular bonding of the materials of the contact partners. In force-fit and form-fit connections with a force-fit component, a minimum contact force $F_{K'min}$ must be maintained during operation. In addition, depending on the operating conditions, static forces, such as the weight force F_G of connected devices or the self-weight of the conductors, and dynamic forces that can occur in the event of a short circuit, must be transmitted. Forces due to thermal expansion F_{TD} in long conductors must be reduced by flexible conductor sections or sliding contact systems to such an extent that transverse forces F_Q in e.g. screwed connections can be transmitted by static friction in the

S. Schlegel et al., *Contact and Long-Term Behavior of Current-Carrying Connections in Electrical Power Engineering*, https://doi.org/10.1007/978-3-662-69644-6_2

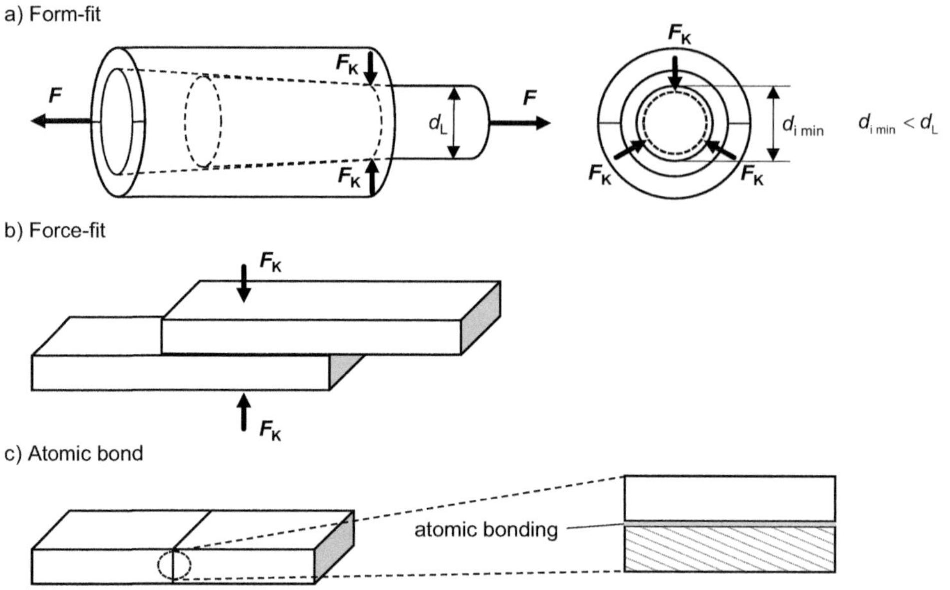

Fig. 2.1 Types of closure **a**) Form-fit, **b**) Force-fit, **c**) atomic bond (schematic)

separation joints of the tensioned conductors (non-sliding connection) [3, 4]. Detachable and conditionally detachable connections must be able to be dismantled after a certain operating time and load at the maximum permissible limit temperature in order to allow opening for revision or maintenance purposes. Key representatives of these types of connections are screwed connections, compression connections, and plug-in connections, which are used to connect current-carrying components such as cables, busbars, switching devices, transformers, battery storage, or motors (Fig. 2.2, Table 2.1).

The conductor material used and the connection must withstand the stresses with the permissible maximum operating temperatures both temporarily and permanently.

For example, in high-voltage engineering, the maximum permissible continuous operating temperature for conductors and connections is currently up to 115 °C [5]. New equipment such as HTLS overhead line conductors (English: **h**igh **t**emperature **l**ow **s**ag) have permissible operating temperatures of up to 240 °C. The connections reach temperatures between 120 °C and 150 °C [6]. In low-voltage engineering, for conductors made of copper temperatures of up to 140 °C are allowed at the busbars and the conductors [7]. For applications in transportation, even higher operating temperatures up to currently 180 °C at the contact may be required [8]. In the event of a fault, short-term currents can occur, leading to temperatures on the conductor of (200 … 270) °C. Even higher temperatures can occur at the a-spots. An excessively high reduction in the mechanical strength of the conductor materials due to the joule heating is not permitted [9–11]. The mechanical effect of short-term currents due to, for example, electromagnetic forces tak-

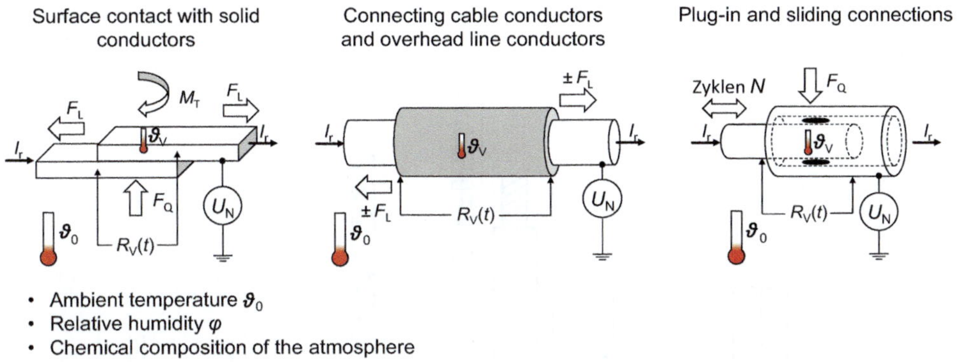

Fig. 2.2 Requirements and function of current-carrying connections depending on the type of connection

ing into account the installation position must also be considered in the design of the current-carrying connections. Another requirement is that the surrounding atmosphere must not unduly influence the functionality of the connection. This means that the strength of the connection must be adapted primarily considering the effect of chemical reactions such as oxidation and galvanic corrosion. The following chapters will therefore focus on contact and long-term behavior. This will include a closer look at conductor and coating materials, the physical principles of contact behavior, the influence of aging on long-term behavior, the design of various types of connections, and the testing procedures for qualification.

Table 2.1 Examples of designs of current-carrying connections

Class	Type of closure	Type of connection	Examples	
detachable/ conditionally detachable	Force-fit	Screwed connection	Screwed connection with busbars or cable lugs, for example in low-voltage switchgear combinations or in the power supply network of electric vehicles	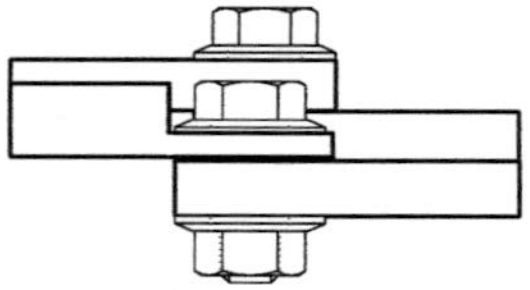
			Clamped connection, e.g. parallel groove clamp in switchgear	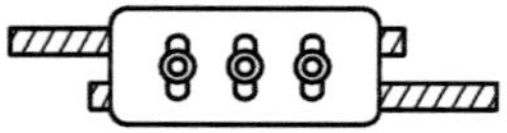
permanent	Force-/ Form fit	Screw connection	Shear-off screw connectors for cables	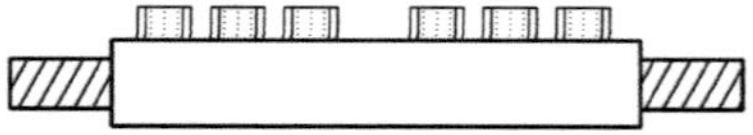
		Compression connection	Circumferential compression connection, e.g. for overhead line conductors	
			Indentation compression connection, e.g. notch connection for overhead line cables	
separable/sliding	Frictional connection	Plug-in connection	Connection in gas-insulated switchgear (GIS) and lines (GIL) with axially loaded coil spring (sectional view)	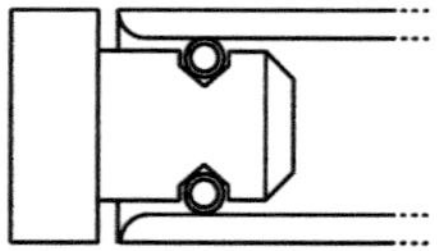

(continued)

Table 2.1 (continued)

Class	Type of closure	Type of connection	Examples	
			Connection in gas-insulated switchgear (GIS) and lines (GIL) with contact element torsion spring (sectional view)	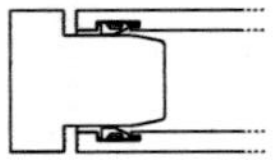
			Finger contacts with separate coil spring usually made of spring steel for generating the contact force (sectional view)	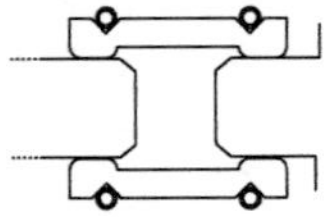
			Contact element with leaf spring (sectional view)	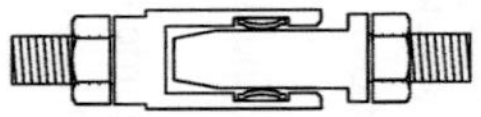
			slotted socket	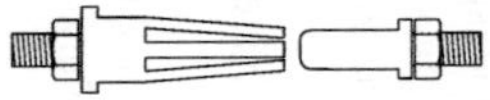

References

1. *Böhme, H.*: Mittelspannungstechnik-Schaltanlagen berechnen und entwerfen. 2. Stark bearbeitete Auflage Berlin: Verlag Technik 2005. – ISBN 3-341-01495-0.
2. *Slade, P. G.*: Electrical Contacts – Principles and Applications. 2. Auflage, CRC Press Taylor & Francis Group 2014. – ISBN 978-1-4398-8130-9
3. *Kloos, K.-H., Thomala, W.*: Schraubenverbindungen – Grundlagen, Berechnung, Eigenschaften, Handhabung. 5. Aufl., Berlin, Heidelberg. 2007. – ISBN: 978-3-540-21282-9
4. VDI-Richtlinie 2230: Systematische Berechnung hochbeanspruchter Schraubenverbindungen – Zylindrische Einschraubenverbindungen, 2003.
5. DIN EN 62271-1 (Ausg.: 2009-08): Hochspannungs-Schaltgeräte und -Schaltanlagen.
6. *Hildmann, C., Schlegel, S., et al.*: Einfluss von Kraft- und Formschlussanteil auf das elektrische Kontaktverhalten von Pressverbindungen. 23. Albert Keil Kontaktseminar, Karlsruhe 2015.
7. DIN EN 61439-1 (Ausg.: 2012-06): Niederspannungs-Schaltgerätekombinationen – Teil 1: Allgemeine Festlegungen.
8. LV 215-1 (2013-02): Elektrik/Elektronik Anforderung an HV-Kontaktierungen.
9. DIN EN 60865-1 (Ausg.: 2012-09): Kurzschlussströme – Berechnung der Wirkung – Teil 1: Begriffe und Berechnungsverfahren.
10. DIN EN 61238-1 (Ausg.: 2004-03): Pressverbinder und Schraubverbinder für Starkstromkabel für Nennspannungen bis einschließlich 36 kV ($U_m = 42$ kV) – Teil 1: Prüfverfahren und Anforderungen.
11. DIN EN 61284 (Ausg.: 1998-05): Freileitungen – Anforderungen und Prüfungen für Armaturen.

3 Contact Materials

In electrical engineering, metals are primarily used as conductors. The specific electrical resistance ρ or the electrical conductivity κ of metals depends on the freely moving electrons in the lattice (represented by ρ_0). Due to temperature-induced crystallographic oscillation (ρ_T) and lattice disturbances, such as defects, dislocations, grain boundaries or foreign atoms (ρ_G), the electrons are scattered and thus the specific electrical resistivity ρ is increased (Eq. (3.1)) [1].

$$\rho = \rho_0 + \Delta\rho_T + \Delta\rho_G \tag{3.1}$$

Through the movement of the electrons, energy and thus also heat is transported. This means that metallic conductors with a low specific electrical resistance ρ also have a high thermal conductivity λ. This indirect proportionality is described with the Wiedemann-Franz-Lorenz law, taking into account the thermodynamic temperature T and Lorenz number L (Eq. (3.2)).

$$\mathrm{L} \cdot T = \lambda \cdot \rho \tag{3.2}$$

In the temperature range in which electrical engineering systems are operated, it is the case that electron movement dominates energy transport. At very low temperatures near absolute zero, crystallographic oscillation is the dominant mechanism of energy transport that must be taken into account in thermal conductivity [1]. By cold forming or alloying pure metals, their mechanical properties are changed. This increases the number of crystallographic defects and thus the specific electrical resistance due to greater scattering of the electrons. This allows the mechanical and electrical properties to be set specifically for the application. As a result, the properties of the contact materials must not change impermissibly during a long operating period and at maximum operating temperature, as this can strongly negatively affect the contact and long-term behavior

S. Schlegel et al., *Contact and Long-Term Behavior of Current-Carrying Connections in Electrical Power Engineering*, https://doi.org/10.1007/978-3-662-69644-6_3

Table 3.1 Standardized aluminum for use in electrical engineering [3–6]

Conductor material	naturally hard/cold formed		precipitation hardened	
	E-Al99.7 (EN AW-1370A)	E-Al99.5 (EN AW-1350A)	E-AlMgSi0.5 (EN AW-6101B)	AlSi7Mg0.3 T6 (EN AC-42100) Sand casting
Electrical conductivity κ in m/(Ω·mm²) at 20 °C	34[3)] … 35.8[1)]	34[3)] … 35.4[1)]	30[5)] … 32[4)]	20 … 27
Temperature coefficient of the electrical resistance α_{20} in 1/K	0.004	0.004	0.0036	0.0022
Hardness H in HBW[6)]	20[1)] … 41[2)]	20[1)] … 41[2)]	55[4)] … 70[5)]	75
Yield strength $R_{p0.2}$ in N/mm²	65[1)] … 130[3)]	50[1)] … 130[3)]	120[4)] … 160[5)]	190
Tensile strength R_m in N/mm²	65[1)] … 165[2)]	60[1)] … 165[2)]	170[4)] … 215[5)]	230
Young's modulus E in kN/mm²	65	65	70	69 … 75
Linear thermal expansion coefficient α in 1/K·10⁻⁶	24	24	23	22
Thermal conductivity λ in W/(mK)	210 … 230	220	190	160 … 180
Specific heat capacity c_p in J/(gK)	0.94	0.92	0.92	0.92
Thermodynamic melting temperature T_S in K	933		903	823 … 898

[1)] Condition O (annealed)
[2)] Condition H28 (cold hardened annealed, 4/4 hard)
[3)] Condition H19 (cold hardened, extra hard)
[4)] Condition T7 (over-hardened)
[5)] Condition T6 (heat treated)
[6)] Brinell hardness (test ball tungsten carbide)

If very high requirements are placed on the mechanical properties, such as the spring action in contact elements for plug-in or screw connections, alloys with a higher proportion of foreign elements must be used. Compared to pure copper, such alloys have significantly lower electrical conductivity and are usually precipitation-hardened. Therefore they have a higher yield strength, meaning they can be deformed more elastically (Table 3.3) [12, 14].

Table 3.2 Standardized copper for use in electrical engineering [7, 9, 10]

Conductor material	Cu-ETP (CW004A) Cu-FRHC (CW005A) Cu-OF (CW008A) CuAg0.1 (CW013A) CuAg0.10(OF) (CW019A) Cu-PHC (CW020A)	CuAg0.1P (CW016A) Cu-HCP (CW021A)
Electrical conductivity κ_{20} in m/(Ω·mm²)	56 … 58	55 … 57
Temperature coefficient α_{20} in 1/K	0.0039	
Hardness H in HV	40 … 110[1)]	
Yield strength $R_{p0.2}$ in N/mm²	100 … 320[1)]	
Tensile strength R_m in N/mm²	200 … 360[1)]	
Young's modulus E in kN/mm²	107 … 118 (annealed)	
Linear thermal expansion coefficient α in 1/K·10⁻⁶	16.8 … 17.3	
Thermal conductivity λ in W/(mK)	385 … 394	
Specific heat capacity c_p in J/(gK)	0.385 … 0.393	
Thermodynamic melting temperature T_S in K	1355 … 1356	

[1)] annealed to cold formed

Table 3.3 Types of copper for mechanically highly stressed current-carrying components (typical examples) [7, 8, 13, 15]

Conductor material	CuBe2	CuCo2Be	CuCr1Zr
Electrical conductivity κ in m/(Ω·mm²) at 20 °C	13 … 16	≤26	≤43
Temperature coefficient α_{20} in 1/K	–	–	≤0,0031
Hardness H in HV	–	265	≤190
Yield strength $R_{p0,2}$ in N/mm²	960 … 1420	500 … 850	≤440
Tensile strength R_m in N/mm²	1130 … 1520	650 … 900	≤570
Young's modulus E in kN/mm²	130	130 … 138	≤130
Thermal conductivity λ in W/mK	approx. 104	–	≤330
Thermodynamic melting temperature T_S in K	1144 … 1255	1303	1343

A crucial criterion for the stability of current-carrying conductors and connections is that the thermal stress caused by current heat losses in operation and by extreme environmental conditions does not lead to the softening of the material. This in combination with the dimensioning and construction of the connection, ensures that no plastic deformation of the conductors and connections occurs due to forces resulting from, for example, short-circuit currents. In addition, the softening of the conductor material leads

Table 3.4 Softening temperature of conductor materials [2, 18–21]

Conductor material	E-Al99.5	AlMgSi0.5	AlMn0.5Mg0.5[1)]	Cu-ETP[2)]	CuAg0.1P[1)]
Steady-state Softening temperature ϑ_{ED} in °C	80 … 93	≈120 (T6, T7)	≥140 °C	≈140 (R300)	>160
Estimated softening temperature for pure metals approx. ((0.3 … 0.4)·T_S) in °C	80 … 100	–	–	134 … 269	–
Short-time softening temperature ϑ_{EK} in °C	160 … 200	220 … 230	–	165 … 170	–

[1)] Material developed within the framework of the funding project with the identifier: 03EK3538D [19]
[2)] Cold formed (permanently permissible load temperature at a forming degree of up to 30%, the higher the degree of forming, the lower the softening temperature)

to a significant reduction in contact force (Sect. 5.1.1). The temperature at which softening can occur depends heavily on the material and its manufacturing process. The softening temperature is often given as the temperature at which the material loses at least 10% of its strength [13]. For pure metals that have gained their mechanical strength through cold forming, softening through recrystallization is likely from a temperature of (0.3 … 0.4)·T_S. The degree of cold forming must be at least (2 … 15)%. The higher the degree of forming, the stronger the tendency for recrystallization [13, 15]. In particular, the proportion of foreign atoms and the type and size of these significantly influence the temperature at which softening occurs [13]. For industrial purposes, such as the production of semi-finished products, the softening temperature is given in annealing diagrams [16, 17]. For very long periods in the range of several months or years and for low temperatures, hardly any information is provided. Few details were compiled from the literature or determined by our own experimental investigations. For this, hardness was determined as an indicator and metallographic examinations were carried out and evaluated depending on the time for some of the most important conductor materials (Table 3.4) [2, 18].

Since current-carrying connections are often frictionally or positively connected with a frictional component, the conductor material must also have a high creep strength and creep limit. Whether aluminium or copper is used as a construction material depends

Table 3.5 Related properties of conductor materials at the same current carrying capacity in %—values from Table 3.1 and 3.2 [23]

Conductor material/property based on Cu-ETP[1)] in %	Cu-ETP	E-Al99.5	E-AlMgSi0.5
Conductor cross-section A_L	100	160	187
Mass of the conductor m_L		49	57
Linear expansion Δl		139	133
Heat ΔQ		114	133

[1)] Current displacement not considered, temperature of 20 °C; temperature change due to short circuit from 80 °C to 200 °C

heavily on the type of connection and the operating conditions. A comparison of related properties at the same current carrying capacity illustrates this and shows that both aluminium and copper are equivalent and indispensable conductor materials (Table 3.5) [23].

Depending on the required current density, the permissible weight, the mechanical requirements, the operating temperature and constructional as well as environmental parameters, a conductor material must be selected that meets the requirements in normal and short-term operation [23].

3.2 Coating of Conductors

To improve the contact and long-term behaviour of current-carrying connections, the conductors or just the contact surfaces are often coated. In electrical energy technology, silver, tin and nickel are used in particular. The coating is mainly done electrolytically but also autocatalytically or by plating, metal spraying, PVD[1] and thermal processes such as hot-dip tinning. The physical properties of the electrolytic and autocatalytic layers depend heavily on the composition of the electrolytes. These significantly determine the contact and long-term behaviour of the current-carrying connections in operation (Table 3.6).

For the contact behavior, it is crucial that no or only very thin, soft film layers form on the surface of the coating materials under normal atmosphere, which can be broken up with low contact forces. In addition, a high electrical conductivity and a low to medium hardness are necessary to create large-scale, well-conductive a-spots. Especially for plug-in connections, additional requirements for the tribological properties of the material must be taken into account [12, 14].

[1] Physical Vapour Deposition—physical vapour deposition.

Table 3.6 Properties of typical coating materials [8, 18, 24, 25]

Coating Material/ Property	Ag (pure[1)])	Sn (matt[1)])	Ni (pure[1)])	NiP[2)] (P content in %)	
				Type 2 (1 … 3)	Type 3 (2 … 4)
Density ρ in g/cm^3	10.49	7.30	8.91	8.5	
Young's modulus E in kN/mm^2	79	47	216	55 … 65	
Hardness H in HV	30 … 90	8	300 … 500	725 … 800	
Linear thermal expansion coefficient α in 10^{-6} K^{-1} (20 °C)	19.7	23	13.3	12 … 15	
Electrical conductivity κ in MS/m (20 °C)	60	8.6	14.6	3.3 … 10	3.3 … 5
Temperature coefficient α_{20} in 1/K 10^{-3}	4.1	4.6	6.75	–	–
Thermal conductivity λ in W/K·m	419	63	92	–	–
Thermodynamic melting temperature T_S in K	1235	505	1726	1523 … 1633	1153 … 1253

[1)] Additives in the galvanic bath can cause the properties to differ from those of the pure metals.
[2)] Precipitation hardenable crystalline layers

Silver hardly forms film layers under normal atmospheric air and is therefore very suitable as a coating material, especially with low contact forces, such as in plug-in connections. In addition, the hardness is low and the electrical conductivity is high (Table 3.6). Tin oxidizes in atmospheric air, but due to its low hardness, it can be plastically deformed very easily during joining, so that large a-spots can be created. The oxide layers are also relatively soft and can be easily broken up. Due to the low hardness, the tribological properties are moderate to poor compared to silver. Nickel is often used in corrosive atmospheres in industrial plants because it is chemically very stable. However, it forms a slowly growing oxide layer in normal air, which is very dense and hard. In addition, nickel itself has a very high hardness, so high contact forces are necessary to achieve good electrical contact behavior. The same applies to nickel-phosphorus, which can be deposited very evenly on the surface of conductors but has an even lower electrical conductivity and higher hardness than pure nickel (Table 3.6). Further information on the properties of coating materials can be found, for example, in [24, 25].

Table 3.7 Recommended thickness of intermediate and top layers according to standards [26–30]

Conductor material → Coating material ↓		Al	Cu
Top layer (TL)	Intermediate layer (IL)	Thickness of the TL/IL in μm	
Sn	Ni[1]	≥5/(10 … 25)	>2 or ≥5/(0,5 … 2)[4]
	Cu		>2 or ≥5/≥2
Ag	Ni[1]	application-specific/20[2]	application-specific/no information[2, 3]
	Cu	no information	
Ni	none	(10 … 35)/no information	(8 … 25)/no information
NiP	none	(0,1 … 60)/no information	application-specific (no information)

[1] Intermediate layer of NiP with a thickness of (2 … 5) μm is possible [30]
[2] Intermediate layer recommended at an operating temperature >150 °C
[3] Intermediate layer of Cu or Ni recommended for alloyed Cu
[4] Intermediate layer recommended at an operating temperature >120 °C

Silver, tin, and nickel can be deposited on copper conductors without an intermediate layer. For conductors made of aluminum, an intermediate layer of, for example, zincate and nickel or copper must be used as an adhesion promoter and diffusion barrier for silver and tin (Table 3.7). The diffusion barrier is intended to prevent the conductor and coating material of the top layer from forming intermetallic phases or mixed crystals with each other and adhering well to each other. Recommended layer thicknesses from different standards are presented below.

The thickness of tin layers on the contact surface of copper and aluminum conductors is standardized. When tinning copper conductors, an intermediate layer is recommended if an operating temperature above 120 °C is required [26]. With regard to aging, a thickness of the intermediate layer of copper of ≥2 μm or of nickel of (0.5 … 2) μm is recommended [26]. For conductors made of aluminum, intermediate layers of copper, bronze or nickel of (10 … 25) μm are specified [27]. The thickness of the top layer of tin is specified for copper conductors with >2 μm [26] or ≥5 μm [27] and for aluminum conductors with ≥5 μm [27] depending on the application and environmental conditions. Lead-free (RoHS-compliant) tin surfaces tend to form whiskers, which manifest as wire-like, thin protrusions from the surface. Influencing factors are mechanical stresses in the coating, the current density, and the electric field. The formation of whiskers can be minimized or prevented by using suitable additives, layer sequences, deposition parameters, and the use of bright instead of matte tin [34, 35].

For silver-plated contact surfaces, no intermediate layer is required, for conductors made of technically pure copper and for alloyed copper, an intermediate layer of technically pure copper or nickel is required. The thickness of the layer is to be determined by

Table 3.8 Typical physical limit temperatures of coating materials exemplarily for contact surfaces of plug connections [31, 32, 35]

Coating material	Au	Sn	Ag	AgPd (Pd content <4 %)[2)]
Physical limit temperature ϑ_{ED} steady-state operation[1)]	Low-alloyed with Ni IL up to max. 180 °C	130 °C Sn alloys up to 150 °C	140 °C Ni IL up to 170 °C	up to 200 °C/1000 h

IL ... Intermediate layer

[1)] Relaxation behavior of the conductor material must be considered

[2)] In development

the user. For conductors made of technically pure aluminum and alloyed aluminum, a layer thickness of 20 µm is required for nickel. Conductors made of copper without an intermediate layer of nickel or conductors made of aluminum with an intermediate layer of copper should not be used permanently above 150 °C [29].

For electrolytically nickel-plated contact surfaces, the following layer thicknesses are required for conductors made of copper and aluminum, especially for decorative coatings. Depending on the stress, which is divided into five stages, a thickness of the nickel layer of (8 ... 25) µm for conductors made of copper and (10 ... 35) µm for conductors made of aluminum is defined [28]. Autocatalytic nickel-phosphorus layers can be divided into five types depending on the phosphorus content. For use in electrical energy technology, high electrical conductivity and partially increased corrosion resistance are required. Therefore, types 2 and 3 are particularly relevant (Table 3.6). Nickel-phosphorus layers can be heat-treated. This affects the hardness and adhesion strength. Depending on the required corrosion resistance, a layer thickness on aluminum of (0.1 ... 60) µm is recommended [29].

The physical limit temperatures of coatings for current-carrying connections depend not only on the temperature but also on the environmental conditions, the conductor material used, and the requirements in operation such as plug-in, sliding cycles, or the solubility after loading. Information on layer thicknesses and limit temperatures is therefore not generally possible and must be considered application-specific. For example, the following limit temperatures are given for the surfaces of plug-in connections (Table 3.8).

To improve tribological behavior, dispersion layers can be used as coating material. Currently, silver layers with graphite particles Ag-C are mainly in use, which are used as self-lubricating coatings for connectors. The particles that are incorporated into the layer affect the electrical properties of the layer to varying degrees (Table 3.9). In experiments, a positive effect on the tribological behavior compared to pure silver layers could be demonstrated for some types of particles (Sect. 4.2.3)

Table 3.9 Electrical properties of silver dispersion layers [37]

Coating material	Ag-C	Ag-hBN	Ag-WS_2			Ag-MoS_2	Ag-SiC	Ag-Bi_2S_3	Ag-SnS
Particle content (vol-%)	6.2	4.8	0.2	2	5	4	3.2	2.5	5
Electrical conductivity κ in MS/m (20 °C)	51	49.5	53.8	47.6	43.4	50.9	56.6	31.5	18.2
Temperature coefficient α_{20} in 1/K 10^{-3}	3.8	3.3	3.4	3.4	2.9	3.8	3.6	3.1	1.2

hBN … hexagonal boron nitride

When selecting a coating, it should generally be ensured that only the same materials touch each other in an electrical contact. If different materials meet, it must be ensured that these bimetallic connections are suitable for permanently conducting electrical current (Sect. 5.3).

References

1. *Hornbogen, E., Warlimont, H.*: Metallkunde – Aufbau und Eigenschaften von Metallen und Legierungen. 4. Auflage, Springer-Verlag, 2001 – ISBN 3-540-67355-5.
2. *Schlegel, S.*: Langzeitverhalten von Schraubenverbindungen mit Stromschienen aus Reinkupfer in der Elektroenergietechnik unter besonderer Berücksichtigung der Temperatur. Dissertation, TU Dresden, 2011.
3. DIN EN 62271-1 (Ausg.: 2009-08): Hochspannungs-Schaltgeräte und -Schaltanlagen.
4. DIN 40501-2 (Ausg. 2005-03): Aluminium für die Elektrotechnik – Teil 2: Rohre, Stangen und Profile, Technische Lieferbedingungen.
5. *Gremmel, H. (ABB)*: Schaltanlagen. 10. Auflage, Cornelsen Verlag, Berlin, 1999. – ISBN 3-464-48235.
6. DIN EN 1706 (Ausg.: 1998-06): Aluminium und Aluminiumlegierungen – Gussstücke – Chemische Zusammensetzung und mechanische Eigenschaften
7. DIN EN 13599 (Ausg. 2014-12): Kupfer und Kupferlegierungen – Platten, Bleche und Bänder aus Kupfer für die Anwendung in der Elektrotechnik.
8. DIN EN 12164 (Ausg. 2016-119): Kupfer und Kupferlegierungen – Stangen für die spanende Bearbeitung.
9. Deutsches Kupferinstitut: Werkstoff-Datenblatt Cu-ETP. DKI, 2005.

10. Deutsches Kupferinstitut: Werkstoff-Datenblatt CuAg0,1P. DKI, 2005.
11. *Schneider, R.*: Langzeitverhalten geschraubter Stromschienenverbindungen in der Elektroenergietechnik. Dissertation, TU Dresden, 2012.
12. *Blumenroth, F.*: Zum Langzeitverhalten von Steckverbindungen mit Schraubenfedern in Anlagen der Elektroenergietechnik. Dissertation, TU Dresden, Shaker Verlag, 2010. – ISBN 978-3-8322-9506-6.
13. *Kammer, C.*: Aluminium Taschenbuch 1 – Grundlagen und Werkstoffe. 16. Auflage, Aluminium-Verlag, 2003 – ISBN 978-3410220282.
14. *Lücke, N.*: Zum Langzeitverhalten elektrischer Steckverbindungen mit Kupfer-Beryllium-Kontaktlamellen. Dissertation, TU Dresden, TUDpress, 2004. – ISBN 978-3944331751.
15. *Bürgel, R.*: Handbuch Hochtemperatur-Werkstofftechnik. Vieweg & Sohn Verlag, 2006. – ISBN 978-3-528-23107-1.
16. *Gremmel, H. (ABB)*: Schaltanlagen. 10. Auflage, Cornelsen Verlag, Berlin, 1999. – ISBN 3-464-48235.
17. *Wieland-Werke AG:* Profile, Profildrähte, Rohre und Profilrohre. Broschüre, 2003.
18. *Fuhrmann,* T.: Kontakt- und Langzeitverhalten von stromführenden Schraubenverbindungen mit vernickelten und versilberten Leitern aus Aluminiumwerkstoffen. Dissertation, TU Dresden, 2019. – ISBN 978-3-86780-636-7.
19. Merkblatt E1: Aluminium in der Elektrotechnik und Elektronik. 1. Auflage, Aluminium-Zentrale e.V. Düsseldorf.
20. IEEE 1283 (Ausg. 2004): IEEE Guide for Determining the Effects of High-Temperature Operation on Conductors, Connectors, and Accessories.
21. *Poudeu, F.*: Langzeitverhalten stromführender Schraubenverbindungen mit Stromschienen kleiner Dicke bei wechselnden Umgebungsbedingungen. Dissertation, TU Dresden, 2022 (in press).
22. *Großmann, S.; Löbl, H.; Schlegel, S.*: Erfahrungen zum Langzeitverhalten von Aluminium-Verbindungen in Starkstromanlagen. Konferenzbd. 21. Albert Keil Kontaktseminar, Karlsruhe, 2011, pp. 129–135.
23. *Lücke, N.; Schlegel, S.; Großmann, S.*: Vergleich von Werkstoffen auf Basis von Cu und Al sowie Trends bei deren Anwendung in der Elektroenergietechnik. Zeitschrift Metall, 67. Jahrgang, 11/2013.
24. *Vinaricky, E.*: Elektrische Kontakte, Werkstoffe und Anwendungen – Grundlagen, Technologien, Prüfverfahren. 3. Auflage, Springer Verlag, 2016. – ISBN 978-3-642-45427-1.
25. *DODUCO GmbH:* Datenbuch der elektrischen Kontakte. überarbeitete 3. Auflage, Stieglitz Verlag, 2012.
26. DIN EN 14436 (Ausg. 2014-11): Kupfer und Kupferlegierungen – Elektrolytisch verzinnte Bänder.
27. ISO 2093 (Ausg. 1986-12): Electroplated coatings of tin – Specification and test methods.
28. DIN EN ISO 1456 (Ausg. 2009-12): Metallische und andere anorganische Überzüge – Galvanische Überzüge aus Nickel, Nickel plus Chrom, Kupfer plus Nickel und Kupfer plus Nickel plus Chrom.
29. DIN EN ISO 4521 (Ausg. 2009-01): Metallische Überzüge und andere anorganische Überzüge – Galvanische Silber- und Silberlegierungs-Überzüge für technische Zwecke – Anforderungen und Prüfverfahren.
30. DIN EN ISO 4527 (Ausg. 2003-10): Metallische Überzüge – Autokatalytisch (außenstromlos) abgeschiedene Nickel-Phosphor-Legierungs-Überzüge – Spezifikationen und Prüfverfahren.
31. *Buresch, I.*: Oberflächenbeschichtungen für Steckverbinderkontakte bei höheren Einsatztemperaturen. Konferenzbd. 24. Albert Keil Kontaktseminar, Karlsruhe, 11.–13.10.2017, pp. 15–23. – ISBN 978-3-8007-4449-7

32. *Buresch, I., Schmidt H., et al.*: Steckverbinderoberflächen für hohe Temperaturen. Konferenzbd. 25. Albert Keil Kontaktseminar, Karlsruhe, 09.–11.10.2019, pp. 32–38. – ISBN 978-3-8007-5083-2.
33. RICHTLINIE 2011/65/EU DES EUROPÄISCHEN PARLAMENTS UND DES RATES vom 8. Juni 2011 zur Beschränkung der Verwendung bestimmter gefährlicher Stoffe in Elektro- und Elektronikgeräten. RL 2011/65/EU
34. *Spörrer, A. et al.:* Untersuchung der Bildung von Druckwhiskern sowie ein Verfahren zu deren Vermeidung. Konferenzbd. 24. Albert-Keil-Kontaktseminar, Karlsruhe, 11.–13.10.2017, pp. 56–62. – ISBN 978-3-8007-4449-7.
35. *Buresch, I.; Horn, J.:* Bleifreie Zinnoberflächen Steigende Anforderungen bei Kontakt- und Lötanwendungen führen zu neuen Lösungen. Konferenzbd. 18. Albert-Keil-Kontaktseminar, Karlsruhe, 5.–7.10.2005, VDE-Fachbericht 61, VDE Verlag GmbH, Berlin, Offenbach, 2005, pp. 89–94.
36. *Adler, U.*: Werkstoffinnovationen für Steckverbinder-Anwendungen im Hinblick auf neuere gesetzliche Vorschriften. Konferenzbd. 22. Albert-Keil-Kontaktseminar, Karlsruhe, 09.–11.10.2013.
37 *Oberst, M.; Israel, T.; Hildmann, C.; Schlegel, S.; Egetenmeyer, A.-K.; Kappl, H.; Willing, H.*: Elektrische und tribologische Eigenschaften von Silber-Dispersionsschichten für die Anwendung in Steckverbindungen. Konferenzbd. 8. Symposium Connectors – Elektrische und Optische Verbindungstechnik 2021, VDE-Verlag, Lemgo, 23.–24.03.2021, pp. 75–85.

Contact Behavior of Current-Carrying Connections

4

4.1 Physical Fundamentals of Point, Line, and Surface Contact

4.1.1 Physical Model

The current-carrying transition of two electrical conductors is referred to as an electrical contact [1]. Depending on the geometry of the conductors, point, line or plane contacts are created (Fig. 4.1). The technical realization of an electrical contact is an electrical connection, which is also called current-carrying connection.

Current-carrying connections in electrical energy technology are implemented in all three variants. The resulting macroscopic apparent contact area A_s is larger than the mechanically supporting contact area A_t, particularly in the case of plane contacts [1, 2]. Due to the surface roughness of the contact partners, both only touch each other at microscopically small areas (Fig. 4.2c).

The size, position, and distribution of the contact points strongly depend on the elasto-plastic deformation of the micro-peaks during joining and are significantly influenced by the surface roughness, the contact force, and the resulting mechanical stress distribution in the contact interface (Sect. 4.3.1). In addition, foreign and especially oxide layers form on most metallic surfaces (Fig. 4.2—blue). These cause not all of the mechanically supporting contact area to contribute to current transmission. Oxide layers thicker than (10 … 20) Å[1] act electrically insulating. Oxide layers with a thickness less than (10 … 20) Å still contribute to current transport and are referred to as quasi-metallic contact area A_{qm} [1, 2]. In the case of the quasi-metallic contact area, the quantum mechanical tunnel effect occurs, in which the electrons can penetrate the oxide layer

[1] 1 Å = 0.1 nm.

S. Schlegel et al., *Contact and Long-Term Behavior of Current-Carrying Connections in Electrical Power Engineering*, https://doi.org/10.1007/978-3-662-69644-6_4

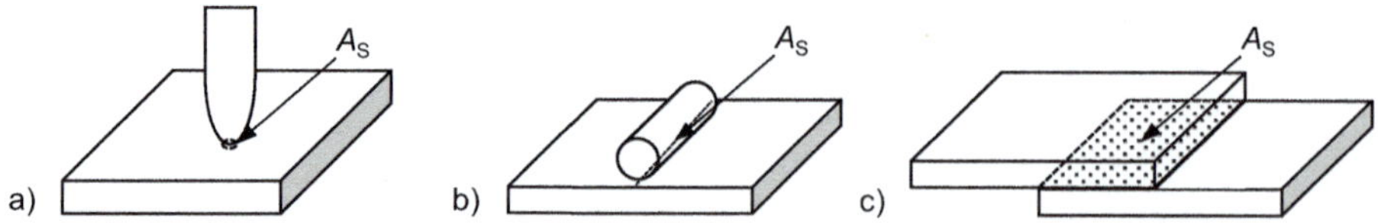

a) Point contact	b) Line contact	c) Plane contact
• Measuring tips • Relay • Plug-in connection with Contact elements	• Plug-in connection with contact elements • Compression connection with stranded	• Screwed connection with busbars • Compression connection • Pipe clamps

Fig. 4.1 **a**) Point, **b**) Line, and **c**) Surface contact and examples of connection types of the respective contact type with drawn-in apparent contact area A_s

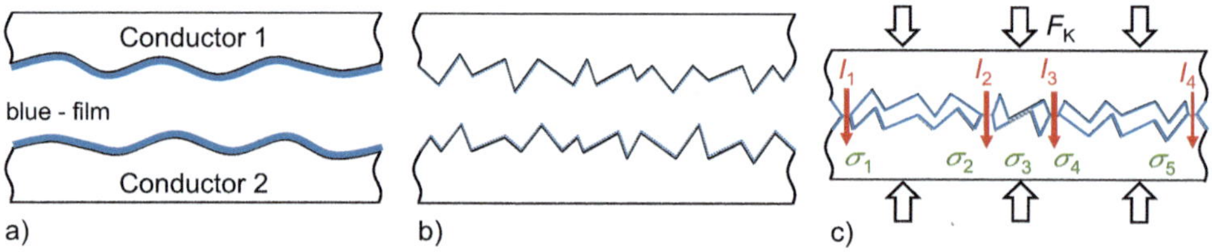

Fig. 4.2 Formation of microcontacts **a**) Conductor initial state, **b**) Conductor abrasively pretreated, **c**) Conductor force-fitted (schematic)

almost without loss [1]. Contact areas free of films are referred to as microcontacts or also a-spots. They carry the majority of the current. The sum of all microcontacts forms the true metallic contact area A_w (Fig. 4.3).

The following applies in general:

$$A_w < A_t << A_s \tag{4.1}$$

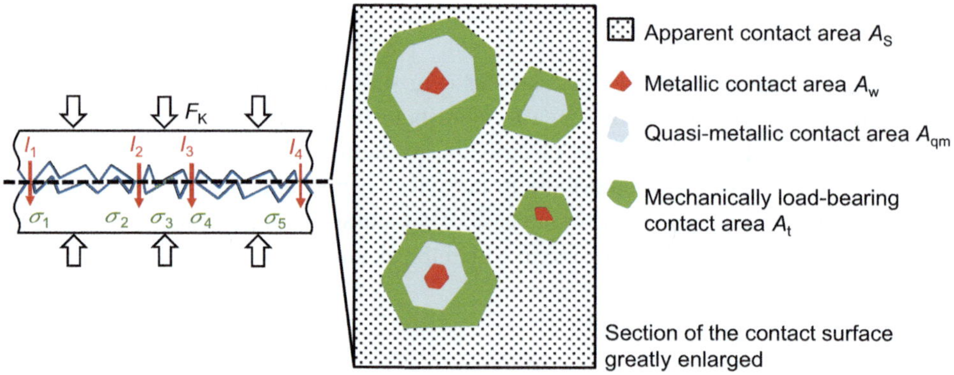

Fig. 4.3 Contact areas between two contact partners (schematic)

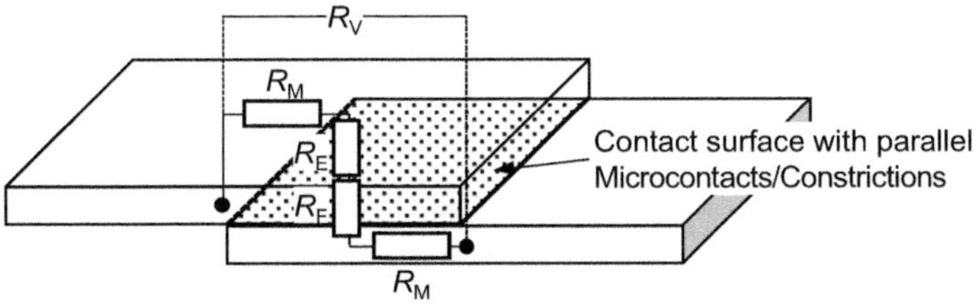

Fig. 4.4 Basic equivalent circuit diagram of a connection using the example of a plane contact

The contact behavior can be simplified with the contact resistance R_K. The contact resistance is the sum of constriction R_E and film resistance R_F. It results in:

$$R_K = R_E + R_F \tag{4.2}$$

Equation (4.2) only applies under the condition that $R_E << R_F$, as the current density distribution changes with increasing influence of films and both resistances are no longer independent of each other [4]. If the material resistance R_M is added to the contact resistance, the connection resistance R_V results as follows:

$$R_V = R_K + R_M \tag{4.3}$$

The material resistance to be considered depends on the current density distribution in the contact partners, which is influenced by the geometry of the connection. In a real-world connection, the connection resistance R_V is often the only size that can actually be measured and used as a basis for evaluating the connection (Fig. 4.4, additionally Sects. 4.1.2 and 4.1.3). A division of the connection resistance into contact and material resistance is valuable for a more precise consideration of the connection and requires an extended model (see Sect. 4.3). The constriction resistance can be described for single-point contacts with the ball or ellipsoid model [2]. For multi-point contacts, the distribution and size as well as the interaction of the electrical flow lines of adjacent microcontacts must be taken into account (Table 4.1). The film resistance is strongly dependent on the thickness of the film and can be described as a disc with a radius r_i and a thickness l_f[2] in the microcontact.

For thick oxide layers, the film resistance R_F must be considered, and for thin oxide layers (less than (10 … 20) Å), the skin resistance R_H must be considered (Table 4.2) [1]. Since the exact location of the microcontacts is not known, especially for plane contacts, the connection resistance can often only be estimated using empirical approaches (see Sect. 4.3).

$$R_{E,E} = \frac{\rho_M}{2r_i} \tag{4.4}$$

[2]This is an estimate, as the thickness of the foreign layer is inhomogeneously distributed and thus the effect is location-dependent.

Table 4.1 Calculating the constriction resistance—single-point contact (based on [2, 5])

	Ellipsoid model
Schematic	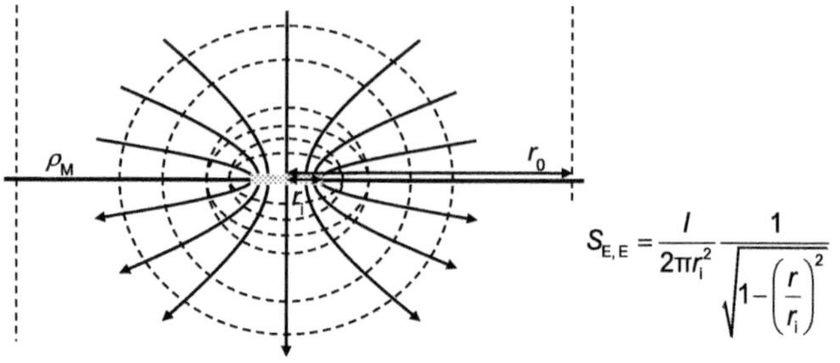
Equation	(4.4) applies when $10 \cdot r_i \leq r_0$ (r_0 … outer radius conductor)
Parameter	ρ_M … spec. electrical resistance, n_E … number of a-spots, r_i … radius of the a-spot, I … current through the constriction $S_{E,E}$ current density in the constriction for the ellipsoid model

Table 4.2 Calculation of the film resistance [15]

	Thin foreign layers (< (10 … 20) Å)	Thick foreign layers (> (10 … 20) Å)
Schematic assumption	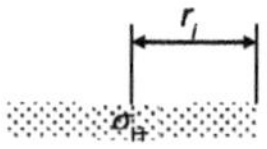	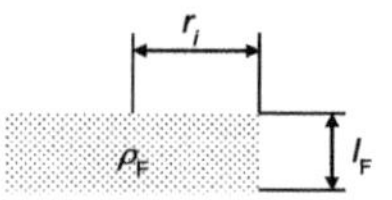
Equation	(4.5)	(4.6)
Parameter	σ_H … specific electrical, ρ_F … specific electrical resistance of the film layer, l_F … thickness of the film layer	

$$R_H = \frac{\sigma_H}{\pi r_i^2} \tag{4.5}$$

$$R_F = \frac{\rho_F l_F}{\pi r_i^2} \tag{4.6}$$

4.1.2 Performance Factor: Evaluating Connections with Elongated Conductors

The evaluation of connections with elongated conductors can be done with the performance factor k_u. The performance factor is the ratio of the connection resistance R_V, measured over the connection length l_V, relative to the resistance R_{BL} of the reference conductor of the same length l_{BL} (Fig. 4.5, Eq. (4.7)) [5]. It is assumed that the current flow is homogeneous in the reference conductor.

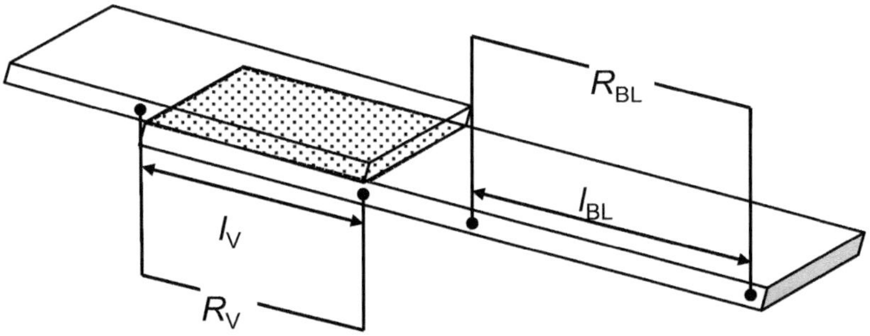

Fig. 4.5 Performance factor—longitudinally elongated connection with conductors of the same geometry

The performance factor also describes the ratio of the power loss in the connection P_V relative to that in the reference conductor P_{BL} with homogeneous current flow.

$$k_u = \frac{R_V}{R_{BL}} = \frac{U_V}{U_{BL}} = \frac{P_V}{P_{BL}} \tag{4.7}$$

A performance factor of $k_u = 1$ means that the same power loss is generated in the connection as in the reference conductor with homogeneous current flow. For connections with elongated conductors, a temperature-equivalent performance factor $k_{uT\text{-}20}$ can be specified, in which the connection and the surrounding conductors of the same geometry have the same temperature. In addition, an upper limit of the performance factor $k_{uS\text{-}20}$ can be specified, at which the melting temperature is reached at the microcontacts during operation. If the connected conductors have a different geometry or are made of different materials, a mean resistance should be used as the reference value, or the resistance of the reference conductor with the lower resistance should be used. For screwed connections with busbars, depending on the geometry of the rails, the number of holes and screws, and the connection length l_V, values for $k_{uT\text{-}20} = (1.41 \ldots 2)$ can be specified (Table 4.21—Sect. 4.5.1 (p. 106) [6].

Example

Reference value R_{BL} of a screwed connection with busbars of dimensions $(40 \times 10)\ \text{mm}^2$ and a connection length $l_V = 40$ mm made of Cu-ETP

$$R_{BL} = 0.0178 \frac{\Omega \cdot \text{mm}^2}{\text{m}} \cdot \frac{0.04\ \text{m}}{400\ \text{mm}^2} = 1.78\ \mu\Omega$$

Reference value R_{BL} of a compression connection with a 400 mm² round conductor made of Al99.5 class 2 and a length of 70 mm.

$$R_{BL} = 0.028 \frac{\Omega \cdot \text{mm}^2}{\text{m}} \cdot \frac{0.07\ \text{m}}{400\ \text{mm}^2} = 4.9\ \mu\Omega$$

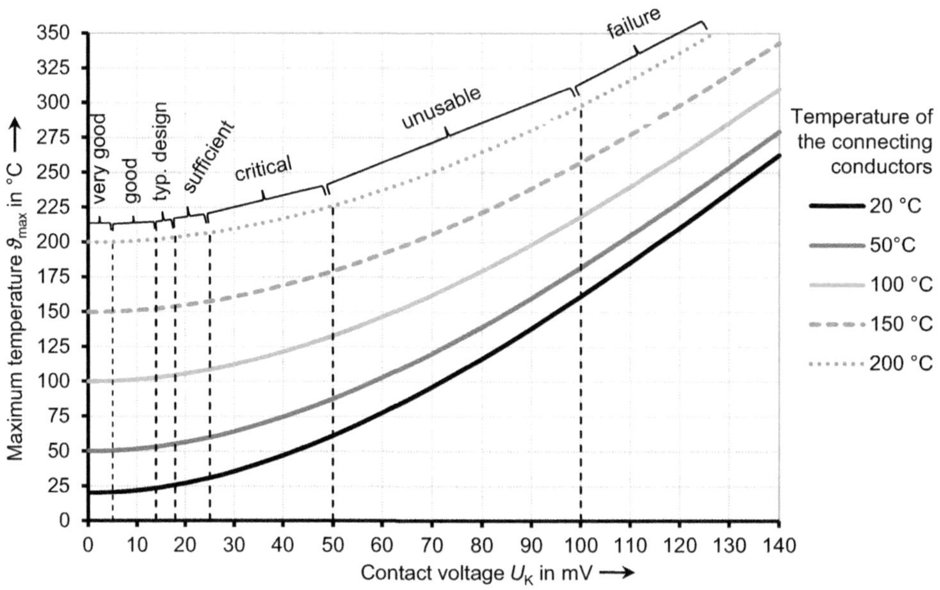

Fig. 4.7 Maximum temperature ϑ_{max} of the contact depending on the contact voltage U_K and the temperature of the connections ϑ_1

4.2 Mechanical Contact Behavior

A long-term stable current-carrying connection must also be a long-term stable mechanical connection. The mechanical function includes absorbing and transmitting operationally acting tensile, compressive and/or shear forces. In addition, the mechanical contact behavior is characterized by the size of the mechanically bearing contact area of the connection as well as the height and distribution of the mechanical stresses in the contact layer. When two or more conductors are joined to form a current-carrying connection, this is usually done by forming the contact partners or by additional joining elements or contact elements. During joining, the conductors are elastically or elasto-plastically deformed. Mechanically bearing contact areas are created, which absorb the mechanical stress during joining and subsequently the remaining mechanical stress. The mechanical stress depends on the elastic deformation and possibly on the hardening of the material during joining. Depending on how the contact force is applied and the geometry of the contact partners, a distribution of the remaining mechanical stress within the apparent contact area results. The maximum permissible operating forces are limited in force-fit connections by the acting normal force and the friction between the conductors and the joining and contact elements. If there is a form-fit component, the transferable forces are determined by the mechanical properties of the materials of the conductor or connector body and the friction between the contact partners. Numerous guidelines describe the

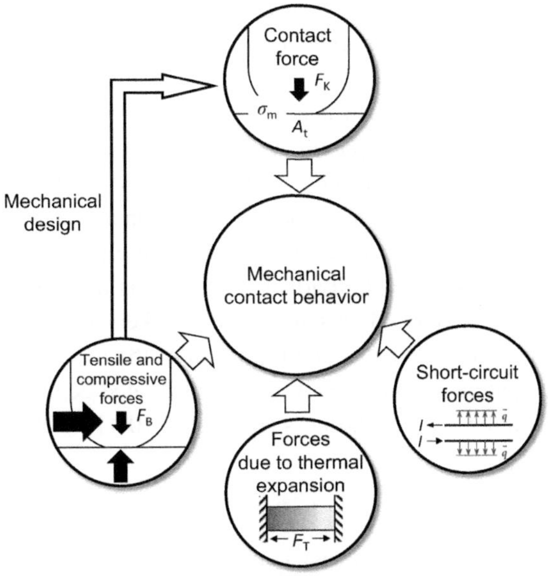

Fig. 4.8 Influencing factors on the mechanical design and the mechanical contact behavior considering operational forces

mechanical design of connections with special consideration of operating forces [26, 27]. The following presents some important relationships for the types of connections considered, which are also essential for the electrical and thermal contact behavior. To ensure the integrity of current-carrying connections and to achieve good mechanical and electrical contact behavior, a mechanical (compressive) stress in the normal direction must act on the contact area permanently. For this purpose, an external mechanical force is applied (Fig. 4.2), which is transmitted via a mechanically bearing contact area between the contact partners. In non-detachable and partly detachable connections such as screw or compression connections, the mechanical stress must be so great that operational forces do not cause relative movements between the contact partners (Fig. 4.8). In moving connections, such as plug-in connections, relative movements may occur. Nevertheless, the mechanical stresses in the normal direction to the contact area must be maintained. Depending on the type of connection, there are various ways to generate the mechanical force in the normal direction and to estimate a mechanically bearing equivalent area.

The most commonly used approach to approximately calculate the mechanically load-bearing contact area A_t is based on the plastic deformation of roughness peaks of the contact materials, depending on the contact force and the contact hardness H_K.[3]

$$A_t = \frac{F_K}{H_K} \tag{4.9}$$

[3]This approximately corresponds to the Brinell hardness.

This approach is not only valid for point contacts but can also be applied to plane contacts, where the contact force and the surface roughness are evenly distributed in the contact plane [20]. It is assumed that with a large apparent contact area, many small mechanically load-bearing contact points are formed, which are less deformed, and with a small apparent contact area, larger but fewer mechanically load-bearing contact points are formed, which are more strongly deformed. Up to a proportion of 10% of the mechanically load-bearing contact area to the apparent contact area, Eq. (4.9) can be applied. If the proportion is higher, more complex models must be used, as the deformation of individual contacts has a feedback effect on the material in the immediate vicinity. If this feedback is taken into account, with high plastic deformation of the microcontacts, the proportion of the mechanically load-bearing contact area A_t to the apparent contact area A_s can be estimated depending on the average mechanical stress σ_m and the contact hardness H_K (Eq. (4.10)) [28, 29].

$$\frac{A_t}{A_s} = \frac{\sigma_m/H_K}{1+\sigma_m/H_K} \tag{4.10}$$

When considering the material proportion curves (Fig. 4.9) of the contact partners in this context, it can be seen that the mechanically load-bearing area in current-carrying connections is formed at the highest elevations of the surface topology [36]. In practical connections, a uniform distribution of the contact force and/or a uniformly distributed

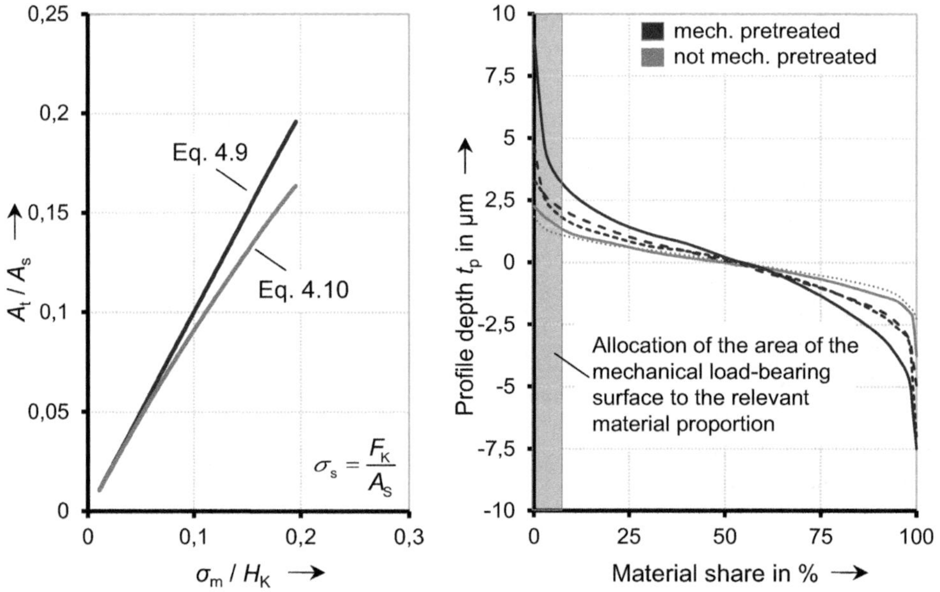

Fig. 4.9 Comparison of the approaches (Eqs. (4.9) & (4.10)) for calculating the mechanically load-bearing contact area and exemplary assignment of the proportion A_t/A_s calculated with Eq. (4.10) to measured height profiles of contact areas [36]

surface roughness is usually not present, so the microscopic mechanically load-bearing contact area is unknown.

Alternatively, depending on the type of connection, the mechanically load-bearing contact area can be estimated using substitute areas in order to be able to determine an average mechanical stress between the contact partners (Eq. (4.11)).

$$\sigma_{\mathrm{m}} = \frac{A_{\mathrm{Ers}}}{F_{\mathrm{K}}} \tag{4.11}$$

The relationships for screwed, compression, and plug-in connections will be considered subsequently. For further literature on this topic, the following books are referred to [5, 26, 27].

4.2.1 Screwed Connections

Screwed connections are force-fit or force-form-fit. The preload force F_{V} and the resulting clamping or contact force F_{K} is applied by the screw, which is elastically deformed during assembly. To optimally utilize the mechanical properties of screwed connections, the force-deformation processes in the connection must be examined more closely. When describing the assembly state, the flexibility of the screw is considered in the case of superimposed, clamped parts. In the case of a centrally, flexure-free clamped screwed connection, the pressure cone emanating from the screw head is symmetrical to the screw axis. This can fully develop or it is limited by the outer diameter D_{A} of the contact plane or the contact partners D'_{A} [26].

The pressure-stressed area widens from the screw head and the nut towards the contact level between the contact partners, when the contact area of the screw head d_{w} or the outer diameter of the flat contact area of a washer to the clamped part d_{wa} is smaller than the transverse dimension D_{A} of the screwed parts. In this case, the mechanical stress decreases in the radial direction. For simplification, in the mechanical design, the pressure cone towards the contact plane is considered as a truncated cone with a circular ring-shaped base. If the pressure cone can fully form (Fig. 4.10), a circular ring-shaped equivalent cross-sectional area A_{Ers} in the contact plane can be calculated taking into account the clamping length l_{k} and the hole diameter d_{h} with Eq. (4.12) . This applies to the condition that the equivalent outer diameter of the base body in the contact plane with D_{A} is larger than the sum of the outer diameter of the flat head contact area of the screw d_{w} and the clamping length l_{k} [26, 27].

$$\begin{aligned} &A_{\mathrm{Ers}} = \tfrac{\pi}{4}\left(d_{\mathrm{W}}^2 - d_h^2\right) + \tfrac{\pi}{8} d_{\mathrm{W}} l_{\mathrm{K}}\left[(x+1)^2 - 1\right] \text{ with } x = \sqrt[3]{\tfrac{l_{\mathrm{K}} d_{\mathrm{W}}}{(l_{\mathrm{K}}+d_{\mathrm{W}})^2}} \\ &\text{für } D_A \geq d_w + l_k \end{aligned} \tag{4.12}$$

If the pressure cone is not fully formed due to the component dimensions, for example, an additional cylindrical sleeve must be taken into account [26]. If the diameter of the

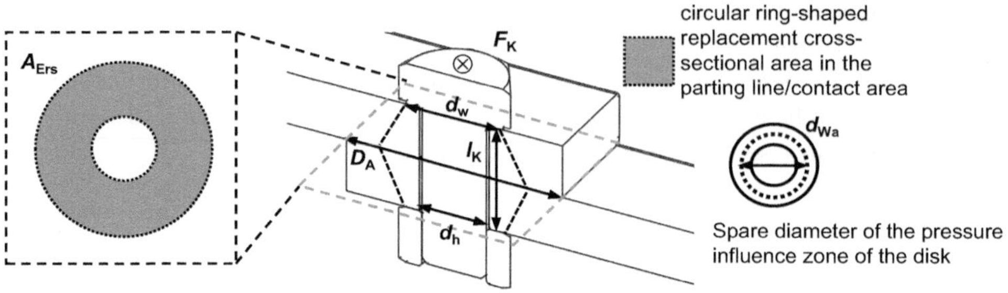

Fig. 4.10 Pressure cone of a centrally clamped through-screwed connection [26]

flat head contact area of the screw is smaller than the equivalent outer diameter of the base body in the contact plane, the following applies:

$$A_{\text{Ers}} = \frac{\pi}{4}\left(D_{\text{A}}^2 - d_h^2\right) \quad \text{für } D_A < d_w \tag{4.13}$$

If the outer diameter D_{A} of the contact plane is larger than the flat head contact area of the screw d_{w} and smaller than the sum of the outer diameter of the flat head contact area of the screw d_{w} and the clamping length l_{k}, the following equation should be used:

$$A_{\text{Ers}} = \frac{\pi}{4}\left(d_{\text{W}}^2 - d_{\text{h}}^2\right) + \frac{\pi}{8} d_{\text{W}}(D_{\text{A}} - d_{\text{W}})\left[(x+1)^2 - 1\right] \text{ with } x = \sqrt{\frac{l_{\text{K}} d_{\text{W}}}{(D_{\text{A}})^2}}$$
$$\text{für } d_w \leq D_A \leq d_w + l_k \tag{4.14}$$

In these area considerations, it is assumed that, unlike in reality, the mechanical stress in the contact plane is evenly distributed and only an elastic deformation of the contact partners occurs. The force ratios in the connection are considered with the equivalent cross-sectional area A_{Ers} [2].

Example

Equivalent cross-sectional area A_{ers} of a through-screwed connection with an M12 screw whose head support diameter d_{W} is 16.63 mm, two washers of the normal series of size 12 with a height of $h_{\text{s}} = 2.5$ mm, two busbars of dimensions (40 × 10) mm with a connection length l_{V} of 40 mm and a hole diameter d_{h} of 13 mm.

$$d_{\text{Wa}} = 16.63\text{ mm} + 1.6 \cdot 2.5\text{ mm} = 20.6\text{ mm}$$

$$A_{\text{Ers}} = \frac{\pi}{4}\left(20.6\text{ mm}^2 - 13\text{ mm}^2\right) + \frac{\pi}{8} 20.6\text{ mm} \cdot 25\text{ mm}\left[\left(\sqrt[3]{\frac{25\text{ mm}\cdot 20.6\text{ mm}}{(25\text{ mm}+20.6\text{ mm})^2}} + 1\right)^2 - 1\right]$$

$$A_{\text{Ers}} = 536\text{ mm}^2$$

◀

In order to approximately calculate the surface pressure at the contact area A_{P} of the screw head and the nut to the contact partner, when using standard washers with a thickness h_{s}, a equivalent diameter d_{Wa} is calculated instead of d_{w}, which takes into account the pressure influence zone between the washer and the contact partner [27].

$$d_{\mathrm{Wa}} = d_{\mathrm{W}} + 1.6 \cdot h_{\mathrm{S}} \tag{4.15}$$

Depending on the requirement, the screwed connection is designed in such a way that the calculated minimum clamping force achieves a sealing function against gases and liquids, shear forces or torques can be transmitted by static friction in the contact plane, and a gaping of the contact plane is prevented [26, 27].

The assembly of current-carrying screwed connections in electrical energy technology is usually carried out torque-controlled. For this purpose, information about the value of the tightening torque M_{A} is given in the assembly instructions. This results from the torque under the screw head M_{K} and the thread moment M_{G}.

$$\begin{aligned} &M_{\mathrm{A}} = M_{\mathrm{K}} + M_{\mathrm{G}} \text{ with } M_{\mathrm{K}} = F_{\mathrm{V}} \cdot \tfrac{D_{\mathrm{Km}}}{2} \cdot \mu_{\mathrm{K}} \text{ and} \\ &M_{\mathrm{G}} = F_{\mathrm{V}}[0.159 \cdot P + 0.58 \cdot d_2 \cdot \mu_{\mathrm{G}}] \\ &M_{\mathrm{A}} = F_{\mathrm{V}}\left[0.159 \cdot P + 0.58 \cdot d_2 \cdot \mu_{\mathrm{G}} + \tfrac{D_{\mathrm{Km}}}{2} \cdot \mu_{\mathrm{K}}\right] \end{aligned} \tag{4.16}$$

The relationship between preload force F_{V} and tightening torque can be calculated taking into account the pitch diameter of the screw d_2, the thread pitch P and the effective diameter for the friction moment in the screw head and nut support D_{Km}.

It should be noted that the minimum friction coefficient in the thread μ_{Gmin} and at the screw head or nut support μ_{Kmin} have a tremendous influence on the calculated tightening torque and depend on the rotational speed during assembly. Precise knowledge of the friction coefficient between the joining elements is therefore absolutely necessary in order to achieve sufficient preload and thus contact forces (Tables 4.4 and 4.5). Detailed explanations can be found in [2]. With repeated tightening, the friction coefficient increases due to increasing adhesive-abrasive wear processes in the contact planes up to seizing or decreases due to leveling processes [26]. This must be taken into account in a reassembly.

Table 4.4 Coefficient of friction μ_{G} for selected material pairings and lubrication conditions [27, 31]

Material pairing		Coefficient of friction μ_{G}
Material 1	Material 2	
Galvanized steel (Zn6) dry	Galvanized steel (Zn6) dry	0.12 … 0.2
Galvanized steel (Zn6) oiled	Galvanized steel (Zn6) oiled	0.10 … 0.18
AlMg bare dry	Galvanized steel or phosphated	0.08 … 0.20
Stainless steel A2 Special lubricant (Chlorinated paraffin base)	Stainless steel A2 Special lubricant (Chlorinated paraffin base)	0.12 … 0.23
Stainless steel A2	AlMgSi dry	0.32 … 0.43
Stainless steel A2	AlMgSi Special lubricant (Chlorinated paraffin base)	0.28 … 0.35

Table 4.5 Coefficient of friction μ_K for selected material pairings and lubrication conditions [27, 31, 30, 33]

Material pairing		Coefficient of friction μ_K
Material 1	Material 2	
Stainless steel A2 dry[1)]	Stainless steel A2 dry[1)]	0.35 … 0.5
Stainless steel A2 with corrosion protection grease[1)]	Stainless steel A2 with corrosion protection grease[1)]	0.25 … 0.35
Stainless steel A2 with special lubricant (chloroparaffin-based)[1]	A2 with special lubricant (chloro-paraffin-based)[1)]	0.08 … 0.12
Stainless steel A2	AlMgSi with special lubricant (chloroparaffin-based)[1)]	0.08 … 0.11
ZnLa[2)]	Cu-ETP	0.12 … 0.16
	galv. Ag	0.12 … 0.15
	NiP (Type 2)	0.12 … 0.19
	galv. Sn (matte)	0.16 … 0.28
	galv. Sn (glossy)	0.13

[1)] Large tightening angles approx. 1000°
[2)] Screw head support external carrier [32, 33]

Example

Screwed connection with a M12 screw and nut made of stainless steel A2 (lubricated) and AlMgSi busbars (lubricated)—desired preload force $F_V = 25$ kN, $\mu_G = 0.12 \ldots 0.23$ (Table 4.4) $\mu_K = 0.08 \ldots 0.11$ (Table 4.5), one flank diameter of the screw d_2 of 10.86 mm, one of the thread pitch P of 1.75 and a screw head and nut support D_{Km} of 14.82 mm.

$$M_{A,\min} = 25000\ \text{N}\left[0.159 \cdot 1.75 + 0.58 \cdot 10.86\ \text{mm} \cdot 0.12 + \frac{14.82\ \text{mm}}{2} \cdot 0.08\right] = 41\ \text{Nm}$$

$$M_{A,\max} = 25000\ \text{N}\left[0.159 \cdot 1.75 + 0.58 \cdot 10.86\ \text{mm} \cdot 0.23 + \frac{14.82\ \text{mm}}{2} \cdot 0.11\right] = 64\ \text{Nm}$$ ◄

In addition to the screw and nut, washers are often used, which generate a defined coefficient of friction between the screw head and washer or nut and are intended to prevent loosening and unscrewing. Various designs of resilient and non-resilient joining elements are available to prevent loosening due to settling and relaxation. A spring effect with tensioned resilient joining elements is only guaranteed if, according to [26]

- the spring force is approximately as high as the screw preload force,
- the surface pressures in the contact areas can be calculated,

- the required preload force when tightening is not affected,
- crevice corrosion is avoided,
- during loading and unloading, there must not be different preload forces (not even with multiple applications).

Spring washers (DIN 137:1970-12, DIN 6904:1990-12), spring rings (DIN 127:1987-10, DIN 128:1968-04 and DIN 6905:1990-12), toothed washers (DIN 6797:1988-07) and fan washers (DIN 6798: 1988-07) do not meet the mentioned requirements, especially the first one, so they should not be used in the design and rightly the standards for them were withdrawn. In tests on electrical contact behavior, tension washers according to DIN 6796:2009-08 have proven effective, taking into account the technical delivery obligations according to DIN 267-26:2005-12 [26]. Measures against self-loosening due to external unscrewing moments, relative movements between the contact areas or breathing of the thread of the nut due to a load in the axial direction are shown as examples and can be read in [26].

- Screw preload force should be chosen sufficiently high
- As large as possible elastic flexibility of the screw
- Use of fit screws to limit shifting by positive locking
- A high head friction
- The use of suitable securing measures such as screws and nuts with locking toothed contact areas

Which measures need to be taken depends on the operational requirements. In long-term tests on screwed connections with busbars made of copper and aluminum, the effectiveness of tension washers could be demonstrated (Fig. 4.11). In addition, other joining elements were comparatively examined. It is clearly visible that the force reduction is strongest with non-resilient washers.

The use of pressure plates leads to a more uniform distribution of mechanical stress and a larger free elongation length of the screw, which results in less force reduction after the same time than when using washers. Similarly stacked disc springs did not lead to any improvement due to the settling amounts in the numerous separating plane and the high coefficient of friction between the individual elements despite a larger elongation length of the screw (Fig. 4.11) [8]. When using aluminum as a conductor material, it is recommended to additionally use washers of the large series to achieve sliding of the tension washers when tensioning and to prevent digging into the softer aluminum, so that the spring effect is maintained [8, 9, 34, 35].

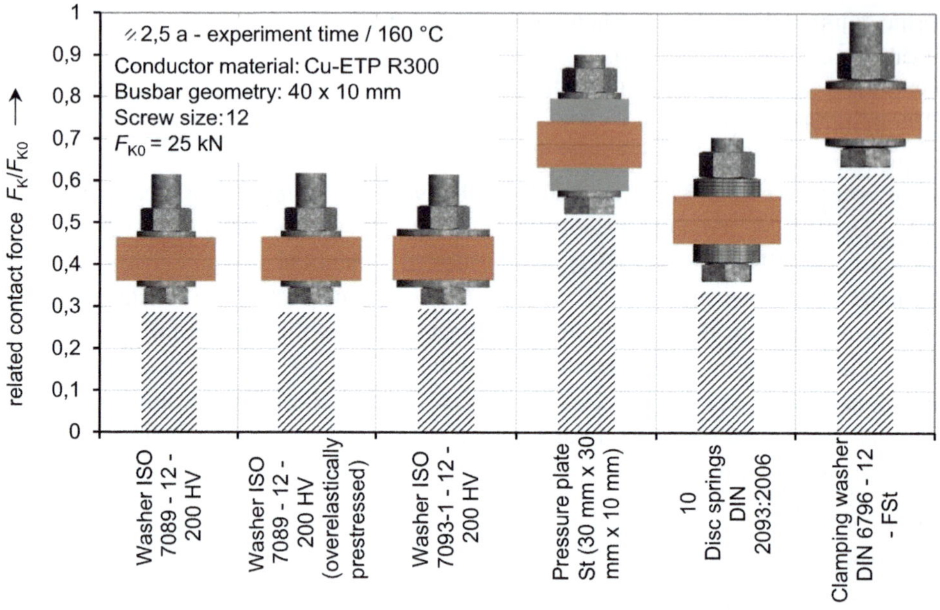

Fig. 4.11 Relative contact force depending on the joining elements used in screw-nut connections with busbars made of Cu-ETP R300 after a test period of 2.5 years at an operating temperature of 160 °C [8]

4.2.2 Compression Connections

A compression connection is made by the form-fitting and force-fitting joining of at least two conductors or a conductor with another connecting element (e.g., compression sleeve, -cable lug or -fitting). In all known methods for joining current-carrying compression connections, a defined shape change of the contact partners is realized. In the process, very high degrees of deformation are achieved in the fitting and in the conductor. The desired shape change always leads to a certain state of strain in the overall system, so that the mechanically load bearing contact area and the force-fit component are resulting quantities and thus, for example, depend on the mechanical properties of the materials used. Compression connections are either joined with mechanical or thermal energy. The methods and the connecting element used for this are adapted to the type, structure, dimensions and material of the respective conductor (Table 4.6).

Individual geometric parameters are used to design the current-carrying connection. The currently known methods for producing compression connections differ in terms of the conductor used, the degree of deformation, and the type of energy input that causes the deformation. Circumferential, indentation compression, and crimp connections for multi-strand conductors, as well as longitudinally compression connections for conductor tubes, are joined by causing elastic and/or plastic deformation through an external

Table 4.6 Classification of methods for making current-carrying compression connections[a]

Type of conductor	*Technology*	*Procedure (examples)*
Solid conductor according to IEC 60228:2005	Circumferential compression	Hexagonal pressing
		Oval pressing
		Square pressing
Multi-/fine and finest stranded conductors according to IEC 60229:2009		Four-prong pressing
		Round pressing
	Indentation compression	Mandrel pressing
		Stamp pressing
		C8 pressing
Conductor tubes (not standardized)	Longitudinal pressing (joining by pressing in)	
	Transverse pressing (joining by stretching)	
Transmission line conductors according to DIN EN 50182:2001	Circumferential compression connection	Hexagonal pressing
	Indentation compression	Notching

[a]The classification of current-carrying compression connections differs from that of the DIN 8593-0:2003 standard series, which considers the underlying joining methods

force. Cross-pressed connections for conductor tubes are joined by changing the thermal expansion of one or both contact partners [37]. The mechanical properties of the material and the possible temperature range are limiting the maximum expansion of the contact partners in this process. The desired change in expansion can be qualitatively characterized for all methods using individual geometric parameters. Therefore, the resulting force effect in a compression connection is a size dependent on the expansion and the properties of the conductor materials. The basic functioning of a compression connection is explained using the example of the hexagonal compression of a overhead line conductor with a compression fitting (Fig. 4.12) [37–39].

The resulting change in the strain of the fitting and the conductor during joining, and thus also the force applied to the conductor, is determined by the unidirectional movement of the compression dies and its geometry. During the closing of the compression dies, both contact partners are continuously deformed elastically and plastically (Fig. 4.12—Phase I).

The inner surface of the fitting and the conductor come into contact with each other and the apparent contact area forms. As soon as the press inserts lie on top of each other, the formation of the contact areas in the connection is completed (Fig. 4.12—Phase II). The deformation of the contact partners is finished and the form fit between conductor and fitting is established. Accordingly, with closed compression inserts, the mechanically

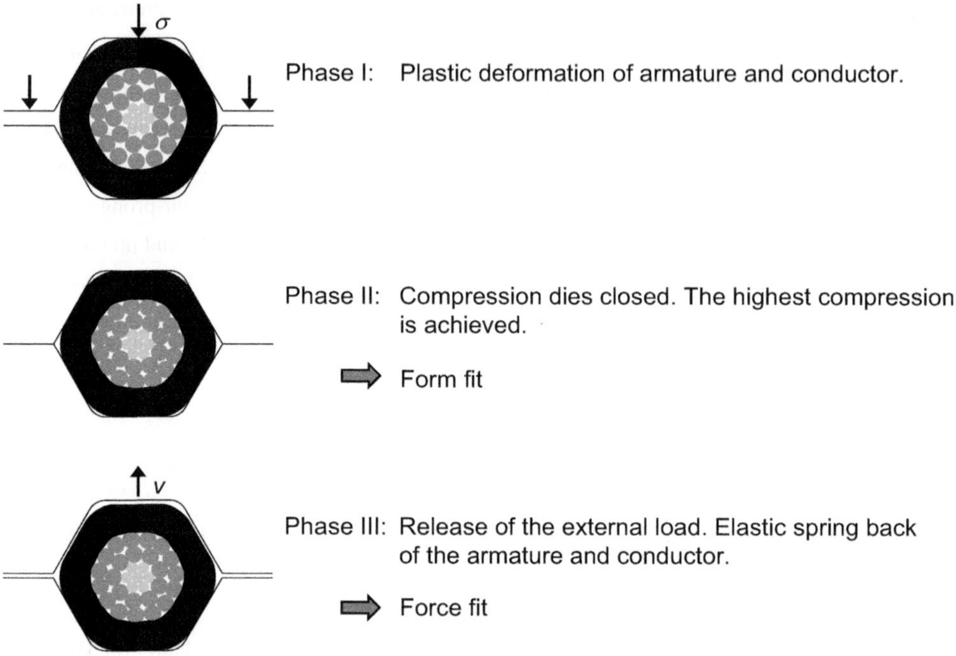

Fig. 4.12 Phases of the compression process using the example of a hexagonal compression [11]

bearing contact areas between the contact partners of a compression connection are the largest. The compression dies then open and the external load is immediately removed (Fig. 4.12—Phase III).

The conductor and the sleeve are relieved and spring back elastically. If the sleeve limits the elastic rebound of the conductor, a mechanical tension or a contact force remains between the two contact partners [37]. The remaining mechanical tension in the contact areas is does not solely depend on the mechanical properties of the materials used. The equivalent area A_{Ers}, in which mechanical compressive stresses act on the microcontacts between the contact partners, results after the joining of the compression connections and cannot be calculated analytically (Fig. 4.13). In a numerical calculation,

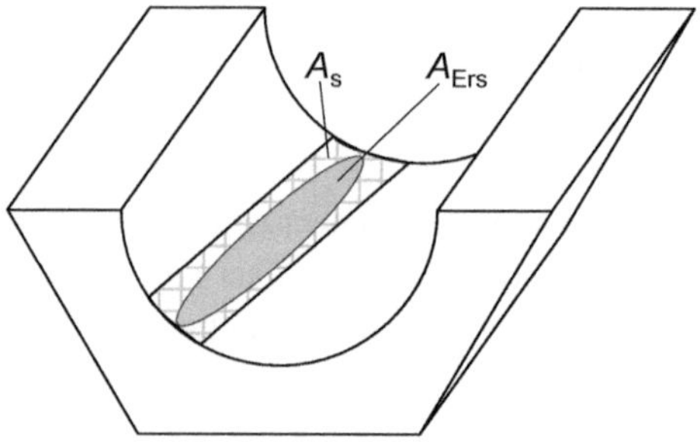

Fig. 4.13 Pressure-loaded equivalent cross-sectional area between a trapezoidal wire and the press fitting in a hexagonal compression (schematic)

the area can be estimated with great effort. A simplified engineering approach is therefore not given for this type of connection.

Example

A fine-stranded 95-mm^2 copper conductor consists of approximately 484 wires with a diameter of 0.5 mm. The hexagonal crimp connection of the conductor between a tubular cable lug ($d_a = 17.5$ mm, $d_i = 13.5$ mm) and a crimp cable lug ($d_a = 19.0$ mm, $d_i = 13.5$ mm) is made with hexagonal compression dies with the key numbers[4] (KZ) 17 ($A_3 = 177$ mm^2) and 18 ($A_3 = 210$ mm^2).

Crimp Inserts	*Tubular Cable Lug (IEC 1238:1993)*	*Crimp Cable Lug (DIN 46235:1983)*
KZ 17	$K_F = \frac{95\,\text{mm}^2 + \frac{\pi}{4}\left[(17.5\,\text{mm})^2 - (13.5\,\text{mm})^2\right]}{177\,\text{mm}^2} = 1.09$	$K_F = \frac{95\,\text{mm}^2 + \frac{\pi}{4}\left[(19.0\,\text{mm})^2 - (13.5\,\text{mm})^2\right]}{177\,\text{mm}^2} = 1.33$
KZ 18	$K_F = \frac{95\,\text{mm}^2 + \frac{\pi}{4}\left[(17.5\,\text{mm})^2 - (13.5\,\text{mm})^2\right]}{210\,\text{mm}^2} = 0.92$	$K_F = \frac{95\,\text{mm}^2 + \frac{\pi}{4}\left[(19.0\,\text{mm})^2 - (13.5\,\text{mm})^2\right]}{210\,\text{mm}^2} = 1.12$
Evaluation	With KZ 18, little to no compression of the conductor → low contact forces between the subconductors	With KZ 17, too much material in the connection point → crimp inserts do not close completely

◀

Empirical approaches are used to characterize the contact and long-term behavior of compression connections with geometric parameters specific to the respective type of connection (Table 4.7). These parameters describe the forming process on the one hand, by considering the change in the cross-sectional areas of the conductor through joining (degree of compression) [40, 41] and the change in the cross-sectional areas by the respective tool (compression factor). Other parameters use the dimensions of the contact partners in the initial state (oversize, fit factor). For hexagonal compression connections as well as longitudinal and transverse compression connections, relationships between the respective geometric parameters and the mechanical contact behavior have been established [11, 36]. A higher elastic strain or a higher degree of forming in the contact partner of a hexagonal compression connection, empirically described by higher compression factors, leads to higher applied and remaining contact forces (Fig. 4.14). In contrast to the other types of connections, the contact force in compression connections can significantly decrease immediately after joining. Also in crimp connections, relationships

[4] Standardized crimp inserts exist for hexagonal crimping, whose dimensions are classified by key numbers. These are specified in DIN 48083-4:1985.

Table 4.7 Geometric parameters of various compression connections (press connections)

Compression connection technology/ process	**Geometric parameter**		**Sketches / Explanation of geometric parameters**	**Application example**
Crimp connections	Degree of compression [41], [42]	$V = \frac{A_{L0}}{A_{LV}}$	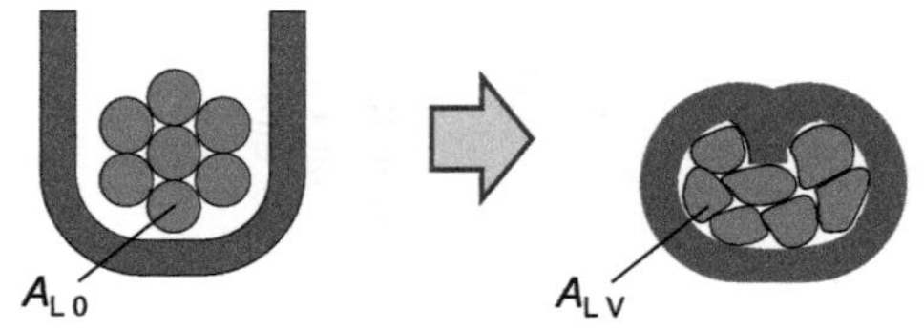	$V \approx 1{,}25$ [43] For crimp connections with bare copper wires; lower values permissible if wires are tinned
Hexagonal compression connections	Compression factor Fit factor	$K_F = \frac{A_L + A_H}{A_3 - A_K}$ $p = \frac{d_L}{d_i}$	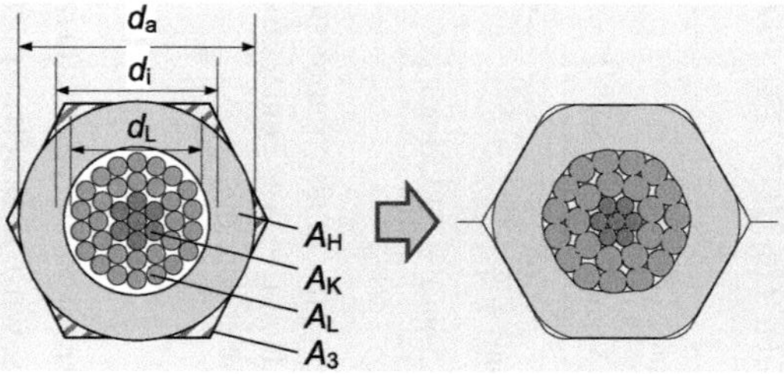	$K_F = 1{,}04...1{,}16$ $p > 0{,}9$ For hexagonal compression joints with conventional and HTLS transmission line conductors
Longitudinal presses/ transverse compression connections	Oversize	$ü = d_{Ia} - d_{Ri}$	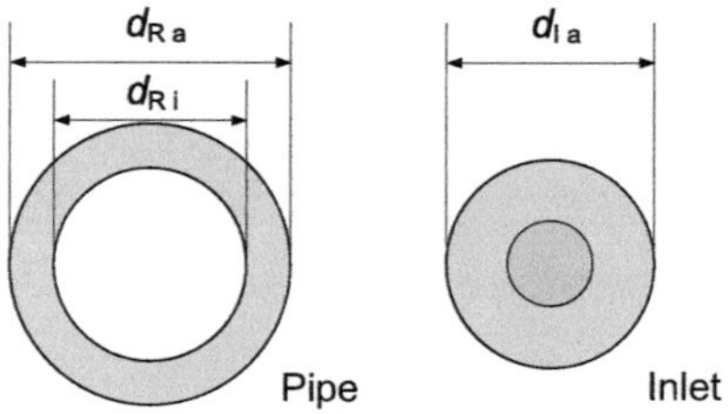	$ü = (0{,}07...0{,}17)$mm Longitudinally pressed connections $ü = (0{,}17...1{,}0)$mm Cross-pressed connections

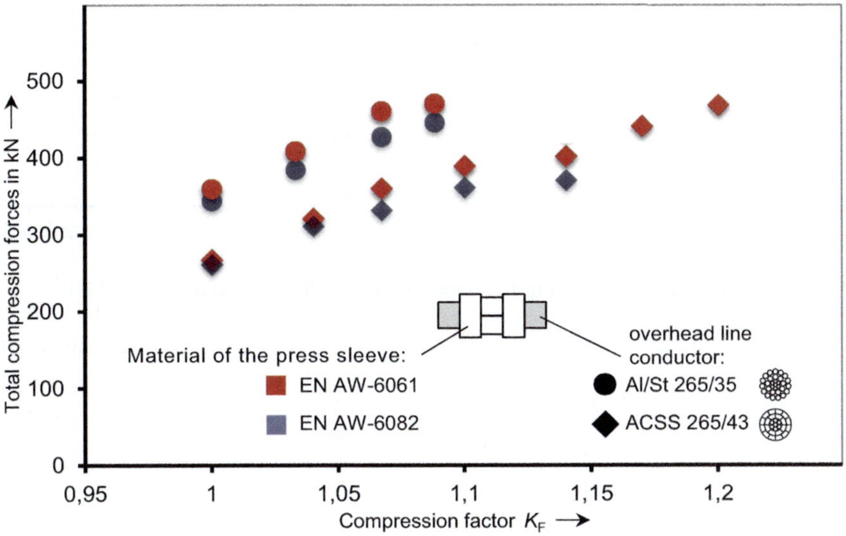

Fig. 4.14 Sum of the highest forces during joining perpendicular to the contact areas of a single hexagonal compression with two conductor wires depending on the compression factor [11]

between the geometric characteristic and the contact or long-term behavior have been established in studies [42]. If the contact partners are formed too much, the mechanical strength of the connection can be reduced under tensile load [1].

4.2.3 Plug-in connections

In plug-in connections, spring elements or slotted, spring-loaded sockets are used, which electrically connect two conductors, usually referred to as plug and socket. During the plugging process, the spring elements or the slotted socket are compressed. This requires a plugging force F_S (Fig. 4.15).

After plugging in, a sliding process begins. The sliding path is predetermined by the construction of the plug, socket, and insulating housing. If the contact is moved from its resting position in the closed state, a friction process occurs. The friction process of a plug-in connection (Fig. 4.16) is characterized by the two phases of static friction and sliding friction. First, to generate a relative movement, the force due to static friction $F_{R\,H}$ must be overcome, which depends on the coefficient of static friction μ_H and the contact force F_K:

$$F_{R\,H} = \mu_H F_K \tag{4.17}$$

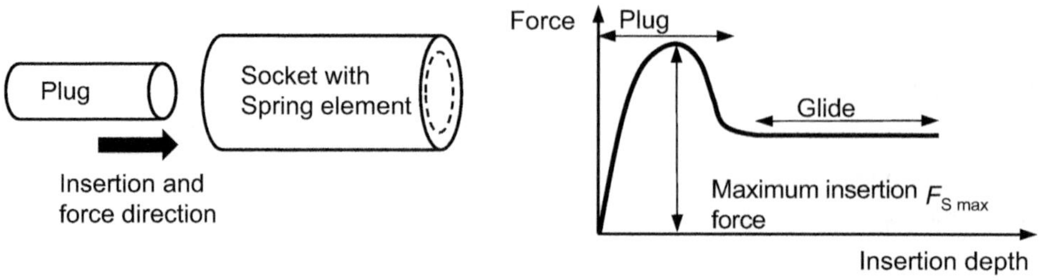

Fig. 4.15 Plugging force F_S of a plug-in connection depending on the insertion depth

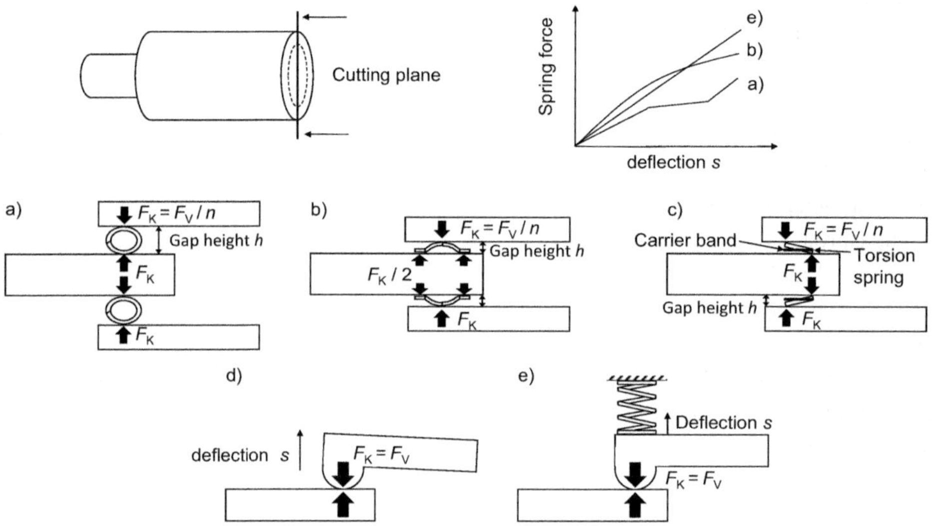

Fig. 4.16 Options for applying a contact force in plug-in connections: **a**) radially stressed helical spring, **b**) contact element according to the laminated spring principle (elliptical or parabolic spring) **c**) contact element according to the torsion spring principle with a carrier band, **d**) Slotted socket with finger contact, **e**) Finger contact with optional compression spring for applying a contact force and associated spring characteristic curve (top right)

he force due to sliding friction $F_{R\ GL}$, which is characterized by the coefficient of kinetic friction μ_{GL} and the contact force F_K ,is:

$$F_{R\ GL} = \mu_{GL} F_K \tag{4.18}$$

In practical plug-in connections, the movement is not pure sliding friction due to the spring elements and the energy stored in them, as well as the inhomogeneous surface structure, but is characterized by discontinuities in the course. In the mentioned simplifications, the elastic behavior of contact points is not depicted (see Sect. 5.1.4). As a result, the metallic contact does not break up during movements in the μm range.

Table 4.8 Values of the sliding friction number of various coatings for connectors

Material pairing	Sliding friction number μ_{GL}	Source
Ag-Ag (galvanic)	0.7 … 0.9 0.5 … 0.8	[87] [1]
Ag-Ag (galvanic) with lubricating grease	0.13 … 0.16 at 25 °C 0.19 … 0.27 at 55 °C 0.35 with "light lubrication"	[86] [73]
Dispersion layers Ag-C against Ag Ag-WS_2 against Ag Ag-MoS_2 against Ag Ag-hBN against Ag Ag-SnS against Ag	0.1 … 0.14 0.2 … 0.33 0.28 … 0.38 0.64 … 0.9 0.41 … 0.73	[88]
Sn-Sn	0.5 … 1.0	[89]

The sliding friction coefficients μ_G of the coating materials for connectors are given in Table 4.8. Spring elements are characterized by a spring characteristic curve. In the extended state, the contact elements have a height h_0. By compression of the spring travel s to a gap height h, a contact force F_K is generated (Fig. 4.16).

$$h = h_0 - s \tag{4.19}$$

For simple geometries, the spring force can be determined analytically based on the spring constant, e.g., for compression springs. For more complex geometries, the spring characteristic curve can be calculated with good approximation using numerical models (Sect. 4.5.3). Depending on the number of turns, louvers or contact fingers n of the spring element, a contact force F_K results for each contact point. To prevent them from shifting during the plugging process, contact elements are placed in a groove or are otherwise secured against displacement.

The contact area of an elastically deformed point contact is calculated based on the Poisson´s ratio v, the Young's modulus E of the material, and the convexity radius of the contact r_B (Eq. (4.21)). For plastically deformed point contacts, the size of the contact area can also be analytically estimated (see also Eq. (4.9)), if the contact hardness H_K is known [23]:

$$\text{Elastic deformation } A_{Ers} = \pi \left(1.5 \cdot (1 - \nu) \frac{F_K \cdot r_B}{E} \right)^{2/3} \tag{4.20}$$

$$\text{Plastic Deformation } A_{Ers} = \frac{F_K}{H_K} \tag{4.21}$$

Table 4.9 Contact hardness H_K of different materials at room temperature

Material	Ag (pure)	Sn (pure)	Cu (pure)	Ni (pure)
Vickers hardness HV	30 … 105	8	40 … 110	250 … 700
Contact hardness H_K in N/mm^2	300 … 750	45 … 60 [24]	470 … 850 [48]	700 … 2000 [24]

The contact hardness of the conductor and the coating materials depends primarily on the local microstructure in the contact area, which in turn depends on the degree of deformation and hardening. Influencing factors include:

- Manufacturing of the conductor or coating: e.g., casting or extrusion, galvanic or chemical coating, heat treatment(s), degree of deformation
- Influence of pretreatment: abrasive pretreatment with wire brush or abrasive fleece, laser texturing, machining post-treatment
- Joining process: contact force, friction path, passivation, lubricant

During the manufacturing of a contact, the contact hardness increases with increasing surface pressure σ_{Sm} due to the increasing cold hardening [48]. Depending on these influencing factors, the hardness at the contact point can deviate from the macroscopic hardness of the unaffected conductor material (Chap. 3). Therefore, only a range of the contact hardness of different materials can be specified (Table 4.9).

In addition, the contact hardness is nonlinearly dependent on temperature and decreases significantly above the softening temperature. The numerous influences on contact hardness, in addition to surface roughness, are a major reason why the determined contact resistances in film-free contacts significantly scatter under seemingly identical boundary conditions (see Sect. 4.3). In point contacts, a typical contact force of (2 … 10) N, which is typical for high-current plug connections, results in a size of the mechanically supporting contact area of

- Silver: $A_{Ers} = (0.003 \ldots 0.03)\ \text{mm}^2$
- Tin: $A_{Ers} = (0.03 \ldots 0.22)\ \text{mm}^2$
- Nickel: $A_{Ers} = (0.001 \ldots 0.01)\ \text{mm}^2$

According to the contact hardness, tin forms the largest and nickel the smallest mechanically load-bearing contact areas (see Sect. 4.5.3).

4.2.4 Current-Induced Forces during Operation

In general, operational forces can arise e.g. due to thermal expansion of conductor materials or housing components, due to electromagnetic forces in the event of a short circuit, the static weight force of connected devices, vibrations as well as ice and wind loads. These mechanical loads must not lead to the failure of the connection within the planned lifetime and must be absorbed by the friction or form-fit between the contact partners. Current-induced forces during operation arise from the thermal expansion of conductors as a result of the operation-related heating due to current heat losses and can act directly on connections. In the case of plug-in and sliding connections, they lead to slowly occurring friction cycles. In connections with solid conductors, thermal expansions lead to transverse and longitudinal forces in the connection. Furthermore, electromagnetic forces can exert a very high force on the connections, especially during a short-circuit current. Also, blow-off forces, due to current deflection at the contact points, lead to a reduction of the contact force and must be taken into account in the design. The following explains the cause and effect on current-carrying connections using practical examples. This chapter only considers the operationally induced and current-induced forces that have a direct effect on the current-carrying connection. More detailed information and findings are published in additional publications [5, 25].

4.2.4.1 Electromagnetic Forces

Forces act on charges in a changing magnetic field, which are referred to as the electromagnetic force F_{em}. These depend on the current I, the magnetic field B, and the conductor element dl. This force is calculated via the effect of a changing magnetic field on a charge, the so-called Lorentz force (Eq. (4.22), Fig. 4.17).

$$\vec{F}_{em} = I \int d\vec{l} \times \vec{B} \xrightarrow{\text{straight conductor}} \vec{F}_{em} = I\left(\vec{l} \times \vec{B}\right) \tag{4.22}$$

The magnetic field B is calculated according to the Biot-Savart law. The line load on two infinitely long, parallel, current-carrying conductors is obtained according to Eq. (4.23). The direction of action of this line load depends on the direction of current flow. Conductors carrying current in the same direction attract each other, while conductors carrying current in opposite directions repel each other (Fig. 4.18). An electromagnetic force

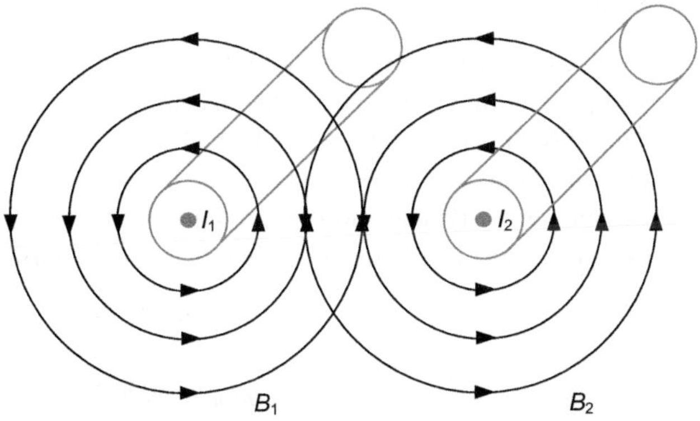

Fig. 4.17 Magnetic flux density around linear conductors

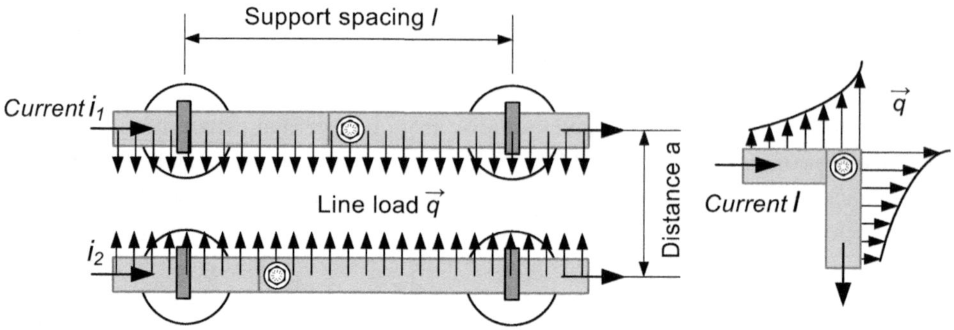

Fig. 4.18 Force effect on parallel, infinitely long conductors, which are traversed in the same direction by a current I (left) and at corners (right)

also acts on corners. This force causes the corner to be driven apart (Eq. (4.24), assumption of line-like conductors).

$$q_{\text{parallel}}(t) = \mu_0\mu_r \frac{i_1(t)i_2(t)}{2\pi a} \tag{4.23}$$

$$q_{\text{Ecke}}(t) = \mu_0\mu_r \frac{i(t)^2}{4\pi x} \tag{4.24}$$

With time-varying currents, the resulting line load is also time-varying (Fig. 4.19). For a sinusoidal load current, the line load consists of a static, time-independent component and a dynamic component that oscillates at twice the network frequency.

With rated continuous current, the resulting line loads are usually very low and do not lead to any significant stress on the structure. In the case of short-circuit currents, which are usually an order of magnitude higher than the rated continuous current, the acting line load is therefore two orders of magnitude higher and leads to considerable mechanical stress on the conductors and connections, which must be designed for this.

Example

Two busbars with a cross-section A of 40 mm × 10 mm and a distance a of 0.1 m are laid parallel in a switchgear. The busbars have a rated continuous current I_r of 850 A. This current results in a static line load of $q_{stat} = 1.4$ N/m. On the conductors, the support insulators and the substructure. This is practically negligible. The switchgear is designed for a typical short-circuit current of 20 kA, at which a peak current of up to 50 kA can act. The static and dynamic line load during the short circuit is then

$$q_{\text{stat}} = 800 \text{ N/m}$$

$$q_{\text{dyn}} = 5000 \text{ N/m}.$$

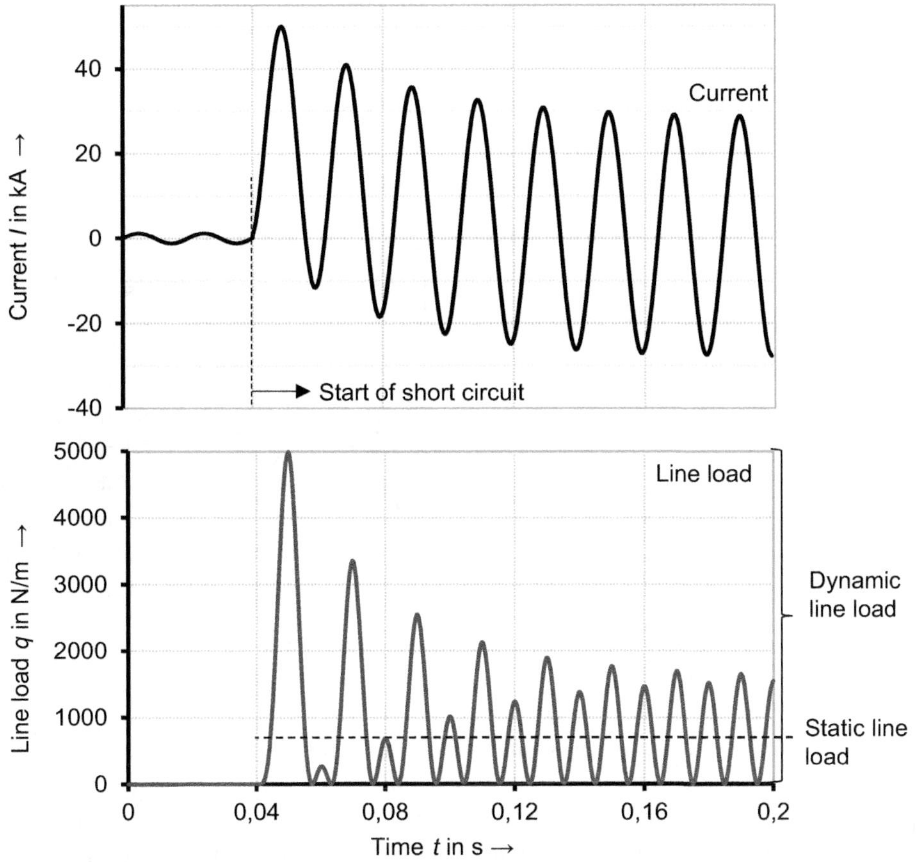

Fig. 4.19 Progression of the line load q after the onset of the short-time current $I_K = 20$ kA with a peak current of $i_p = 50$ kA for parallel conductors at a distance of $a = 0.1$ m

This line load must be absorbed by the support insulators and the substructure without exceeding the breaking force or causing the conductor to deform plastically. This force also acts on connections that lie in the current path as shear or tensile force. ◀

The electromagnetic force F_{em} on the conductor causes a deflection from the resting position (Fig. 4.20). To determine the deflection of the conductor, the deflection must be calculated from the line load, the geometry, and the material used. Depending on the cross-section and position of the conductors to each other (horizontal, vertical), their section modulus against bending differs. Whether the conductor arrangement follows the dynamic line load depends on its mechanical natural frequency. If this is significantly greater than twice the network frequency, the movement follows the dynamic line load. The construction must be designed accordingly to be resistant. If the natural frequency is much smaller than twice the network frequency, the system is too sluggish to follow

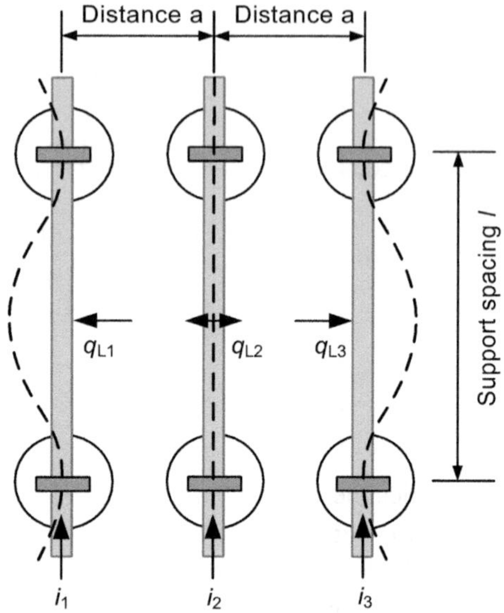

Fig. 4.20 Deflection of the conductor in a horizontally arranged, three-phase busbar system (static deflection, exaggerated representation)

the dynamic component. In this case, the construction must be resistant to the lower static line load. A detailed description of the mechanical design of busbar systems can be found in [5].

In three-phase systems, this mechanical deflection can unfavorably overlap with the current distribution due to the proximity effect. As a result, in connections, contacts that are already heavily mechanically stressed may additionally have to carry a comparatively high proportion of the total load current. This is particularly critical in the short-term loading of plug-in connections, for example in switchgear applications.

4.2.4.2 Blow-off Forces

In addition to the electromagnetic forces F_{em}, which depend on the geometry of the surrounding conductors, a blow-off force F_A acts at contact points due to the constriction of the current lines. This can be calculated analytically for point contacts depending on the radius of the constriction r_i and the radius of the conductor R [2]:

$$F_{\mathrm{A}} = \frac{\mu_0 \hat{i}^2}{4\pi} \ln \frac{R}{r_{\mathrm{i}}} = 10^{-7} \hat{i}^2 \ln \frac{R}{r_{\mathrm{i}}} \propto k\,\hat{i}^2 \tag{4.25}$$

In general, the resulting blow-off force depends on the square of the current and a geometry-dependent constant k. Numerous experimental tests have determined a range of values for the constant k of $k = (4 \ldots 5)\ 10^{-7}$ N/mm² [15, 49]. The resulting contact force $F_{K\,res}$ for short-term currents results from the sum of the contact force F_K, the blow-off force F_A and the electromagnetic force F_{em} (Fig. 4.21 and Eq. (4.26)).

$$F_{\mathrm{K,res}} = F_{\mathrm{K}} - F_{\mathrm{A}} \pm F_{\mathrm{em}} \tag{4.26}$$

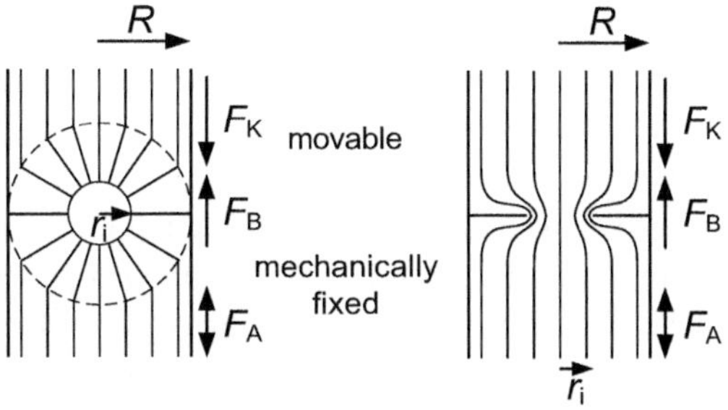

Fig. 4.21 Distribution of the flow lines and the acting forces in a point contact according to the sphere model (left) and the ellipsoid model (right, schematic)

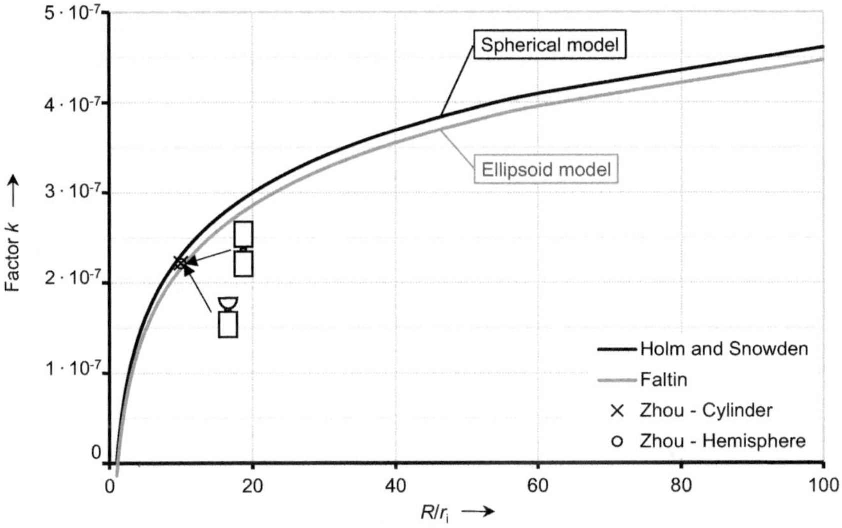

Fig. 4.22 Factor k for determining the blow-off force F_A in a point contact calculated according to different models

For larger ratios of R/r_i, the blow-off force F_A increases. If the simplified sphere model is replaced by the ellipsoid model to represent a more realistic distribution of current density in the contact area, slightly lower blow-off forces result (Eq. (4.27), Fig. 4.22) [81].

$$F_A = \frac{\mu_0}{4\pi} i^2 \left(\ln\left(\frac{R}{r_i}\right) - 2\ln(2) - \frac{5}{4} \right) \approx \frac{\mu_0}{4\pi} i^2 \left(\ln\left(\frac{R}{r_i}\right) - 0.1363 \right) \qquad (4.27)$$

The model according to *Holm* and *Snowdon* was checked with an FEM calculation with the same geometry of the arrangement. This resulted in approximately 3.5% lower blow-off forces than with Eq. (4.26) [82]. In addition, the influence of the shape of the contact area was considered. Assuming the same current-carrying contact area, a reduction of the blow-off force by about 10% was determined for very elongated ellipses with a semiaxis ratio of 1:8. The number of contact points in a multi-point contact has a major influence on the blow-off force. The greater the number of parallel contact points, the lower the blow-off force due to the current distribution [80, 82]. The empirical relationship between the surge current and the electromagnetic force F_{em} on a model arrangement was also investigated [83]. The acting electromagnetic force F_{em} is proportional to a constant multiplied by the square of the surge current (Eq. (4.25)). The factor k was empirically determined for point contacts to be (4.45 … 4.8) 10^{-7} N/A^2 [78, 82, 83]. The force due to the surrounding conductors must be considered separately and depends heavily on the geometry of the conductors and their routing. This affects the joining or spring elements in the current path and also the size, bending stiffness, and position of the connected conductors relative to each other [80, 83]. Recent publications suggest that the total effective blow-off force of a real connection is smaller than that indicated by *Slade and Taylor*. It was found that the blow-off forces in a real contact geometry with a different current deflection than in the point contact model can be lower than the analytically calculated values [80].

Example

A separable contact in a switchgear has 12 contacts. The contact force is applied externally via a steel coil spring and is 150 N per contact F_K. Point contacts are formed with a contact area diameter of approximately 0.8 mm. The current-carrying cross-sectional area of the conductors leading to each individual contact has a diameter of 8 mm. The switch is supposed to be able to carry a short-circuit current with a surge current $i_p = 160$ kA without the contacts lifting off. The maximum acting blow-off force F_A needs to be determined.

$$F_A = 4.47 \cdot 10^{-7} \frac{\mathrm{N}}{\mathrm{A}^2} \left(\frac{160000\,\mathrm{A}}{12} \right)^2 = 79.5\,\mathrm{N}$$

Since the contact force F_K is greater than the blow-off force F_A, the contact should not open during the short circuit. A melting of the contacts needs to be checked accordingly. ◀

4.2.4.3 Thermal Expansion

Metals expand when heated, depending on the temperature change $\Delta\vartheta$ and the linear thermal expansion coefficient α. In the case of firmly clamped conductors, this leads to compressive stresses, which result in elastic or plastic deformation of the conductors (Fig. 4.23).

If there are electrical connections in the current path, the resulting force F_{TD} also acts on these.

$$l_1 = l_0(1 + \alpha \Delta T) \tag{4.28}$$

$$F_{TD} = A\,E\,\alpha\Delta\vartheta \tag{4.29}$$

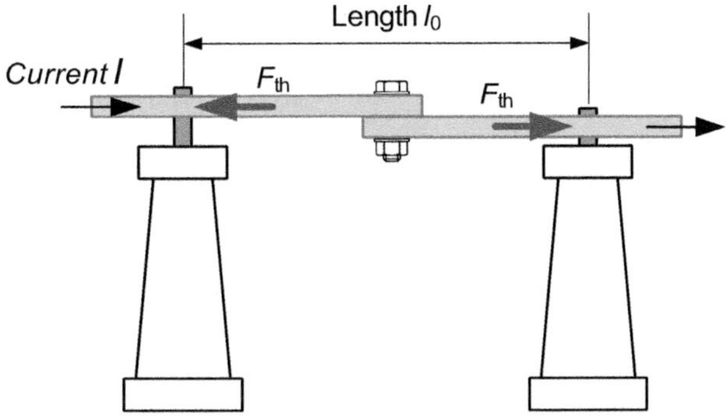

Fig. 4.23 Effect of thermal expansion on a rigid conductor clamped at both ends

To avoid these thermal expansion forces, plug-in connections or flexible expansion elements can be installed in the current path. These absorb the expansion of the conductors through relative movement and compensate for it. If this is not the case, e.g., with short conductors, the thermal expansion leads to a shear force F_Q in the current-carrying connection and must be estimated for the design.

Example

In a switchgear, busbars made of Cu-ETP are laid on support insulators. The distance between the support insulators l_0 is 600 mm. The maximum conductor temperature is 80 °C at an ambient temperature of 35 °C. The busbars have a cross-section of (50 × 10) mm². The resulting force on the support points is therefore:

$$F_{TD} = 500\,\text{mm}^2 \cdot 127\,\text{kN/mm}^2 \cdot 16.5 \cdot 10^{-7}\text{K}^{-1} \cdot 45\,\text{K} = 47.2\,\text{kN} \blacktriangleleft$$

4.2.4.4 Effect on Connections

The presented current-induced forces (Table 4.10) have different effects depending on the type of connection. The following will explain the effect in more detail using selected examples.

In screwed connections with busbars, electromagnetic forces act as a shear force on the connection and thus shear force in the contact plane when laid horizontally (Fig. 4.24). When the conductors are laid vertically, the resulting force is a composite force. A displacement of the conductors must be prevented by a sufficiently high contact force. If this is not the case, in the worst case, an arc can form between the conductors due to the breaking of the microcontacts. This can occur, for example, in screwed cross-connections of lightning protection devices. Thermal expansions also lead to shear forces in screwed connections and must be absorbed as operating forces by friction and the force fit between the conductors. This must be ensured by a sufficient contact force during the design [27]. If this is not possible, expansion sections in the form of expansion bands or sliding connections must be installed.

Table 4.10 Selection of the most common current-induced operational forces on electrical conductors and connections

	Force on current-carrying conductors	Blow-off force	Thermal expansion
Schematic	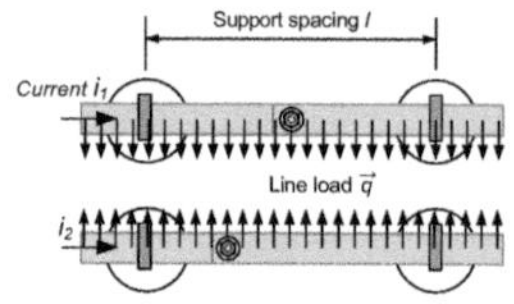	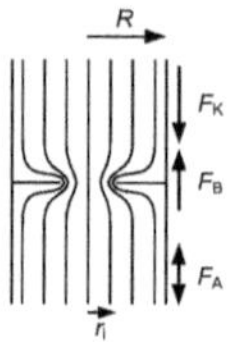	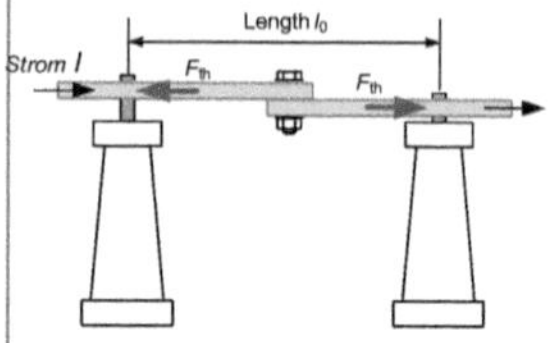
Equations	$q = \frac{\mu_0 i(t)}{2\pi a}$	$F_A = \frac{\mu_0 i^2}{4\pi} \ln \frac{R}{r_K} = 10^{-7} i^2 \ln \frac{R}{r_K} \propto k\, i^2$	$l_1 = l_0(1 + \alpha_{therm}\,\Delta T)$ $F_{th} = A\, E\, \alpha_{therm}\, \Delta\vartheta$
Note	Generally only relevant for high short-term currents	Generally only relevant for high short-term currents	Compensation possible through flexible bands or sliding connections
Parameters	q_h … Main conductor stretch load due to electromagnetic forces q_t … Partial conductor stretch load due to electromagnetic forces F_{TD} … Forces due to thermal expansion	Contact area and shape, shape of surrounding conductors, current intensity	α_{therm} … Coefficient of thermal expansion Load current, mechanical boundary conditions
Examples	Screwed connections with busbars, overhead lines	Plug-in connections, Switch contacts for e.g. relays and circuit breakers	Cables, gas-insulated switchgear (GIS) and lines (GIL), switchgear busbars

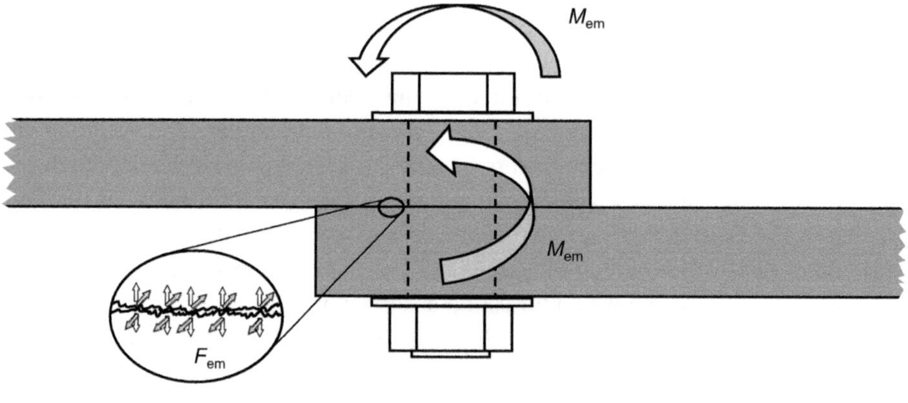

Fig. 4.24 Operating forces and moments on a screwed connection

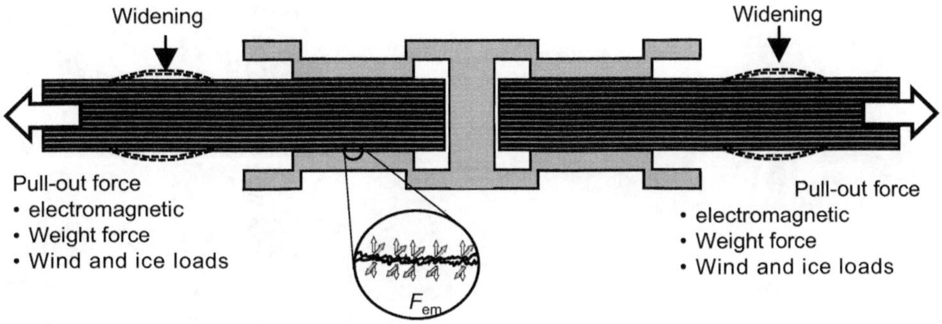

Fig. 4.25 Operating forces on a compression connection

The compression connections used in overhead lines or cable systems must absorb the tensile loads caused by weight forces, wind and ice loads, or electromagnetic forces that occur on conductors in the event of a short circuit (Fig. 4.25). These forces must not lead to the breaking of the microcontacts in order to not change the quality of the connection. In particular, the conductors must not slip in the connection. The forces that occur must therefore be taken into account in the design and testing [43, 44].

In multi-strand conductors, in addition to a compression connection, the outer layers can deform in the event of a short circuit, which causes an expansion of the outer layers of the conductor and can thus influence the connection mechanically and electrically.

In round plug-in connections, but also in flat plug-in connections and disconnect switches in busbar systems, transverse forces on the connected conductors lead to asymmetric compression (Fig. 4.26). On part of the circumference, the spring elements are compressed by the deformation of the plug and the socket and are relaxed on the opposite side. In the case of very large deformations, the connection can be destroyed. This

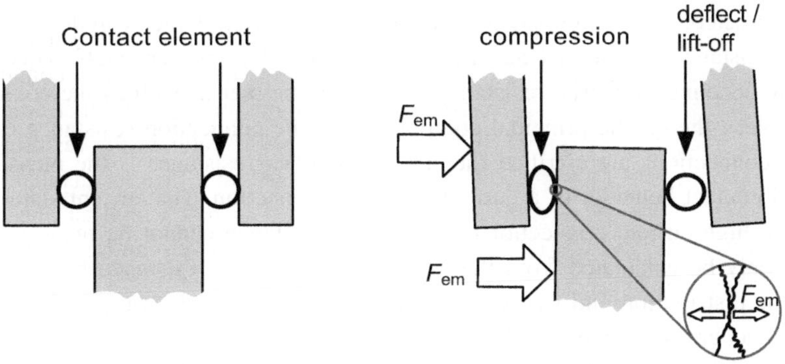

Fig. 4.26 Effect of the electromagnetic force during a short circuit on a plug-in connection (longitudinal section)

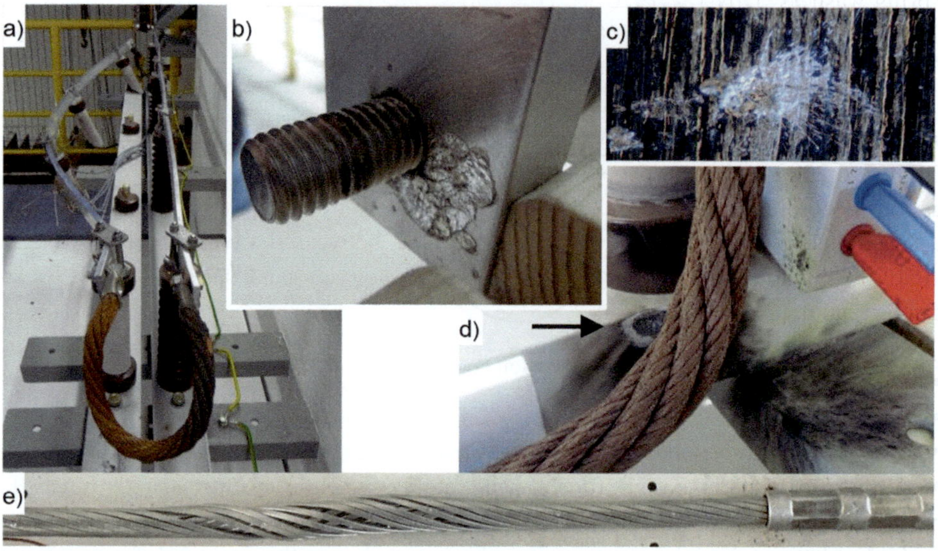

Fig. 4.27 Effect of short-time currents: **a**) Deformation of parallel busbars **b**) poorly assembled, partially melted screwed connection **c**) Point contact with lifting and spraying out of molten material **d**) unintentional contacting of a stranded conductor with a grounded component **e**) Deformation of the outer layers of an overhead line before a compression connection

can be counteracted by mechanically supporting elements that prevent excessive radial displacement between the plug and the socket.

Examples of the effect of short-circuit currents are shown in Fig. 4.27.

4.3 Electrical Contact Behavior

As an abstracted, very simplified equivalent circuit diagram, which can describe the connection resistance, the model according to Eq. (4.3) has proven itself. This generally takes into account a material resistance per contact partner as well as the contact resistance between them. The principal components of the connection resistance of current-carrying connections are clarified therein. The contact resistance is the measure of the electrical contact behavior of a current-carrying connection. The current state of the art is that for high-current connections the contact resistance cannot be measured directly and can only be calculated from the measured connection resistance for a few special cases. The distribution and size of the microcontacts of rough, wavy and usually film-laden contact areas is unknown. Accordingly, the contact resistance of plane contacts can only be calculated directly for certain boundary conditions. Furthermore, it should be noted that the feedback of the contact resistance on the distribution of the current density in the contact partners is specific for the different types of connections. The material portion through which current flows in the connection is not independent of the contact resistance, this applies especially to elongated connections. In particular, in current studies on the electrical modeling of screwed connections and compression connections, these aspects were increasingly taken into account [8, 9, 11]. The aim of these considerations was always to compare the contact behavior of connections independently of the material resistances of the contact partners. For this, parameters are required that can be determined based on empirical studies with simplified considerations as well as electrical models of current-carrying connections. The following will go into more detail on this.

4.3.1 Empirical Relationships

To achieve good electrical contact behavior in non-material connections, the force fit is usually decisive. This means that during assembly, the applied contact force must achieve sufficient elasto-plastic deformation of the micro-peaks on the contact area to create many parallel and large-area microcontacts. Furthermore, there must always be a sufficiently high remaining contact force between the contact partners after joining. The basic force-resistance behavior can generally be described with a hysteresis curve (Fig. 4.28). The hysteresis behavior is strongly dependent on the conductor and coating material, the pretreatment of the contact areas, the mechanical stress distribution in the contact plane and the contact and joining elements used. For surface and line contacts, the true metallic and the mechanically supporting contact area cannot currently be deter-

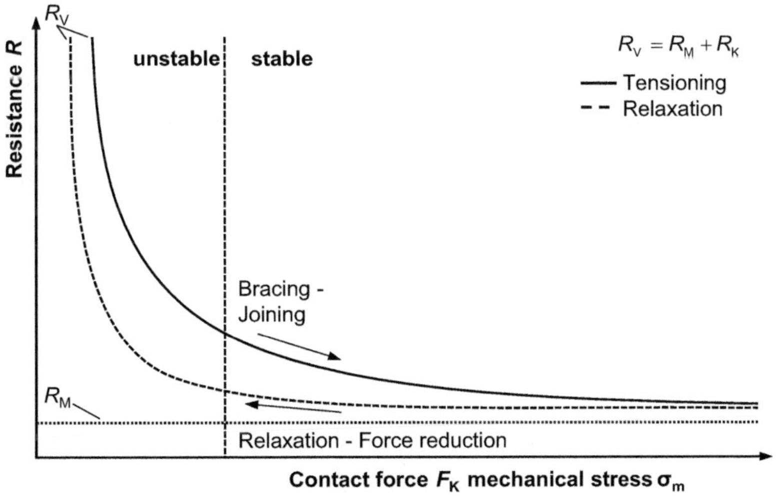

Fig. 4.28 Connection resistance R_V depending on the contact force or average mechanical tension σ_m (schematic)

mined or calculated with FEM programs, as the surface roughness and the films cannot be accurately taken into account. Values of 3% for the ratio of A_w to A_s are generally given for plane contacts in the literature [1]. Even film-laden point contacts cannot currently be fully described [3], so empirical approaches are often used.

Based on a very frequently used, extended approach according to Babikow (Eq.(4.30) [12, 14]), the contact resistance can be calculated for the constructive design of current-carrying connections with the average mechanical tension from the equivalent cross-section model or the contact force (Sect. 4.2.1) [52]. The parameters k_V and m_V must be determined experimentally (Fig. 4.28) and depend on the geometry of the contact partners and the condition of the contact area. The condition of the contact area is considered via the size k_V. The smaller the value for k_V, the less the influence of foreign layers. The size m_V characterizes the shape of the contact area. The larger m_V is, the more flatly the contact partners lie on each other and the mechanical tension is more evenly distributed than, for example, when using pressure plates or clamping washers as joining elements. Small values of m_V are to be used with an uneven tension distribution as it occurs, for example, when using an inner carrier screw without additional washers.

$$R_{K0}(\Omega) = \frac{k_V}{\left(\frac{F_{K0}(\text{N})}{A_{Ers}(\text{mm}^2)}\right)^{m_V}} = \frac{k_V}{\left(\sigma_m(\text{N/mm}^2)\right)^{m_V}} \quad (4.30)$$

For some material pairings, values for k can be exemplified from studies on screwed connections with busbars and plug-in connections.$_v$ and m_v are given, with which the contact resistance R_{K0} in Ω can be calculated (Table 4.11).

For point contacts, Eq. (4.30) can be adapted, as the relationship between apparent contact area and contact resistance is not a meaningful design size.

$$R_{K0}(\Omega) = k_V\,(0.1\,F_K)^{-m_V} \quad (4.31)$$

The parameters k_V and m_V were determined empirically and depend on the material, the surface condition (shape, cleaning, roughness) and the temperature (Tables 4.12 and 4.13).

Table 4.11 Values for k_v and m_v for plane contacts

Material pairing WS1/BS1-BS2/WS2	k_v	m_v	Remark
Al99.5-Al99.5	(1.3 … 2.1) 10^{-5}	0.43 … 0.75	Abrasive pretreatment of the contact area (wire brush)
AlMgSi/Ag[1)]-Ag[1)]/AlMgSi	(0.3 … 0.6) 10^{-5}	0.43 … 0.65	Without pretreatment of the contact area
AlMgSi/Ni[1)]-Ni[1)]/AlMgSi	(2.3 … 3.5) 10^{-5}	0.72 … 0.83	Abrasive pretreatment of the contact area (abrasive fleece)
AlMgSi/Ni[1)]-Al99.5	(3.7 … 4.7) 10^{-5}	0.89 … 0.95	Abrasive pretreatment of the contact area (Ni abrasive fleece, Al wire brush)
Cu-ETP-Cu-ETP	(0.2 … 0.6) 10^{-5}	0.43 … 0.67	Abrasive pretreatment of the contact area (abrasive fleece)
Cu-ETP/Ni[1)]-Ni[1)]/Cu-ETP	(2.2 … 7.9) 10^{-5}	0.83 … 0.95	Abrasive pretreatment of the contact area (abrasive fleece)
Cu-ETP/Ag[1)]-Ag[1)]/Cu-ETP	0.3 10^{-5}	0.52	Without pretreatment of the contact area
Cu-ETP/Sn[1)]-Cu-ETP	(0.4 … 0.5) 10^{-5}	0.59 … 0.67	Without pretreatment of the contact area[2)]
Cu-ETP/Sn[1)]-Sn[1)]/Cu-ETP	(0.7 … 1.1) 10^{-5}	0.56 … 0.75	Without pretreatment of the contact area
Cu-ETP-Ag[1)]/AlMgSi	0.6 10^{-5}	0.63	Abrasive pretreatment of the contact area (Cu abrasive fleece)
AlMgSi/Ag[1)]-Sn[1)]/Cu-ETP	(0.5 … 0.9) 10^{-5}	0.61 … 0.72	Without pretreatment of the contact area
AlMgSi/Ni[1)]-Cu-ETP	(0.6 … 0.8) 10^{-5}	0.54 … 0.59	Abrasive pretreatment of the contact area (abrasive fleece)
Al99.5-Ni[1)]/Cu-ETP	(4.6 … 4.9) 10^{-5}	1.02 … 1.04	Abrasive pretreatment of the contact area (Ni abrasive fleece, Al wire brush)
AlMgSi/Ni[1)]-Ni[1)]/Cu-ETP	(0.6 … 1.6) 10^{-5}	0.54 … 0.73	Abrasive pretreatment of the contact area (abrasive fleece)

WS … Material
BS … Coating
[1)] all coatings applied electrolytically
[2)] Oxide layer on the copper surface can lead to higher contact resistances

Table 4.12 Values for the factor k_V for point contacts [12]

Material pairing	Surface condition	k_V	Remark
Cu-Cu	Cleaned	$(0.08 \ldots 0.14)\ 10^{-3}$	[12]
Al-Al	Cleaned	$(3 \ldots 6.7)\ 10^{-3}$ $0.18\ 10^{-3}$	[12] TUD experiments
Sn-Sn	Unlubricated	$0.1\ 10^{-3}$	[12]
Ag-Ag	Cleaned	$0.11\ 10^{-3}$	TUD experiments

Table 4.13 Values for the factor m_V [12, 13]

Geometry of the contact area/material	Surface condition	m_V	Remark	
Plane-Plane	–	1	[12]	
Sphere-Sphere		0.5		
Sphere-Plane		0.5		
Busbars		0.5 … 0.7 (0.6)		
Cu-Cu	Cleaned	0.6	20 °C	[13]
		0.55	75 °C	
		0.3	200 °C	
	Oxidized	1	20 °C	
		0.2	75 °C	
		0.1	200 °C	
		0.08	500 °C	

Example

The contact resistance of a screwed connection with two busbars made of Al99.5 with a cross-section A of $(40 \times 10)\ \mathrm{mm}^2$, a connection length l_V of 40 mm, a screw and washer of size 12, and a replacement cross-sectional area A_{Ers} of 536 mm² is to be calculated for a contact force of 25 kN. Choice of parameters $k_V = 1.3$ (very well pre-treated contact area with a wire brush) and $m_v = 0.7$ (average contact area of the joining elements)

$$R_{K0} = \frac{1.3 \cdot 10^{-5}}{\left(\frac{25000\,\mathrm{N}}{536\,\mathrm{mm}^2}\right)^{0.7}} = 0.88\,\mu\Omega$$

From this, the connection resistance R_{V0} and performance factor k_{U0} are to be determined.

$$R_M = k_{ul}(A_B) \cdot R_{Sch} = 0.73 \cdot 2.78\,\mu\Omega = 2.03\,\mu\Omega \text{ (Eq. (4.66))}$$

$$R_{V0} = R_{K0} + R_M = 2.03\,\mu\Omega + 0.88\,\mu\Omega = 2.91\,\mu\Omega$$

$$k_{\mathrm{U0}} = \frac{R_{\mathrm{V0}}}{R_{\mathrm{BL}}} = \frac{2.91\,\mu\Omega}{2.78\,\mu\Omega} = 1.05 \qquad \text{(Eq. (4.7))}$$ ◀

4.3.2 Electrical Models of Current-Carrying Connections with Line or Plane Contacts

Electrical models aim to represent the resistance behavior of current-carrying connections and to evaluate the influence of the construction or the materials used. An electrical model is a component for calculating the electrical-thermal operating behavior. The contact behavior of the current-carrying connection can be an input size of the model or it can be determined with the model. Since the contact behavior of point contacts has already been thoroughly investigated, this type of contact will not be further discussed here (Sect. 4.1.1 especially Table 4.1) [3, 15–17]

(a) Resistance networks

Electrical models with linear resistance networks describe the current paths in current-carrying connections with discrete or differential electrical resistances. These networks can represent current-carrying connections in two or three dimensions. The contact partners are represented in these networks by their material resistances, possibly taking into account the deformation occurring during joining. The contact resistances are simplistically calculated from the descriptive size (specific transverse resistance R'_{q}) [11] or can

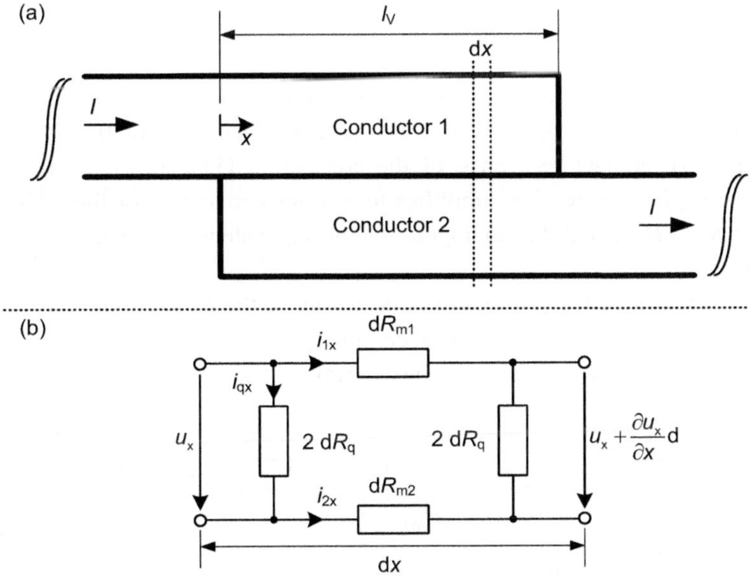

Fig. 4.29 Current-carrying connection of two linear electrical conductors (**a**) and the equivalent circuit diagram of a length element dx (**b**)

be determined with empirical studies of the individual contact behavior [46]. Network calculation programs (e.g., LTSpice) are suitable for solving simple discrete resistance networks. For networks with a high number of resistances, methods of network calculation, such as node voltage analysis, can be used to calculate the potential and current distribution in the networks. A resistance network with differential electrical resistances of a current-carrying connection between two identical contact partners was analytically described by [47] (Fig. 4.29). For this resistance network, the connection resistance

$$R_V = \frac{R'_{m1}}{R'_{m1} + R'_{m2}} \left[R'_{m2} l_V + \frac{(R'_{m1} - R'_{m2})}{\alpha_{el}} \tanh\left(\frac{\alpha_{el} l_V}{2}\right) \right] + \frac{R'_{m1} + R'_{m2} \cosh(\alpha_{el} l_V)}{\alpha_{el} \sinh(\alpha_{el} l_V)} \tag{4.32}$$

with the reciprocal electrical location constant

$$\alpha_{el} = \sqrt{\frac{R'_{m1} + R'_{m2}}{R'_q}} \tag{4.33}$$

can be calculated analytically. If the contact partners in the resistance network with differential elements are connected without contact resistance ($R'_q \to 0$), the limit consideration shows

$$\begin{aligned} \lim_{R'_q \to 0} R_V &= \lim_{R'_q \to 0} \left(\frac{R'_{m1}}{R'_{m1} + R'_{m2}} \left[R'_{m2} l_V + \frac{(R'_{m1} - R'_{m2})}{\alpha_{el}} \tanh\left(\frac{\alpha_{el} l_V}{2}\right) \right] + \frac{R'_{m1} + R'_{m2} \cosh(\alpha_{el} l_V)}{\alpha_{el} \sinh(\alpha_{el} l_V)} \right) \\ \lim_{R'_q \to 0} R_V &= \frac{R'_{m1} R'_{m2} l_V}{R'_{m1} + R'_{m2}} \end{aligned} \tag{4.34}$$

that the resistance of the parallel-connected contact partners $R_{||}$ results (Eq. (4.34)). It follows that the redirection of the flow field in the connection is not taken into account in this resistance network.

From the linear resistance network with differential resistances, a simplified equivalent circuit diagram can be derived with one resistance each of the contact partners and the effective contact resistance of the connection [11]. The equivalent circuit of the current-carrying connection simplifies to a series connection of the three equivalent resistances R_{M1}, R_{M2}, and R_K analogous to the equivalent circuit diagram in Fig. 4.4 (Fig. 4.30).

For this, the sum of the power losses occurring in the differential resistances is calculated. From this, an equivalent resistance, through which the total current flows and in which the same power loss occurs, can be determined (Eqs. (4.35)–(4.37)).

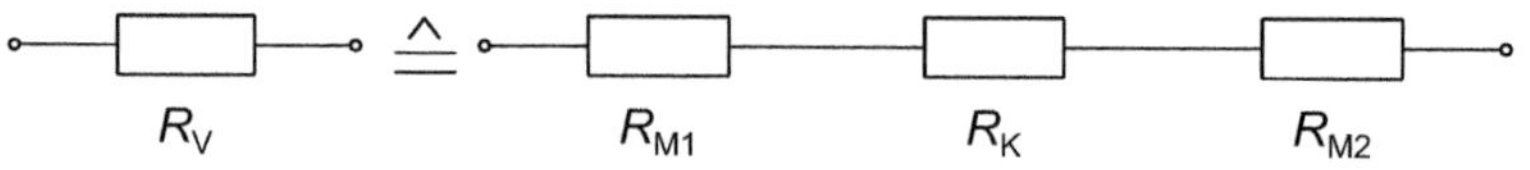

Fig. 4.30 Equivalent circuit diagram of the resistances determined from the power loss balance

$$P_{V1} = \int_0^s dP_{V\,1} = R'_{m1} \int_0^s i_{1x}^2 dx = I^2 R_{M1} \tag{4.35}$$

$$P_{V2} = \int_0^s dP_{V\,2} = R'_{m2} \int_0^s i_{2x}^2 dx = I^2 R_{M2} \tag{4.36}$$

$$P_{V\,q} = \int_0^s dP_{V\,q} = \int_0^s \frac{u_x^2}{R'_q} dx = I^2 R_K \tag{4.37}$$

(b) Numerical electrical models

With numerical electrical models of current-carrying connections, which are solved with the finite element method, the static or dynamic distribution of the electrical flow field in and the electrical potential at the contact partners can be calculated. The contact partners are represented in the models two- or three-dimensionally with the exact or a simplified geometry by surface or volume elements. In addition, the electrical conductivity of their materials is taken into account. The coupling of the flow fields between two contact partners is achieved by additional, pairwise contact elements on two contact areas each. Depending on the user-defined specific areal conductance G', the potential difference between the contact elements results depending on the local current density S (Eq. (4.38)).

$$\Delta\varphi = \frac{S}{G'} \tag{4.38}$$

The specific areal conductance as a model size can be assigned to an infinitesimally thin layer between the contact partners. The specific areal conductivity cannot be directly calculated unrestrictedly from the contact resistance. This is only possible if the current density distribution in the infinitesimally thin layer between the contact partners is homogeneous and results in ((4.40)):

$$G' = \frac{1}{R_K A_S} \tag{4.39}$$

This condition is not met, especially in connections with good contact behavior, where the current density in the layer is determined by the material resistances of the contact partners. In these cases, the specific areal conductivity can be determined by an indirect method (Sect. 4.3.3).

4.3.3 Determining the model sizes of the electrical contact behavior

Model sizes are the material resistances (R_M) and those that describe the electrical contact behavior of a connection (R_K, R'_q, G'). They serve the electrical modeling of the connection and can be determined and used for the simulation of the operating behavior or for constructive improvement. Methods are to be distinguished in which the model sizes are calculated from input parameters (material, geometry, force) (direct) or in which the model sizes are determined from measured values (indirect). Another possibility to describe the electrical contact behavior is the analogy to the thermal contact behavior [42] (Sect. 4.4.1).

(a) Direct solution path

The direct calculation of the contact resistance or a substitute size describing the contact behavior of plane contacts requires, for example, defined structured contact areas, where assumptions can be made for the size and distribution of the microcontacts and the influence of foreign layers can be neglected. Furthermore, for plane contacts, the contact resistance can be directly determined by estimating the effective material resistance of a connection. Such an approximation is well applicable when the contact resistance is the predominant part of the connection resistance. In this case, uncertainties in estimating the material resistance have little effect.

Example

Consider a screwed connection with two copper busbars with a cross-section of $(50 \times 10)\ \mathrm{mm}^2$ and a connection length of $l_\mathrm{V} = 50\ \mathrm{mm}$. A very high connection resistance of $R_\mathrm{V} = 41.2\ \mu\Omega$ was measured at this connection. The material resistance can be estimated with Eq. (4.42). In a first approximation, this results in the specific areal conductivity G' with:

$$R_{\mathrm{M\ w}} = 34.5\frac{\mu\Omega}{\mathrm{m}}0.05\,\mathrm{m}\left[0.639\left(\frac{50\,\mathrm{mm}}{10\,\mathrm{mm}}\right)^{-1.35} + 0.5\right] = 0.99\,\mu\Omega$$

Since the material resistance is small compared to the connection resistance, the areal conductivity G' can be approximated as a model size with

$$R_\mathrm{K} = R_\mathrm{V} - R_{\mathrm{m\ w}} = 41.2\,\mu\Omega - 0.99\,\mu\Omega = 40.2\,\mu\Omega$$

$$G' = \frac{1}{R_\mathrm{K} A_\mathrm{S}} = \frac{1}{0.00004021\,\Omega \cdot 0.0025\,\mathrm{m}^2} = 9.95\,\frac{\mathrm{MS}}{\mathrm{m}^2}$$

can be estimated. ◀

(b) Indirect Approach

The contact resistance of current-carrying connections can be determined in a simplified manner by estimating the effective material resistance in the connection and subtracting it from the connection resistance (Eq. (4.40)).

$$R_\mathrm{K} = R_\mathrm{V} - R_\mathrm{M} \tag{4.40}$$

First, two contact partners with the length-related material resistances R'_{M1} and R'_{M2} are considered, which are connected without contact resistance over the connection length

l_V. The resulting material resistance $R_{\parallel}$ of the connection in an electrical parallel connection of the contact partners over the connection length l_V is

$$R_{\parallel} = \frac{R'_{m1} R'_{m2}}{R'_{m1} + R'_{m2}} l_V \tag{4.41}$$

While in a straight electrical conductor traversed by a direct current, the magnitude and direction of the current density are the same at every point of the conductor, the magnitude and direction of the current density in almost all types of current-carrying connection are distributed depending on the location. The current commutes from one electrical conductor to another. This increases the effective material resistance traversed by the current compared to the resistance of the parallel connection of the contact partners. The effective material resistance $R_{m\,w}$, which is due to the actual flow through the contact partners over the connection length l_V, was empirically determined for current-carrying screwed connections with busbars of height h (Eq. (4.42)) [6].

$$R_{m\,w} = \frac{R'_{m1} + R'_{m2}}{2} l_V \left[0.639 \left(\frac{l_V}{h} \right)^{-1.35} + 0.5 \right] \tag{4.42}$$

For more complex geometries or arrangements with more than two contact partners, the effective material resistance can usually only be determined with numerical electrical models. To determine the current-carrying material resistance, it is advisable to model the contact partners of the relevant current-carrying connection as one body or to make reasonable assumptions for the specific areal conductivity G'. In particular, its influence on the effective material resistance of a current-carrying connection should be considered.

(c) Determining the specific transverse resistance for resistance networks

If the contact behavior of different current-carrying connections is to be compared independently of the geometry and the electrical properties of the contact partners, the specific transverse resistance can be used. This can only be calculated directly from the connection resistance R_V using Eq. (4.44) if the condition

$$R'_q << l_V{}^2 (R'_1 + R'_2) \tag{4.43}$$

is fulfilled.

$$R'_q = \left[\frac{l_V R'_{m1} R'_{m2} \sqrt{R'_{m1} + R'_{m2}}}{R'_{m1} + R'_{m2}} \left(\frac{R_V - R}{R} \right) \right]^2 \tag{4.44}$$

As a rule, the specific transverse resistance can only be calculated iteratively (indirectly) since without the condition in Eq. (4.43) no analytical solution exists. The flow chart in Fig. 4.31 serves as a guide.

(d) Determining the specific areal conductivity for FEM models

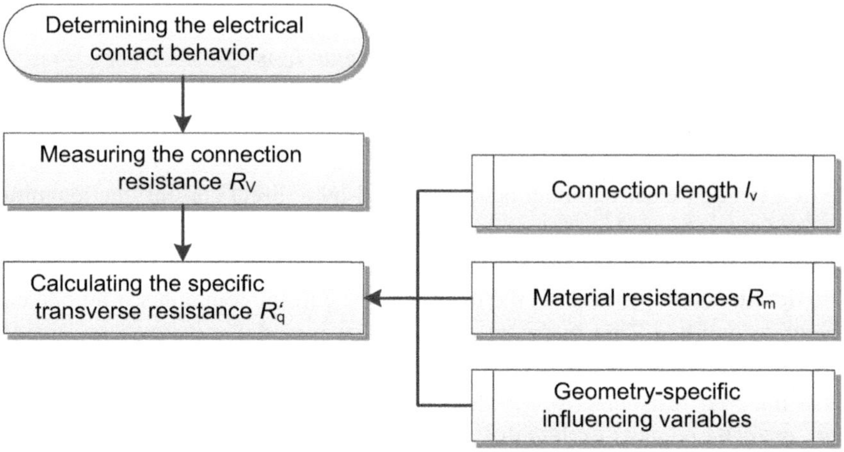

Fig. 4.31 Procedure for determining the specific transverse resistance

First, the boundary conditions under which the specific areal conductivity for surface and line contacts should be determined must be defined. A distinction can be made between arrangement-specific procedures and those that take into account the mechanical contact behavior of a specific material combination under defined, average mechanical stresses. In the second case, the specific aral conductivity G' can be determined independently of the geometry of the contact partners. From the assumption of evenly distributed, average mechanical stress, it follows that the microcontacts in the contact plane are also of equal size and evenly distributed. The specific areal conductivity G' as a model size is not location-dependent in this special case. The starting point for determining the specific areal conductivity G' depending on the mechanical stress are the hysteresis characteristics, which are determined at current-carrying connections (Sect. 4.3.1). The experimental design must ensure an evenly distributed mechanical stress in the contact plane. The effective, current-carrying material resistance R_M can change depending on the contact resistance, as this influences the current density distribution in the contact partners (Sect. 4.3.4). For the considered line or surface contact, the function $R_M(R_V)$ must first be determined (see Fig. 4.32), in order to capture the effective material resistance at low connection resistances in particular. A suitable electrical model can be used for this purpose (Sect. 4.3.2). The corresponding material resistance is subtracted from the respective connection resistance measured in the hysteresis test to calculate the contact resistance depending on the average mechanical stress σ_m (Eq. (4.45)). Subsequently, the coefficients of a relationship describing the contact resistance (e.g., k_V and m_V in Eq. (4.30) [12]) can be determined.

$$R_K(\sigma_m) = R_V(\sigma_m) - R_M(R_V) \tag{4.45}$$

Furthermore, the effective contact resistance of the corresponding current-carrying connection is determined in parameter studies with an FEM model describing the real arrangement, depending on the specific areal conductivity G'. The effective contact

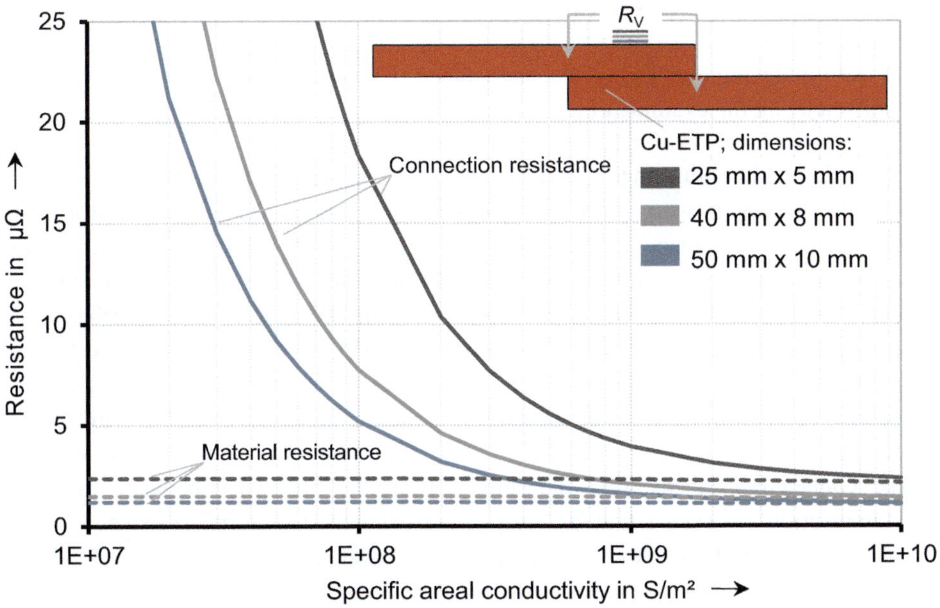

Fig. 4.32 Connection and material resistance of the surface contact between busbars made of CU-ETP with various cross-sections depending on the specific areal conductivity

resistance is the one that can be determined from the power balance of the contact partners (Eq. (4.37)). For this purpose, the power loss densities P'_V of the elements of each contact partner are summed from the results of the FEM calculation. The power loss generated in it is calculated with the volume of the respective contact partner (Eq. (4.46)) and from this the effective material resistance of the contact partner.

$$P_V = V \sum_{i=1}^{n} P'_{V\,n} \tag{4.46}$$

The contact resistance is ultimately the difference between the calculated connection resistance and the effective material resistances of the contact partners. If Eq. (4.30) is now rearranged for the mechanical stress (Eq. (4.47)), a direct relationship between the areal conductivity and the average mechanical stress σ_m can subsequently be established.

$$\sigma_m \left(\mathrm{N/mm^2}\right) = \frac{1}{A_S \left(\mathrm{mm^2}\right)} \left(\frac{k_V}{R_K(\Omega)} \right)^{\frac{1}{m_V}} \tag{4.47}$$

This relationship can be described empirically with a power function (Eq. (4.48)).

$$G' = k_\sigma (\sigma_m)^{m_\sigma} \tag{4.48}$$

4.3.4 Application Examples

(a) Determining the specific areal conductance depending on the average mechanical tension using the example of a surface contact

The specific areal conductance G' was determined for plane contacts with contact partners made of copper with abrasively pretreated contact areas. For this purpose, the contact resistance of plane contacts between busbars made of CU-ETP with various cross-sections was measured using an experimental setup, which can be assumed to have an approximately uniform average mechanical stress in the contact plane. Furthermore, a FEM model was set up, with which the connection and material resistances of the contact partners of the investigated surface contact were calculated (Fig. 4.32). The material resistance is up to 12% lower for a specific areal conductivity of 10^{10} S/m^2 than for 10^7 S/m^2.

The contact resistances were determined from the difference between the measured connection resistance and the respective calculated material resistance $R_M(R_V)$ (Eq. (4.45)). From these data, the parameters k_V and m_V for the empirical function according to Babikow were determined (Fig. 4.33). The function approximates the relationship between the average contact resistance and the mechanical tension well.

In the area of very low contact resistances on the order of 0.1 μΩ, the relative scatter of the determined contact resistances was considerable, which is insignificant for

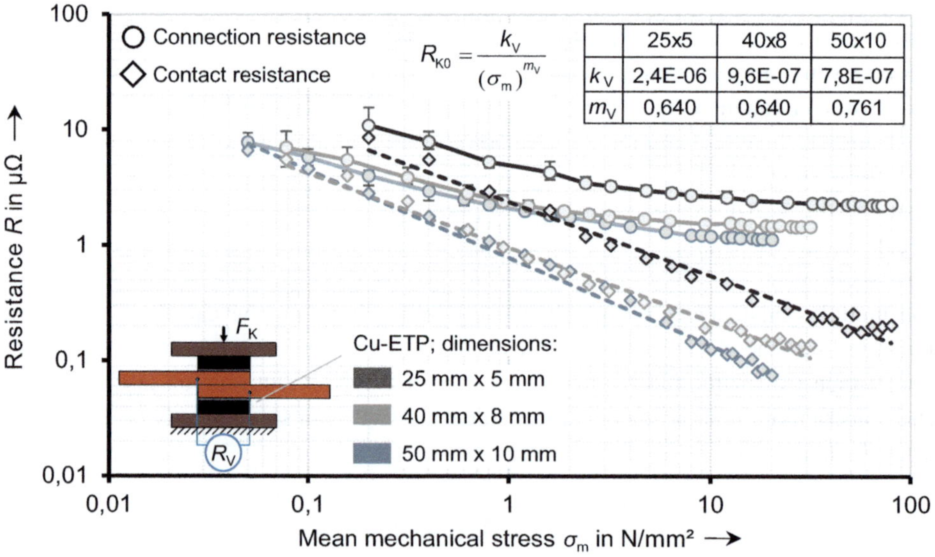

Fig. 4.33 Connection and contact resistance of a surface contact between busbars made of CU-ETP with various cross-sections depending on the surface pressure

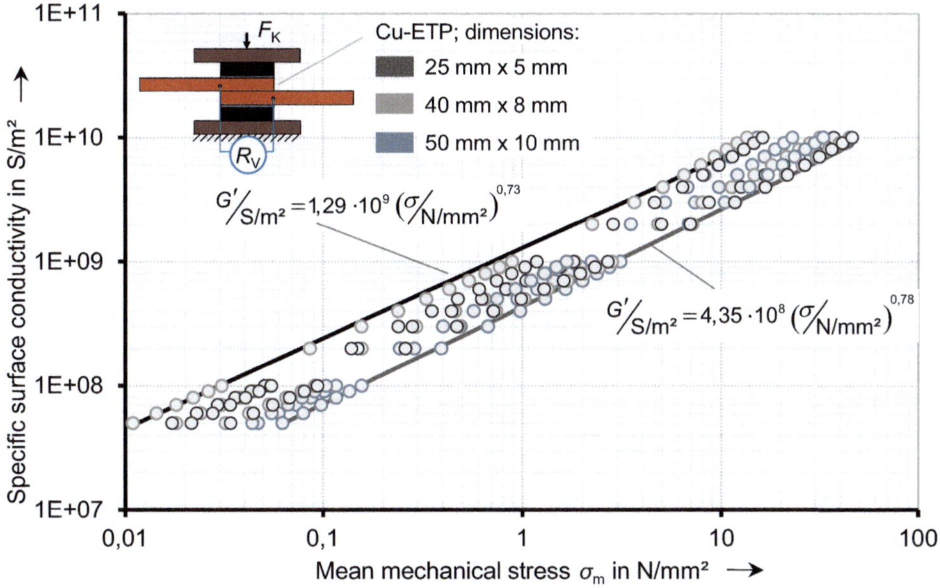

Fig. 4.34 Specific areal conductivity of plane contacts between copper contacts with abrasively pretreated contact areas depending on the surface pressure

practical applications but relevant for calculating the areal conductivity. In the next step, the contact resistances determined with the FEM model can be assigned a mean mechanical stress calculated with Eq. (4.47). This allows an empirical correlation between the mean mechanical stress and the specific areal conductivity to be established. If the range of resistances in the hysteresis experiments is taken into account, a parameter field results (Fig. 4.34). For this, power functions can be specified as adapted quantity equation, with which the optimal or the most unfavorable contact behavior for the material combination copper–copper in surface contact can be estimated.

(b) Optimizing the length of a current-carrying compression connection

As an example, any compression connection consisting of a press sleeve and conductor wire is considered. It is assumed that the quality of the electrical contact does not change along the connection. Thus, the specific transverse resistance is assumed to be independent of the connection length. With this assumption, the connection resistance and its components can be calculated with the loss power balance in the connection (Eqs. (4.32), (4.35)–(4.37)) [11]. The connection resistance is the sum of the three equivalent resistances. In the case study, a minimum of the connection resistance is obtained at a connection length l_V ($R_{V\ min}$) of approx. 0.066 m (Fig. 4.35). If the connection length l_V in the range $l_V < l_V$ (R_{Vmin}) is continuously increased, the connection resistance decreases significantly until reaching the local minimum. In this area, the equivalent contact resist-

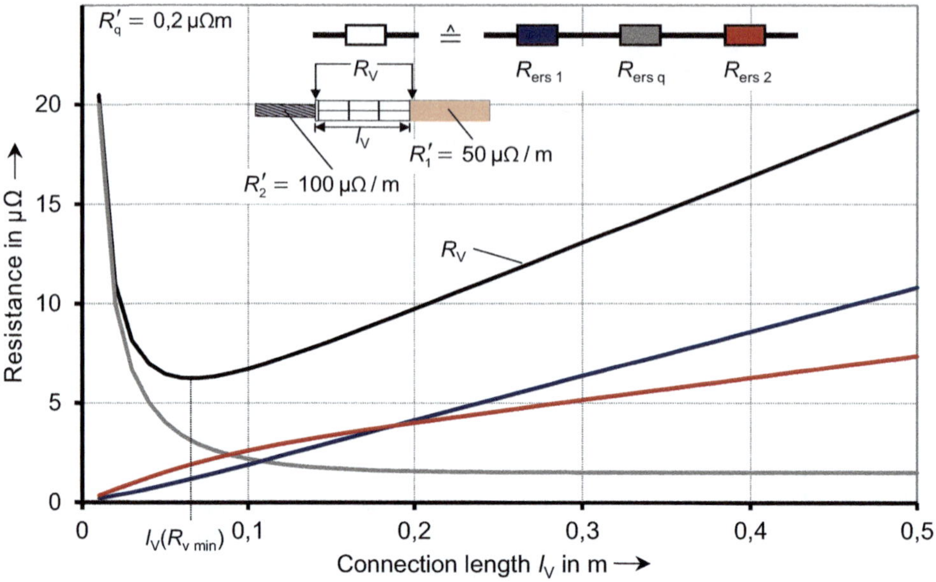

Fig. 4.35 Components of the connection resistance depending on the length of the compression connection assuming an equal specific transverse resistance

ance $R_{ers\ q}$, which decreases by paralleling additional contact points, dominates. In long connections ($l_V > l_V\ (R_{V\ min})$), the equivalent material resistances of the contact partners $R_{ers\ 1}$ and $R_{ers\ 2}$, which change linearly with the connection length in this area as a first approximation, dominate. For very long connections ($l_V >> l_V\ (R_{V\ min})$), the equivalent contact resistance asymptotically reaches a limit value. If additional microcontacts are created by extending a current-carrying connection, these are initially partially not current-carrying in the initial state, as the current already commutes in the axial edge areas from one contact partner to the other in order to fully utilize the available conductor cross-section. Only in the further lifetime after aging of the current-carrying connection can these contact points be effectively current-carrying. This happens when the contact resistance of current-carrying microcontacts has significantly increased.

In general, it has been shown that the position of the minimum connection resistance is independent of the specific transverse resistance and only depends on the ratio of the material resistances [11]. It is recommended to choose the connection length to be at least four times the reciprocal of the electrical spatial constant α_{el} (Eqs. (4.33), (4.36) and (4.49)), in order to keep the equivalent transverse resistance of a current-carrying connection in the initial state low.

$$l_V > 4/\alpha_{el} \tag{4.49}$$

As a result, the power loss in continuous operation is predominantly in the contact partners and the microcontacts are thermally relieved.

4.4 Thermal Contact Behavior

Two basic cases can be distinguished in thermal contact behavior:

- The limit temperature on the surface of the connection as the upper and lower limit of the operating range, especially with regard to long-term stable operation (Sects. 5.2, 5.3 and 6.3.4)
- The thermal contact for transferring a heat flow from one contact partner to another (Sect. 4.4.1)

In current-carrying connections, there is an interaction of the electrical and thermal flow field due to the current heat losses, which can be mathematically described with the heat balance equation or often with the voltage-temperature relationship (Sect. 4.4.2)

4.4.1 Thermal Contact Behavior

4.4.1.1 Basics

Electrical and thermal flow fields can be described mathematically in an analogous way. The cause for the current flow in an electrical conductor, the electrical flow field $\vec{S}$, are potential differences grad φ in the conductor itself. The cause for the heat transport, thus the thermal flow field $\vec{q_t}$, are temperature differences grad ϑ [50].

$$\begin{aligned} \vec{S} &= -\kappa \; grad \; \varphi \\ \vec{q_t} &= -\lambda \; grad \; \vartheta \end{aligned} \tag{4.50}$$

In an electrical contact, the effective transport paths to transfer electrical power or a heat flow are not identical. Electrically effective are especially the purely metallic contact area and to a lesser extent the quasi-metallic contact area. Thermally effective are additionally films and gas-filled gaps, in which heat conduction and heat radiation occur (Fig. 4.36) [45]. The reason for the additional transport paths in thermal contact is the ratio of the electrical and thermal conductivities of the materials in the connections. Oxide layers and gases typically have an electrical conductivity that is 10 to 20 orders of magnitude lower than that of metals and are therefore electrically insulating from a contact technical point of view.

The thermal conductivity, on the other hand, is only about 2 to 4 orders of magnitude smaller than that of metal (Table 4.14). Since the surfaces of the contact partners separated by oxide layers, lubricants, or insulating gases are often much larger than the metallic contact areas (see Sect. 4.1.1), a significant part of the heat flow can be transported via these paths.

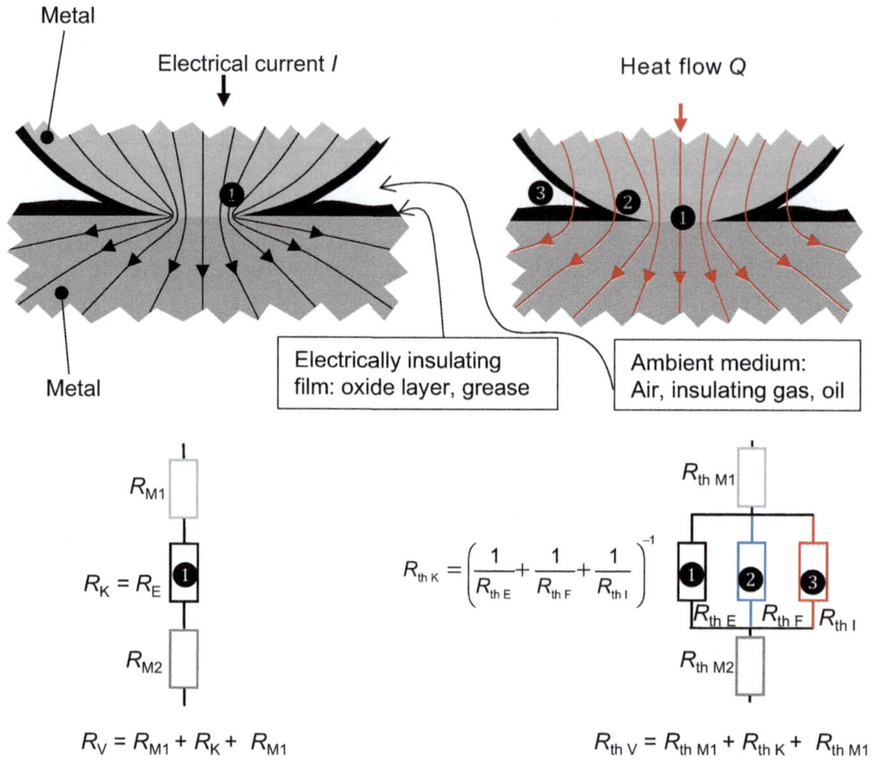

Transport paths in contact					
	Replacement		Designation	Effectiveness	
	electric	thermal		electrically	thermally
1	R_E	$R_{th\,E}$	Metallic and quasi-metallic contact surface	Yes	Yes
2	-	$R_{th\,F}$	Gap filled with film (e.g. oxide layer, grease)	no	Yes
3	-	$R_{th\,I}$	Gap filled with insulating medium (e.g. air, oil)	no	Yes

Fig. 4.36 Transport path for the electrical current I and the heat flow Q as well as the equivalent circuit diagrams for a simplified model connection with a microcontact

For metals, the electrical and thermal conductivity are linked via the Wiedemann-Franz-Lorenz law (Eq. (3.2)), which can be used to calculate the thermal resistance R_{th} of a metallic body from the equivalent electrical resistance R calculate:

$$R_{th} = \frac{\kappa}{\lambda} R_{el} \approx \frac{1}{LT} R_{el} \tag{4.51}$$

Equation (4.51) can also be used for connections, but is not applicable to all elements of the equivalent circuit diagram:

Table 4.14 Ratio of electrical and thermal conductivity of materials in current-carrying connections [59]

Conductor material/Foreign layer/Ambient medium	Electrical conductivity κ		Thermal conductivity λ	
	Value in S/m	Ratio to conductor	Value in W/(K · m)	Ratio to conductor
Aluminium (EAl99.5)	$3.4 \cdot 10^{7}$	–	220	–
Aluminium oxide Al_2O_3	10^{-12}	**$2.9 \cdot 10^{-20}$**	19 … 30	**$8.6 \ldots 13.6 \cdot 10^{-2}$**
Air	$5 \cdot 10^{-14}$	**$1.5 \cdot 10^{-21}$**	$2.6 \cdot 10^{-2}$	**$1.8 \cdot 10^{-4}$**
Copper (Cu-ETP)	$5.6 \cdot 10^{7}$	–	385	–
Copper oxide Cu_2O	$1.8 \cdot 10^{-5}$	**$2.7 \cdot 10^{-13}$**	6.3	**$1.6 \cdot 10^{-2}$**
Transformer oil	$2 \cdot 10^{-13}$	**$3.6 \cdot 10^{-21}$**	0.12	**$3.1 \cdot 10^{-4}$**

- R_M, R_E → applicable
- R_K, R_V → limited applicability

The effective thermal contact resistance $R_{th\,K}$ is always smaller than that calculated from the electrical contact resistance R_K using Eq. (4.51) due to the heat conduction through films and the insulating medium (paths ❷ and ❸ in Fig. 4.36). The following applies:

$$R_{th\,K} \leq \frac{1}{LT} R_{el\,K} \text{ and } R_{th\,V} \leq \frac{1}{LT} R_{el\,V} \tag{4.52}$$

4.4.1.2 Point and Line Contacts with Noble Metal Coating

The thermal and electrical contact resistance of silver-plated high-current plug-in connections with spring contact elements and lubricating grease were determined experimentally. In the operating state, the thermal connection resistance was only (3 … 18)% smaller than that calculated with Eq. (4.51) (Fig. 4.37) [50].

The true and quasi-metallic contact areas are decisive for the electrical and thermal contact resistance (path ❶ in Fig. 4.36). The heat flow through films (lubricating grease in [50]) and insulating medium (air in [50]) is, however, low (paths ❷ and ❸ in Fig. 4.36). If the electrical contact resistance is deliberately increased in plug-in connections due to too much lubricating grease and new contacting in the “cold” state, the thermal contact resistance does not increase to the same extent, as the lubricating grease carries a larger part of the heat flow. In this case, the thermal contact resistance is up to 58% smaller than that calculated with Eq. (4.51). Similar results were experimentally determined for blade contacts for fuse inserts (silver-plated brass on silver-plated copper). The determined thermal resistance is (0 … 34)% smaller than that calculated according to Eq. (4.51) [62].

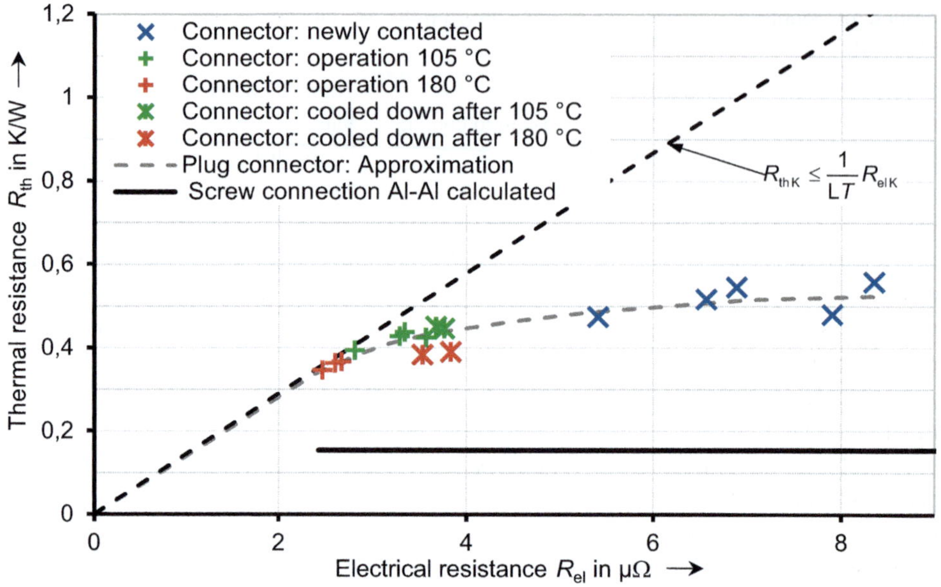

Fig. 4.37 Thermal and electrical connection resistances of a plug and a screwed connection at different operating states and pretreated contact areas

4.4.1.3 Plane Contacts

In the case of plane contacts in screwed connections, the additional transport paths for the thermal flow field are significantly more important, as the contact partners touch each other over a relatively large area via the foreign layers (mechanically supporting contact area, path ❷ in Fig. 4.36) and the contact partners are also connected only by a few µm wide gap filled with insulating medium (apparent contact area, path ❸ in Fig. 4.36). Accordingly, the two parallel heat transport paths in the resulting thermal contact and connection resistance must be taken into account. This means that the resulting thermal contact resistance is significantly lower than that which can be calculated with Eq. (4.51). The thermal resistances of the additional paths can be determined with (Eq. (4.53)

$$R_{\text{th}} = \frac{l_{\text{F}}}{\lambda A} \tag{4.53}$$

Here, s is the thickness of the film or the insulating medium, λ is the thermal conductivity of the foreign layer or the insulating medium, and A is the respective effective cross-sectional area. It is assumed that a gas gap is only a few micrometers large and no molecular movement is possible, so the pure thermal conductivity of the air must be taken into account. The thermal contact resistance from the electrical contact resistance and the thermal resistance of the foreign layers is calculated using the example of a 10

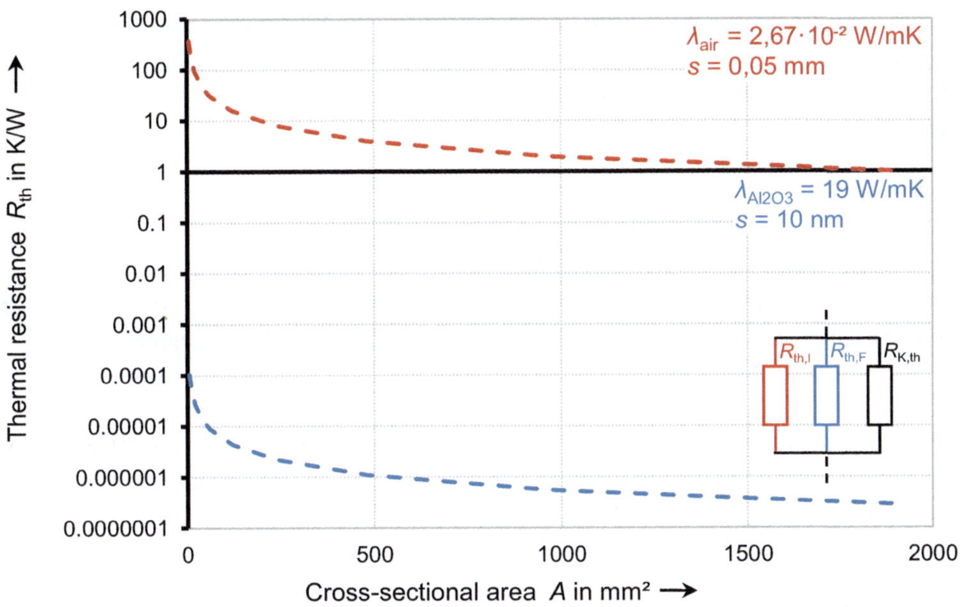

Fig. 4.38 Thermal contact resistance of a 0.05 mm thick air gap and a 10 nm thick oxide layer of Al_2O_3 depending on the respective effective cross-sectional area

nm thick aluminum oxide layer and a 50 μm large air gap. Since the film is three orders of magnitude thinner than the air gap (10 nm vs. 50 μm) and the thermal conductivity is also three orders of magnitude larger, the thermal contact resistance is six to seven orders of magnitude smaller with a similar effective area and therefore dominant in a parallel circuit (Figs. 4.36 and 4.38).

Example

From extensive investigations into the influence of the pretreatment of the contact area, it is known that the electrical connection resistance can vary enormously at identical connections. In the case of uncoated screwed connections with aluminum busbars and an apparent contact area $A_S = 30\ \text{mm} \times 40\ \text{mm} = 1200\ \text{mm}^2$, performance factors in the range $k_{u\,0} = 1.1 \ldots 90$ were measured [61]. This corresponds to good to catastrophic assembly quality. The electrical material resistance is approximately constant $R_M \approx 1.1\ \mu\Omega$. The associated electrical narrow resistances were in the range $R_E = (1.3 \ldots 200)\ \mu\Omega$. With Eq. (4.51), the associated thermal narrow resistance $R_{th\,E}$ and the thermal material resistance $R_{th\,M}$ can be calculated:

$$R_{th\,E} = (0.19 \ldots 27)\ \text{K/W}$$

$$R_{th\,M} = 0.15\ \text{K/W}$$

To this, the thermal resistances through the film layer of aluminum oxide and the insulating medium air are connected in parallel. The area shares are each approximated with 10% ($A_{Al2O3} = 0.1 \cdot A_S = 120\ \text{mm}^2$) and 90% ($A_{Air} = 0.9 \cdot A_S = 1080\ \text{mm}^2$) of the apparent contact area A_S. With Eq. (4.53), the following results:

$$R_{th\ F} = 4 \cdot 10^{-6}\ \text{K/W}$$

$$R_{th\ I} = 0.02\ \text{K/W}$$

Due to the very small thermal contact resistance through the film $R_{th\ F}$, the thermal narrow resistance $R_{th\ E}$ is short-circuited (see Fig. 4.36) and the following applies:

$$R_{th\ K} \approx R_{th\ F} \approx 0 \rightarrow R_{th\ V} \approx R_{th\ M} = 0.15\ \text{K/W}$$

Thus, $R_{th\ V}$ of this exemplary screwed connection is independent of the electrical contact quality determined by the pretreatment. Although k_{u0}, R_V and R_E change in a wide range, $R_{th\ V}$ is practically constant (Fig. 4.37). ◀

4.4.2 Voltage-Temperature Relationship

Steady Load

In the case of thermally steady loading of a contact with a current I, the contact (Fig. 4.39) reaches the steady final temperature. Generally, this is the case when the load duration is many times longer than the thermal time constant of the conductor or contact. Kohlrausch and Diesselhorst showed that in a current-carrying, metallic conductor in the steady state, the equipotential surfaces are also isothermal surfaces if no heat is dissipated to the surroundings [54, 55].

Holm, as well as Greenwood and Williamson, used this knowledge to calculate the temperature distribution from the voltage drop across a contact point U_K [2, 56]. They adapted the so-called voltage-temperature relationship to the electrical contact (Fig. 4.40).

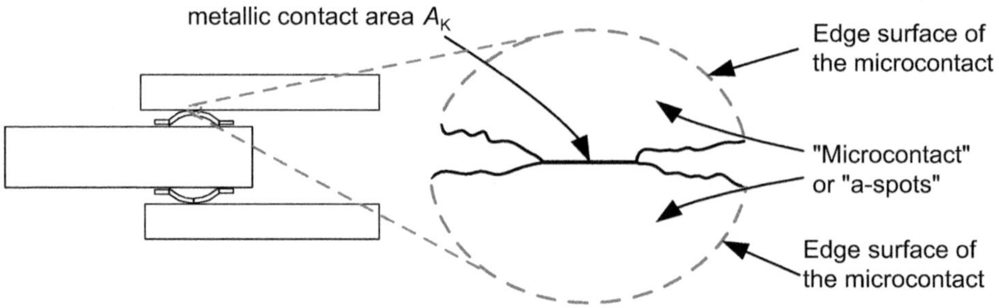

Fig. 4.39 Microcontact using the example of a plug-in connection

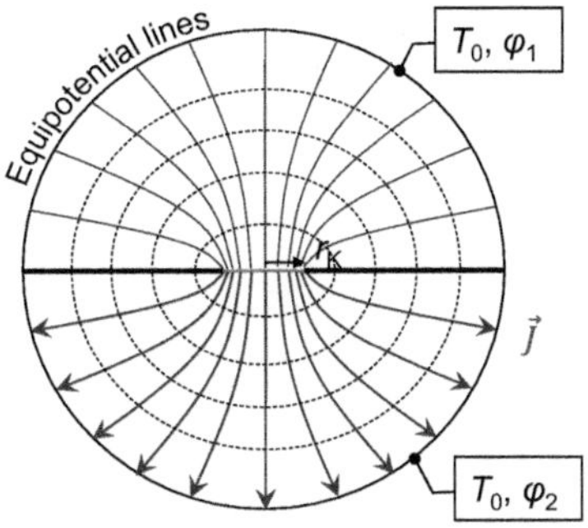

Fig. 4.40 Electrical-thermal model concept for the ellipsoid model

The assumption is that heat flow occurs exclusively through heat conduction, so neither convection nor radiation occur. If the Wiedemann-Franz law is valid for the contact and the opposing, outer boundary surfaces of the point contact have the absolute temperatures T_1 and T_2, the maximum temperature T_{max} in the contact can be calculated analytically:

$$T_{\max} = \sqrt{\frac{U_K^2}{4\,L} + \frac{1}{2}\left(T_1^2 + T_2^2\right) + \frac{L}{4\,U_K^2}\left(T_1^2 - T_2^2\right)^2} \tag{4.54}$$

For symmetrically heated connecting conductors ($T_1 = T_2$), Eq. (4.54) applies in a simplified form (see Sect. 4.1.3):

$$T_{\max} = \sqrt{\frac{U_K^2}{4\,L} + T_1^2} \tag{4.55}$$

This relationship is independent of the actual shape and size of the contact area. For a contact described with the ellipsoid model, the temperature distribution can be calculated numerically (Fig. 4.41). In symmetrically shaped contacts, the maximum temperature occurs at the point of contact between the contact partners when they are heated to the same temperature.

The maximum temperature of the contact depends on the voltage drop and the temperature of the connecting conductors ϑ_1 and ϑ_2 (Figs. 4.41 and 4.42).

Compared to the connections, an excess temperature ($\vartheta_{max} - \vartheta_1$) occurs in the contact. This excess temperature depends on the contact voltage but also on the temperature of the connecting conductors. With the same contact voltage U_K, warmer connecting conductors lead to a higher maximum temperature in the contact, especially at low contact voltages that occur in practical applications (Fig. 4.42). The excess temperature is lower for heavily heated connecting conductors than for connecting conductors at room temperature. In electrical power engineering, contacts have the task of carrying a certain current I. The instantaneous value of the current i is determined by the network volt-

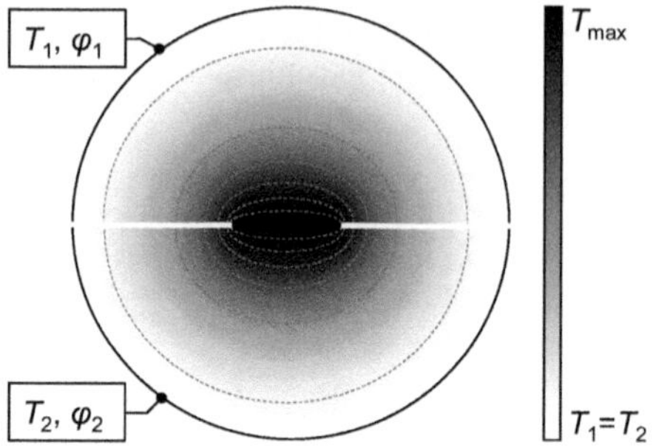

Fig. 4.41 Temperature distribution in a stationary electrical contact (ellipsoid model)

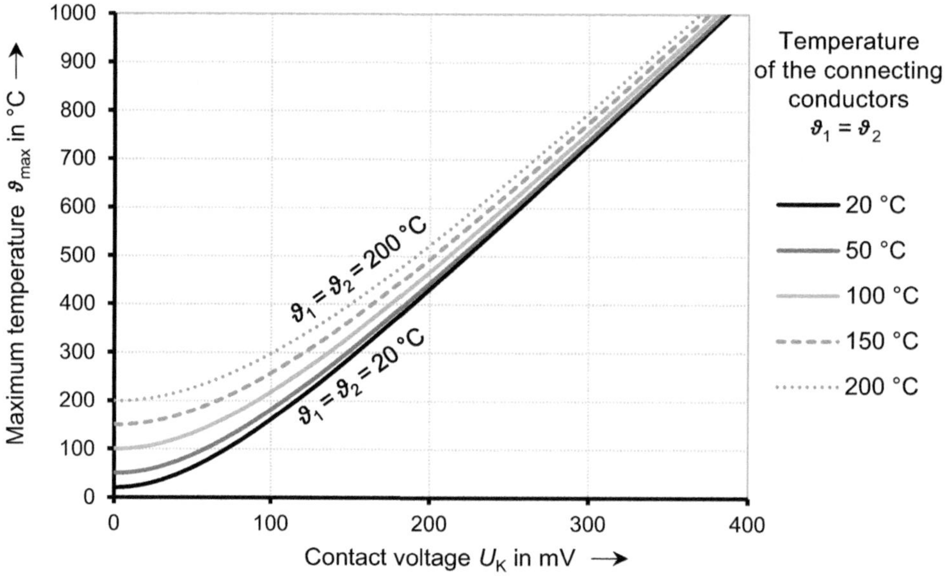

Fig. 4.42 Maximum temperature of a contact depending on the contact voltage and temperature of the connections

age and the network and load impedances and is not influenced by the contact resistance itself. The following considers a film-free contact, so that the constriction and contact resistance are identical. The previously used contact voltage u_K as a control variable is determined according to the temperature-dependent(fromϑ_{max} and u_K itself) constriction resistance R_E and the current i:

$$u_K = R_E \cdot i \tag{4.56}$$

Under certain boundary conditions (maximum temperature within the contact, high current load), an explicit expression for $T_{max} = f(i)$ can be given [54]:

$$T_{\max} = \sqrt{\frac{2T_1T_2\cos\left(i\frac{R_{\mathrm{E\,T1}}}{\sqrt{\mathrm{L}T_1}}\right) - T_1{}^2 - T_2{}^2}{\cos^2\left(i\frac{R_{\mathrm{E\,T1}}}{\sqrt{\mathrm{L}T_{\mathrm{T1}}}}\right) - 1}} \tag{4.57}$$

As the current i increases, the contact voltage u_K also increases (Eq. (4.56). This leads to a higher maximum temperature ϑ_{max} inside the contact (4.57). The stronger internal heating of the contact leads to an increase in the constriction resistance (Fig. 4.44). If i is increased even further, u_K and ϑ_{max} increase disproportionately due to the also increasing resistance. Consequently, a contact can only be loaded with a certain maximum current, even with very good cooling, until the melting temperature is reached inside. Since the contact area is heated inhomogeneously (Fig. 4.41), the electrical constriction resistance $R_{E\,max}$ increases compared to a homogeneously heated contact $R_{K\,T1}$ with $T_{max} = T_1 = T_2$. Greenwood and Williamson [56] calculated this resistance increase for the case that a local temperature maximum exists in the contact. Based on this, the equation was transformed as follows [50] (Fig. 4.43):

$$\frac{R_{\mathrm{E\ max}}}{R_{\mathrm{E\,T1}}} = \frac{1}{T_1}\frac{\sqrt{T_{\max}^2 - T_1^2} + \sqrt{T_{\max}^2 - T_2^2}}{\arccos\frac{T_1}{T_{\max}} + \arccos\frac{T_2}{T_{\max}}} \tag{4.58}$$

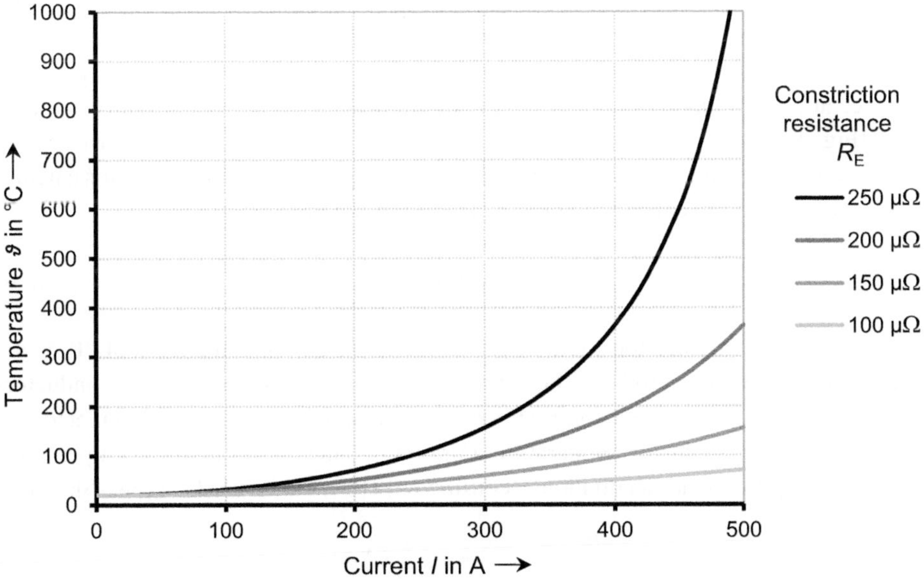

Fig. 4.43 Maximum temperature in the contact as a function of the current depending on the constriction resistance R_E, calculated with Eq. (4.57), $\vartheta_1 = \vartheta_2 = 20$ °C

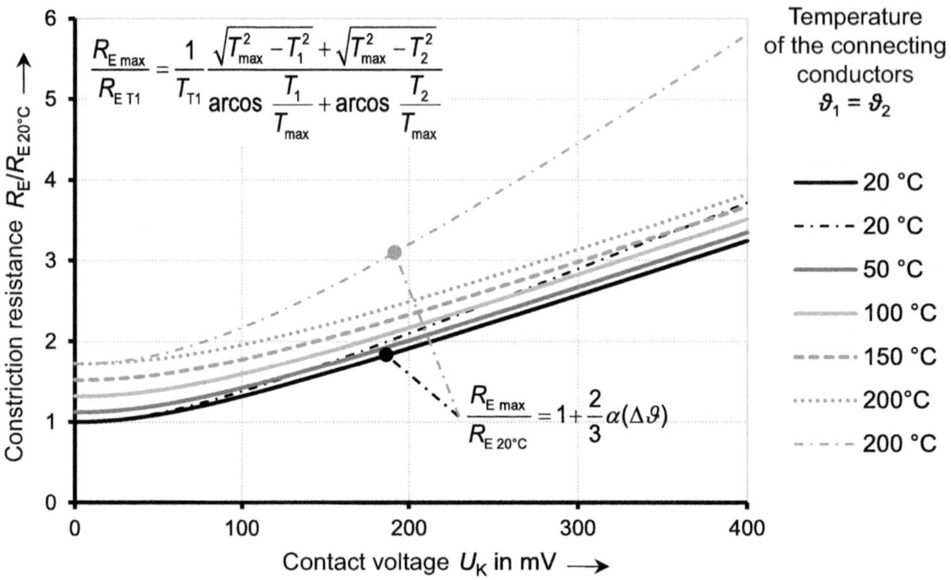

Fig. 4.44 Constriction resistance R_E based on the constriction resistance at 20 °C depending on the contact voltage and the temperature of the connections

It should be noted that in Eq. (4.61) for the calculation of the effective constriction resistance $R_{E\ max}$, the resistance of the contact heated to the temperature of the connection T_1 $R_{E\ T1}$ must be used. If this effective constriction resistance is related to the constriction resistance at room temperature $R_{E\ 20°C}$, it becomes clear that otherwise a significant error occurs (Fig. 4.44).

The well-known and often used approximation equation according to [2] for calculating the contact resistance depending on the maximum temperature

$$\frac{R_{E\quad max}}{R_{E\quad 20°C}} = 1 + \frac{2}{3}\alpha(\Delta\vartheta) \tag{4.59}$$

should be used exclusively for a temperature of the connecting conductors of 20 °C and contact voltages less than 100 mV. For other temperatures of the connecting conductors, large deviations from the exact solution occur, as the approximation was calculated only for this connecting conductor temperature. For other connecting conductor temperatures, the following approximation equation according to [2] should be used (Fig. 4.45):

$$\frac{R_{E\ max}}{R_{E\ T1}} = 1 + \frac{2}{3}\frac{1}{T_1}(\Delta\vartheta) \tag{4.60}$$

In plug-in and sliding connections for transmitting high power, the voltage drop at rated current $U_{K\ r}$ can be used as a nearly geometry-independent criterion (Table 4.3).

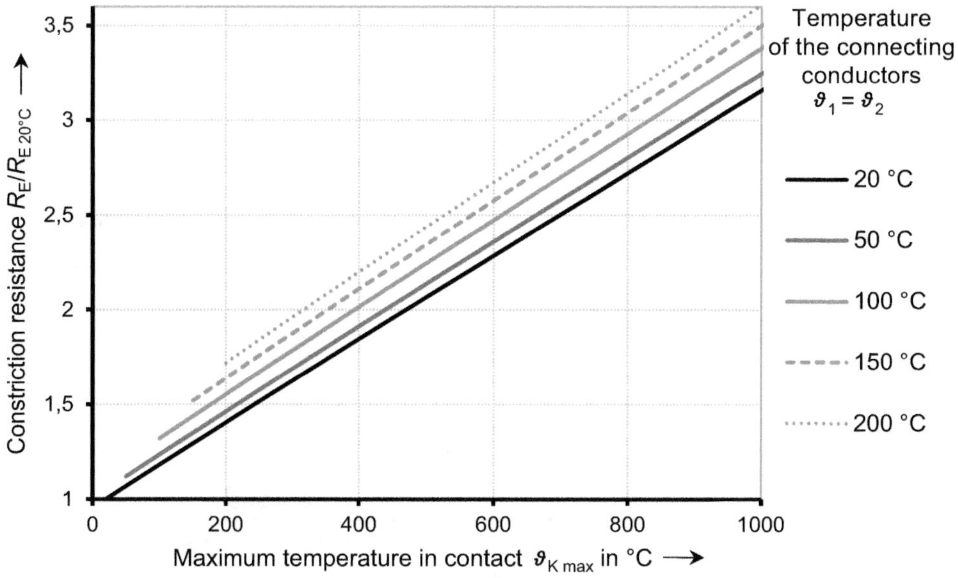

Fig. 4.45 Contact resistance depending on the maximum temperature in the contact and the temperature of the connecting conductors

Transient Load

A contact is in a thermally transient state during transient current load if the load duration or the period of the load current is less than or in a similar time range as the thermal time constant of the contact. The thermal time constant of the contact correlates with the size of the contact area. A calculated minimum load duration can thus be specified, at which, depending on the size of the metallic contact area, a stationary heated contact can still be assumed (Fig. 4.46).

In the transient state, the resulting power loss during current flow is not fully distributed by heat conduction in the contact area. In contrast to the stationary state, the heat capacity of the contact is therefore crucial for its heating during a transient load.

The larger this is, the less the contact heats up during very short current load. In contrast to stationary heating, the equipotential lines of the electrical flow field no longer lie on the isotherms (Fig. 4.47).

The relationship between contact voltage and contact temperature thus depends on the size of the contact area. An extreme case of transient load is the assumption that the generated power loss in the current area cannot be distributed during the load duration. The maximum temperature in the contact is then no longer dependent only on the contact voltage U_K, but is a function of the contact material, the size of the metallic contact area A_W, and the load duration t_K. Hilgarth states for contacts that are loaded briefly, that no

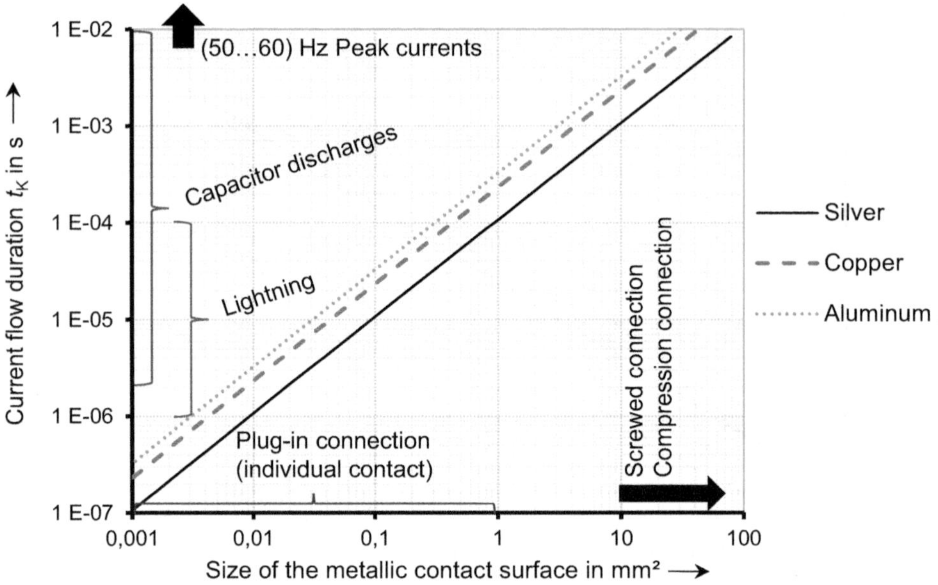

Fig. 4.46 Applicability of the stationary approximation for calculating the temperature of the contact points depending on the duration of current flow and the size of the metallic contact area according to [22, 57], rectangular currents—For current flow durations above the curve, the stationary approximation can be used

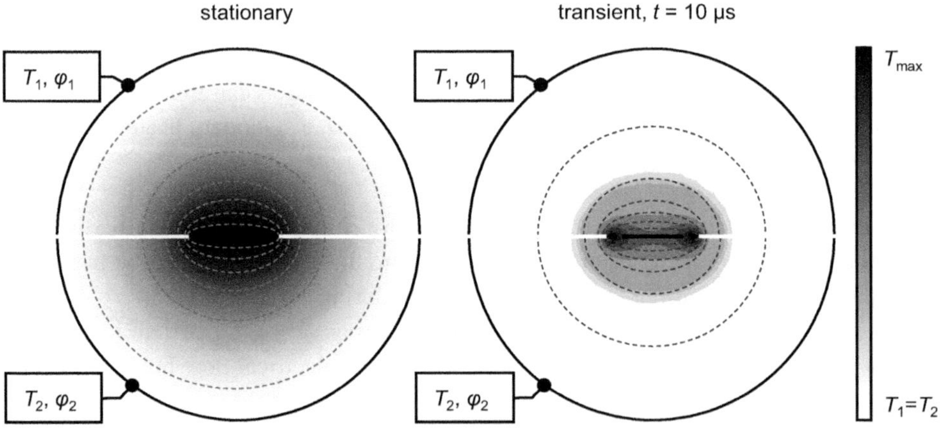

Fig. 4.47 Temperature distribution in the electrical contact during stationary load (left) and short-term, adiabatic heating (right) with the same maximum temperature and equipotential lines (dashed)

heat balance takes place in the contact and it heats up adiabatically [58]. The following relationship between contact voltage in V and maximum temperature in the contact is given for this.

$$U_K = \frac{4}{\pi^2}\sqrt{\frac{c_p \rho \pi \ \ln(1+\alpha\vartheta_{max})}{\alpha}}\left[1+\frac{1}{5}\ln(1+\alpha\,\vartheta_{max})\right]\sqrt{\frac{A_W}{t_K}} \tag{4.61}$$

Here, c_p is the heat capacity at 20 °C in Ws/cm³, ρ is the specific electrical resistance at 20 °C in Ωcm, α is the temperature coefficient of electrical resistance, A_W is the metallic contact area in cm², ϑ_{max} is the maximum temperature in the contact in °C, and t_K is the load duration in s. There is no complete analytical solution for this case, only an approximation, to calculate the change in contact resistance. Hilgarth [58] gives the following approximation for the resistance of a contact that is loaded briefly:

$$\frac{R_{E\ max}}{R_{E\ hom}} = 1 + \frac{1}{5}\ln(1+\alpha(T_{max} - T_{hom})) \tag{4.62}$$

The increase in contact resistance due to contact heating T_{max} is much smaller in these short-term loads than in the steady state, as the surrounding contact material of the contact heats up less than in the steady state (Figs. 4.47, 4.48 and 4.89).

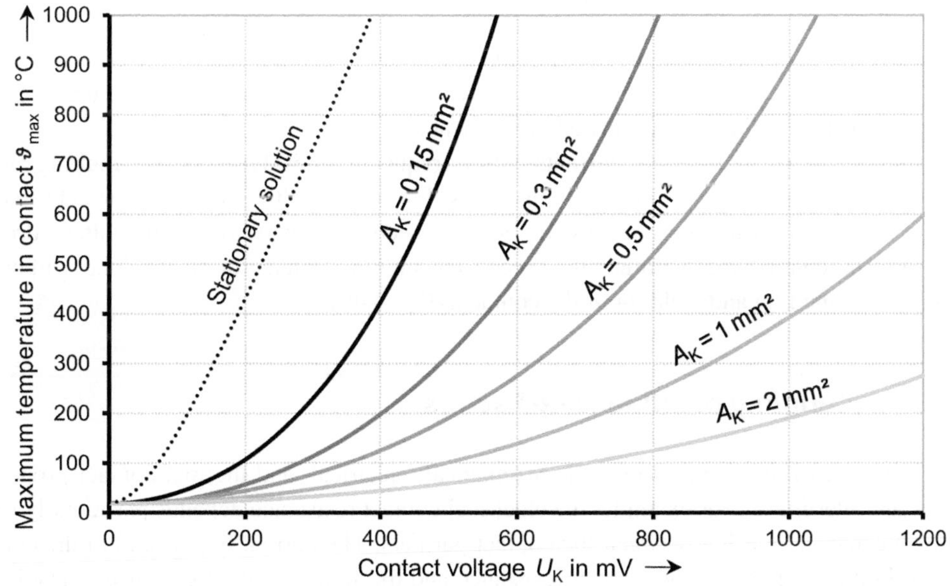

Fig. 4.48 Temperature in a symmetric contact point with copper conductors depending on the size of the contact area for a current flow duration of 10 μs

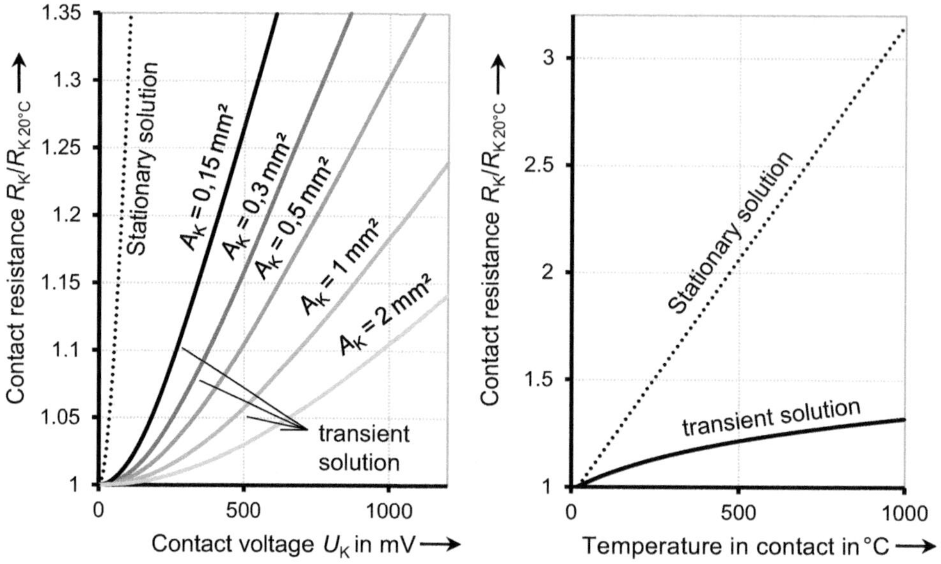

Fig. 4.49 Contact resistance in relation to the unheated contact depending on the contact voltage (left) and the maximum temperature (right) each transiently calculated with Eq. 4.62 for a current flow duration of 10 μs and in the steady state

In relation to the maximum temperature in the contact, the resistance ratio according to Eq. (4.62) is independent of the current flow duration and the size of the contact area, as the temperature balance in the current area is neglected. This transient solution thus represents the limit case in purely adiabatic heating.

In the case of short-circuit currents with a frequency of 50 Hz or 60 Hz, melting and subsequent welding by fusing welding is possible if the melting voltage at steady load U_{KS} is reached. With a current load in the microsecond range or lower, the voltage drop across the contact can significantly exceed the melting voltage at steady load without leading to melting and welding of the contacts (Fig. 4.49).

4.5 Coupling of Influencing Factors

In current-carrying connections, the mechanical, electrical, and thermal contact properties are interlinked. For example, the electrical contact resistance depends particularly on the mechanical tension between the contact partners. The current heat losses in the contact, in turn, depend on the contact resistance and the square of the load current, leading to the heating of the contact and the connection (Fig. 4.50).

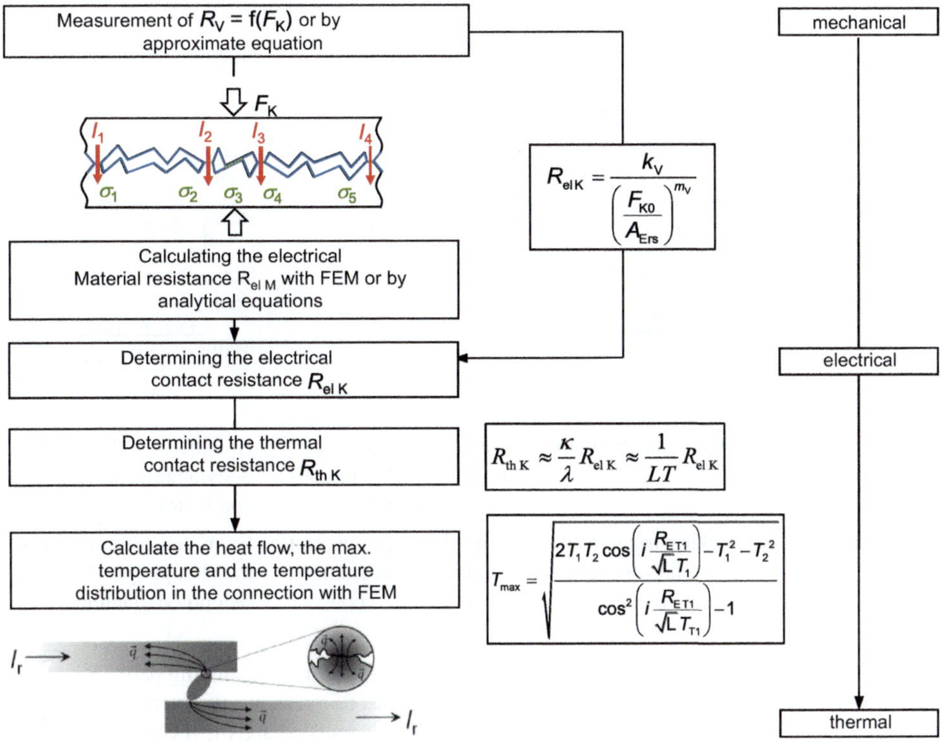

Fig. 4.50 Principle coupling of physical quantities

4.5.1 Screwed connection

The frictional screwed connection is used in electrical energy technology in different designs. The most important representatives are presented below and examined more closely, particularly with regard to their electrical-thermal properties. (Table 4.15). The basic relationships are explained using the example of the screw-nut connection with busbars, which can be transferred to all other types. Specific points are considered separately.

The current flow in classical current-carrying screwed connections in electrical engineering takes place between the conductors via microcontacts (Fig. 4.51a). In some applications, e.g., for potential equalization in vehicle construction, the conductors can be designed with an electrically insulating or very poorly conductive coating for corrosion protection reasons, so that the current flow is only possible via the joining elements (Fig. 4.51b). This creates contact areas between the individual components such as the clamp part/conductor 1 and the screw, the thread of the screw and the nut, and between the nut and the clamp part/conductor 2. If washers are also used, the number of contacts

Table 4.15 Types of screwed connections in electrical energy technology

Design	Screw-nut connection with two clamping parts[1]	Nut connection with clamping part and threaded bolt	Clamping connection
Example connection between	Two busbars 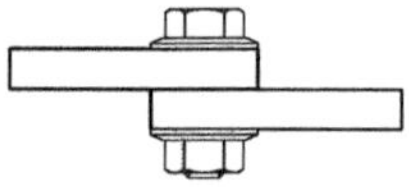 Cable lug/busbar 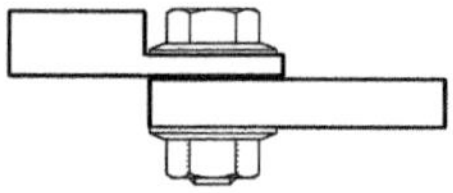 Device connection	Transformer connection 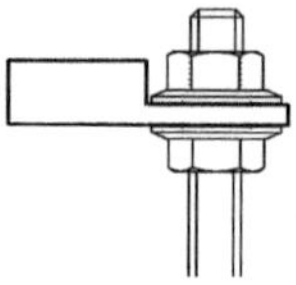	Parallel groove clamp, switchgear terminals 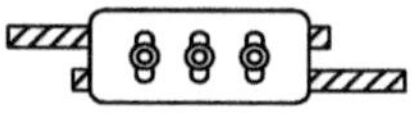 Screw connectors for cables 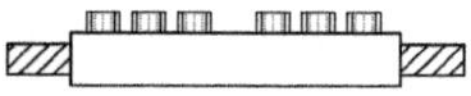
Force effect	Bracing of the clamping parts	Bracing of the clamping part and the threads against the bolt	Bracing of a clamping body against an electrical conductor
Current flow	From clamping part to clamping part, small proportion via the joining elements[2]	From the clamping part via the joining elements to the threaded bolt	From one conductor via the clamping body to the second conductor

[1)] This also includes screwed connections where the thread is cut in the clamping part
[2)] With insulated conductor surfaces e.g. for corrosion protection reasons, the current flows via the joining elements, in this case the current carrying capacity is reduced and the use of special joining elements must be considered [69]

increases accordingly. The connection resistance becomes larger and depends not only on the properties of the conductor materials but particularly on the materials of the joining elements. Therefore, a distinction is made between current-carrying screwed connections with and without insulated clamp parts/conductors. The focus is clearly on screwed connections without insulating clamp parts/conductors, as these are much more common in practice.

Screw-nut connection with two non-insulated clamp parts/conductors
In the case of plane contacts, such as the screwed connection with busbars, the current flow apparently takes place over the entire connection length of the contact partners. In fact, however, the flow lines are constricted in parallel microcontacts, which only occur at the points between the contact partners where a contact force is also transmitted and a mechanical tension is present. The distribution and size of the microcontacts are thus directly dependent on the mechanical stress distribution (Fig. 4.52).

The relationship between the performance factor, the connection or contact resistance, and the contact force or average mechanical tension can so far only be determined experimentally (Sect. 4.3). Approaches to determine this relationship based on experi-

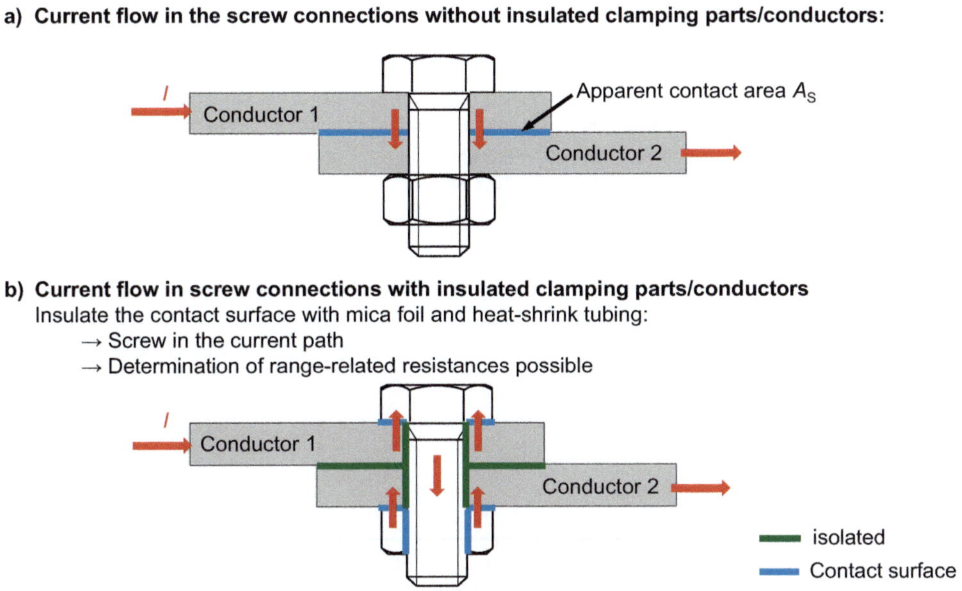

Fig. 4.51 Screwed connection **a**) with and **b**) without insulated contact area between the busbars [69]

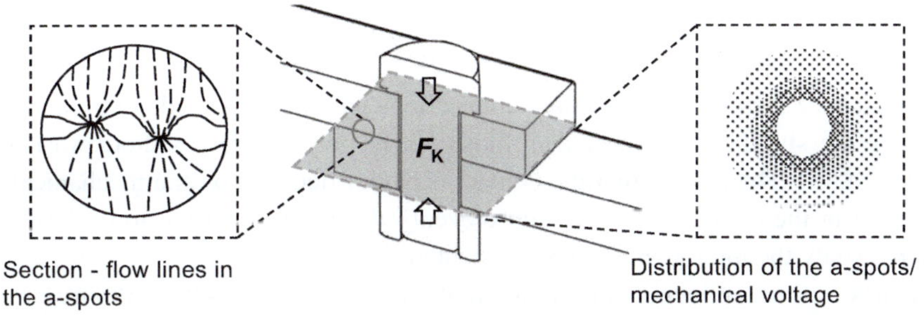

Fig. 4.52 Constriction of the flow lines in the microcontacts using the example of a screwed connection with busbars

mentally determined data have already been presented (Sects. 4.3.1 and 4.3.3). The basis for this are tension tests on contact partners and joining elements, in which the contact force is gradually increased. Subsequently, the average mechanical tension can be calculated (Sect. 4.2.1, Fig. 4.53). It becomes clear that the gradient and the absolute value of the performance factor strongly depend on the joining elements used. The more evenly the force is introduced into the contact partners, the more parallel microcontacts are created and the performance factor decreases significantly even at low average mechanical stresses. At the same average mechanical stress, especially in the range $\sigma_m < 20$ N/mm^2,

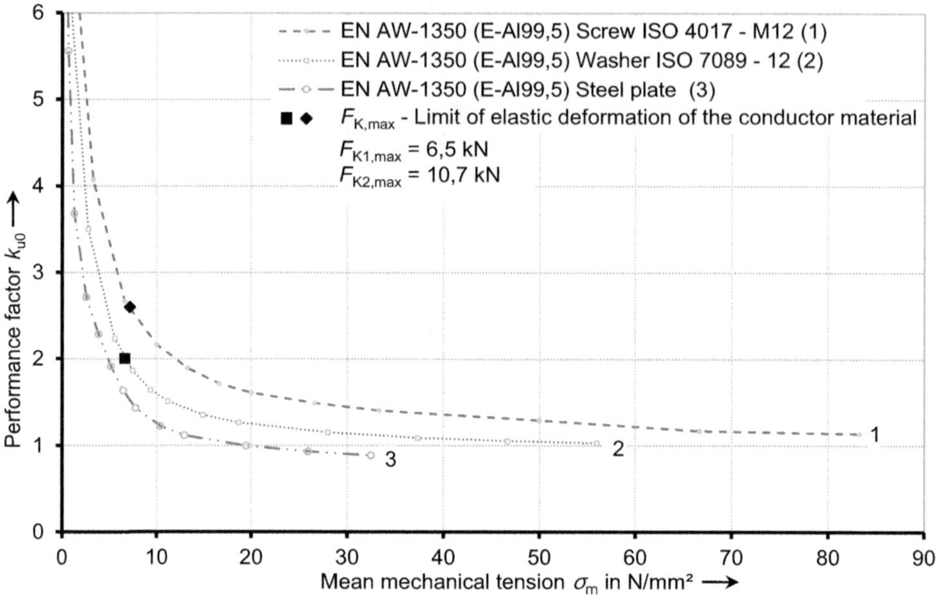

Fig. 4.53 Initial value of the performance factor k_{u0} dependent on the average mechanical tension σ_m exemplarily for busbars with a cross-section of (40 × 10) mm^2 made of E-Al99.5 and different joining elements (average mechanical tension calculated with Eqs. (4.11) and (4.12)). (Maximum contact force $F_{K'max}$ calculated with a yield point $R_{p0.2\text{-}E\text{-}Al99,5\text{-}O} = 20$ N/mm^2 [26, 27])

the performance factors of the screwed connections with busbars made of aluminum and steel plate are significantly lower than those where only a screw without washers was used (Fig. 4.53). This shows that the contact behavior depends on the mechanical stress distribution in the contact area. The more evenly the mechanical tension is distributed, the more evenly the microcontacts are distributed.

This has a direct influence on the distribution of current density in the true metallic contact area. With a uniform distribution of microcontacts, the entire size of the contact area is well utilized and a low connection resistance is created (Fig. 4.54, Table 4.16).

If the contact area and the conductor are homogeneously permeated, i.e., the existing conductor material is well utilized, the material resistance and thus the connection resistance is at its lowest. The material resistance and optimal performance factor can then be estimated (Eq. (4.67)). Furthermore, Fig. 4.53 shows that a minimal mechanical stress is necessary to elasto-plastically deform the micro-peaks of both contact partners, break open thin foreign layers, and thus achieve a low contact resistance. For connections with additional pressure plates, the mechanical stress is more homogeneously distributed on the contact area. The generated micro-contacts are more evenly distributed, the flow lines are hardly influenced before commutation to the other contact partner and only constricted in the micro-contacts. As a result, the current density in the contact partners is

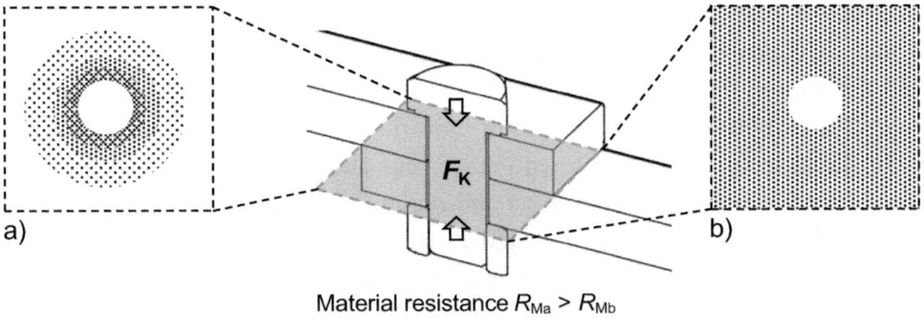

Fig. 4.54 Distribution and size of microcontacts in inhomogeneous as well as homogeneous mechanical stress distribution **a**) screw-nut connection and **b**) additional pressure plates (schematic). (Material resistance $R_{Ma} > R_{Mb}$)

Table 4.16 Calculation of the constriction resistance for the multipoint contact (based on [5])

Multipoint contact	Microscopic surface roughness	Structured surface	
Assumption	R_E with interaction 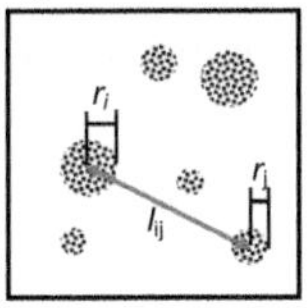	R_E without interaction 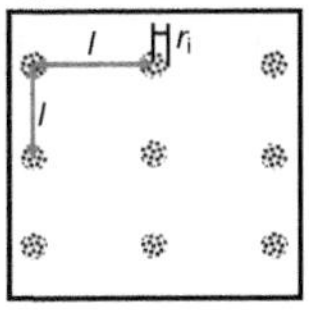 $\bar{l} > 20 \cdot r_i$	R_E with interaction 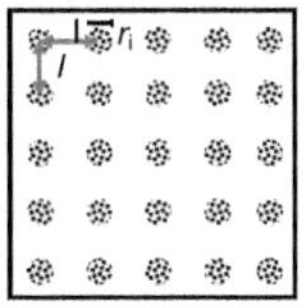 $\bar{l} < 20 \cdot r_i$
Equation	(4.63)	(4.64)	(4.65)
Parameter	ρ_M … spec. electrical resistance, n_E … number of microcontacts, r_i … radius of microcontact, l … distance between microcontacts		

more homogeneously distributed than in the other two connection variants, and the proportion of material resistance is minimal (Fig. 4.54).

$$R_E = \frac{\rho_M}{2\sum_{i=1}^{n} r_i} + \frac{\rho_M}{\pi} \frac{\sum_{i=1}^{n}\sum_{j\neq 1}^{n} \frac{r_i r_j}{l_{ij}}}{\left(\sum_{i=1}^{n} r_i\right)^2} \tag{4.63}$$

$$R_{E,E} = \frac{1}{n_E}\frac{\rho_M}{2r_i} \tag{4.64}$$

$$R_{E,E} = \frac{1}{n_E}\frac{\rho_M}{2r_i}\sqrt{\frac{\left(\frac{l}{2}\right)^2 - r_i}{r_i^2}} - 0.6\frac{\rho_M}{A_S}\sqrt[2]{\left(\frac{I}{2}\right)^2 - I^2} \tag{4.65}$$

The interaction between the micro-contacts also needs to be considered, which has an additional influence on the constriction resistance (Table 4.16). A prerequisite for good current distribution through evenly distributed micro-contacts is careful pre-treatment of the contact area before joining, so that a uniform topography is created on the contact areas. In the case of connections with screw-nut and screw-nut with additional washers, the difference between the average and maximum mechanical stress is significantly larger. This behavior has been repeatedly demonstrated in FEM calculations [8, 9]. Large-scale micro-contacts are generated, particularly in areas of high mechanical stress. With decreasing mechanical stress, for example, at a greater distance in the contact plane from the screw, the size and number of micro-contacts decrease again. No micro-contacts are generated outside the pressure cone. The flow lines are thus constricted more strongly in the conductor and subsequently in the micro-contacts. The conductor cross-section is used less efficiently, and thus the proportion of material resistance and constriction resistance in the connection resistance is higher. For the busbars made of Al99.5 examined in [52], it also becomes clear that the maximum permissible contact force calculated in accordance with the design guidelines [26, 27, 51] is too low to achieve an initial value of the performance factor $k_{u0} < 1.5$ (Fig. 4.53). For similar connections with busbars made of AlMgSi0.5 or Cu-ETP, on the other hand, the contact forces are sufficiently high, as the mechanical strength of the material is higher [52]. Due to the higher strength, a higher surface pressure can be generated without reaching the yield point of the material and a macroscopic plastic deformation occurs.

If the contact area is carefully pre-treated and the connection is made immediately afterwards, values of the performance factor in the new condition close to the optimal performance factor k_{uI} can be achieved for plane contacts with approximately evenly distributed mechanical stress in the contact plane (Fig. 4.55, Eq. (4.67)). To achieve this, suitable joining elements, such as pressure plates, clamping washers or screws with a large head contact area, can be used. Experiments on screwed connections with different sizes of screws and geometries of the busbars but the same construction show that similar relationships apply and these can be practically considered the same (Fig. 4.56).

Furthermore, the contact resistance can be evaluated. To do this, the material resistance must be subtracted from the measured connection resistance (Fig. 4.5, Eq. (4.3)). In order to determine the material resistance, a FEM model of the screwed connection is created and calculated with this or determined with the approximation equation (Eq. (4.66)). The approximation equation takes into account the evenly distributed change in current density over the apparent contact area [6]. The pressure cone is not considered separately.

$$R_M = k_{uI}(A_B) \cdot R_{BL} \tag{4.66}$$

Furthermore, the optimal performance factor can be estimated as follows:

$$\begin{aligned} k_{uI}(A_B = 0) &= 0.639 \cdot \left(\frac{l_V}{h_L}\right)^{-1.35} + 0.5 \\ k_{uI}(A_B) &= C_B \cdot \frac{\sum A_B}{A_S} \cdot k_{uI}(A_B = 0) + k_{uI}(A_B = 0) \text{ with } A_B = \frac{\pi}{4} \cdot d_B^2 \end{aligned} \tag{4.67}$$

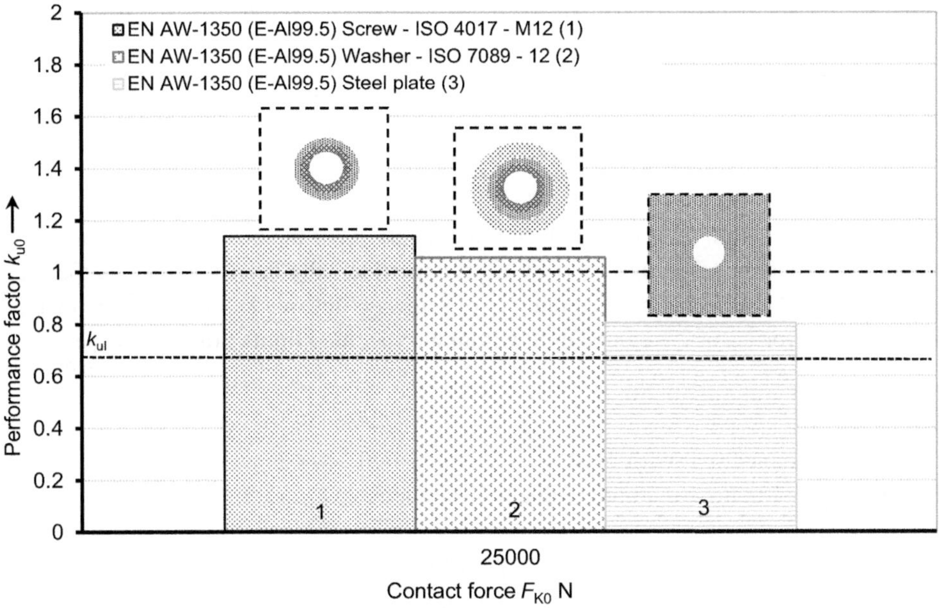

Fig. 4.55 Performance factor k_{u0} compared to the optimal performance factor k_{uI} (screw-nut connection)

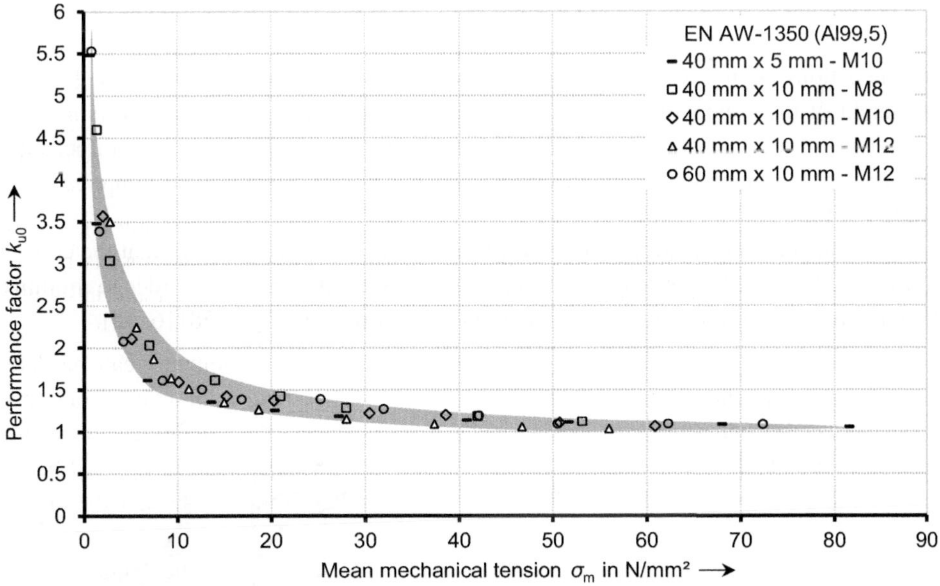

Fig. 4.56 Initial value of the experimentally determined performance factor k_{u0} depending on the average mechanical stress σ_m, size of the screw (ISO 4017), washers of the normal series (ISO 7089) and geometry of the busbars made of E-Al99.5 (connection length = width of the busbar)

Table 4.17 Parameters for the approximation factor C_B [6]

C_B for screwed connections with busbars	Value
(100 × 5) mm four holes	1.8
Intermediate size of the conductor	1.81 … 2.59
(40 × 5) mm (one/two holes)	2.6

The connection length l_V of the conductor, the height h_L of the busbar and the area of the borehole A_B must be taken into account. C_B is experimentally determined and depends on the geometry of the busbar and the number of screws (Table 4.17). For a first estimate, an average value for C_B of 2.3 can be used within the specified limits (Table 4.17) [6].

Example

Screwed connection with busbars of dimensions 40 mm × 10 mm, a connection length $l_V = 40$ mm and a bore with a hole diameter of $d_B = 13$ mm

$$k_{uI}(A_B = 0) = 0.639 \cdot \left(\frac{40\ \text{mm}}{10\ \text{mm}}\right)^{-1.35} + 0.5 = 0.598$$

$$k_{uI}(A_B) = 2.59 \cdot \frac{\pi \cdot \left(\frac{13\ \text{mm}}{2}\right)^2}{40\ \text{mm} \cdot 40\ \text{mm}} \cdot 0.598 + 0.598 = 0.73$$

The optimal performance factor k_{uI} depends on the geometry of the connection. ◀

In connections with overlapping conductors, the ratio of the connection length l_V to the height of the conductor h_L Eq. (4.67) must be considered, as the current is not only constricted in the microcontacts, but also redirected (Fig. 4.57). This has a direct influence on the flow field and the connection resistance and requires a minimum connection length l_V of the contact partners [6, 64].

Experimental investigations have been carried out, which have shown that the connection length l_V should be at least 4 times the thickness of the busbar in order to minimize the influence of the redirection of the flow lines in particular (Fig. 4.58) [6, 65].

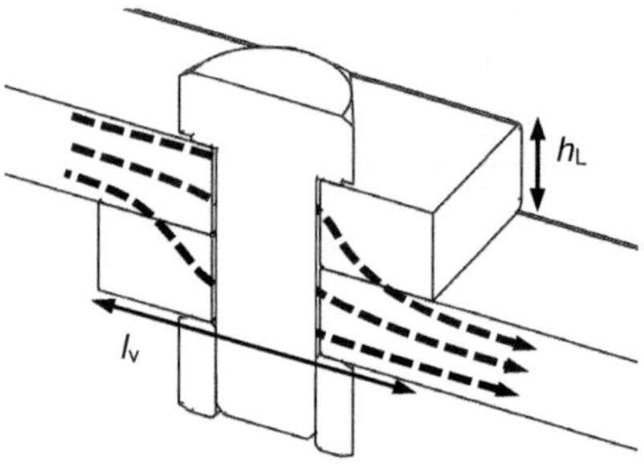

Fig. 4.57 Current redirection in a screwed connection with busbars (schematic)

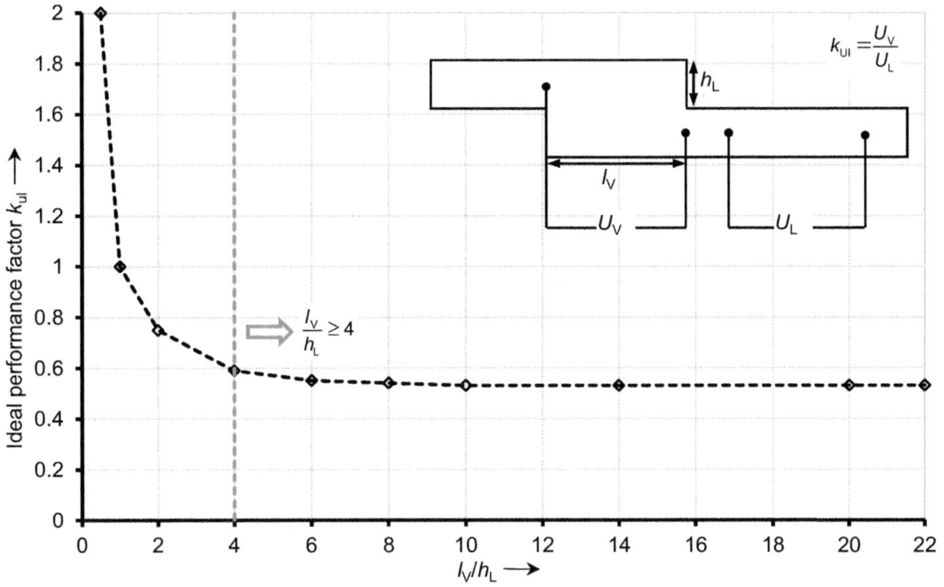

Fig. 4.58 Influence of the connection length l_V to the conductor thickness h_L on the optimal performance factor k_{uI}, milled busbar without hole made of solid material [6]

As a rule, this ensures that for conductor geometries, as used in electrical energy technology, the diameter of the pressure cone in the contact plane between the conductors of the screwed connection fully develops. This creates many parallel microcontacts. Screwed connections with elongated holes in the conductors must be considered separately. These are used to compensate for positional tolerances of the conductors to be joined. The contact area and thus also the number of microcontacts is reduced compared to screwed connections with round holes in the conductors. Investigations on screwed connections with busbars in which elongated holes were introduced resulted in approximately 30% higher connection resistance. These result from a higher contact resistance and the changed current density distribution influenced material resistance of the connection.

Experiments on screw-nut connections with coated busbars made of CW004a (Cu-ETP) show that with silver-plated contact partners, a performance factor of $k_{u0} = 1$ can already be achieved at low average mechanical tensions of $\sigma_m \geq 10$ N/mm^2 (Fig. 4.59). This can be explained by the fact that silver forms only very loosely adhering films, has a moderate hardness and is therefore very well contactable (Sect. 3.2). Connections with tin-plated contact partners achieve a performance factor of $k_{u0} \leq 1.4$ at the same average mechanical stress. No significant difference could be detected whether the contact areas were pretreated with a scouring pad or not. Tin is very soft and can be easily deformed. The oxide layer is also soft and can be easily broken. However, the conductivity of tin is

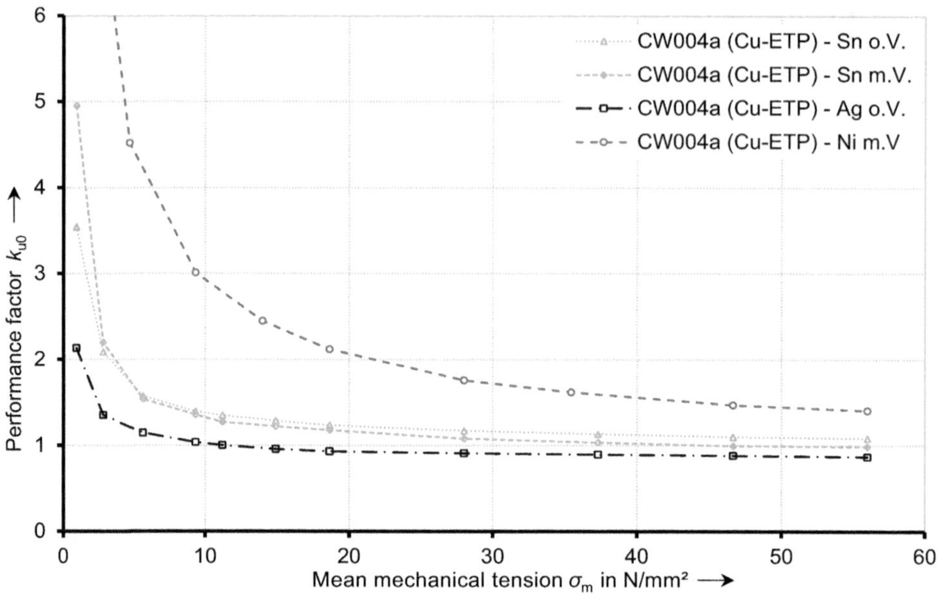

Fig. 4.59 Initial value of the performance factor k_{u0} depending on the average mechanical tension σ_m (Cu-ETP coated without intermediate layers, o. V.—without abrasive pretreatment and m. V.—pretreatment with a grinding fleece) [52]

significantly lower than that of silver, which leads to slightly higher performance factors at the same average mechanical tension as silver (Sect. 3.2). For connections with nickel-plated contact partners, the gradient of the performance factor is significantly larger at low average mechanical stresses. Due to the high hardness of nickel, a much higher average mechanical tension of $\sigma_m > 45$ N/mm^2 must be generated to achieve performance factors of $k_{u0} \leq 1.5$ (Fig. 4.59). Despite the same pretreatment of the contact area, the contact behavior with nickel is significantly worse compared to the coating materials silver and tin (Table 4.18).

Table 4.18 Minimum average mechanical stress/minimum average surface pressure depending on the contact material for static and dynamic load

Material	Load	Al	Cu	Ni	NiP	Sn	Ag	Source
$\sigma_{m\,min}$ [1)]in N/mm^2	Static	20 … 40	20 … 30	45	–	15	10	[52]
		10 … 26	10 … 17	–	33 … 50	2 … 6	3 … 20	[53]
	0.1 … 0.14							
	Dynamic	–	13 … 17	–	29 … 34	16 … 30	13 … 20	[33, 53]

[1)] The values serve as a guide, as they strongly depend on the contact hardness and pretreatment of the contact area. The basis is a performance factor k_{u0} between 1 and 1.5

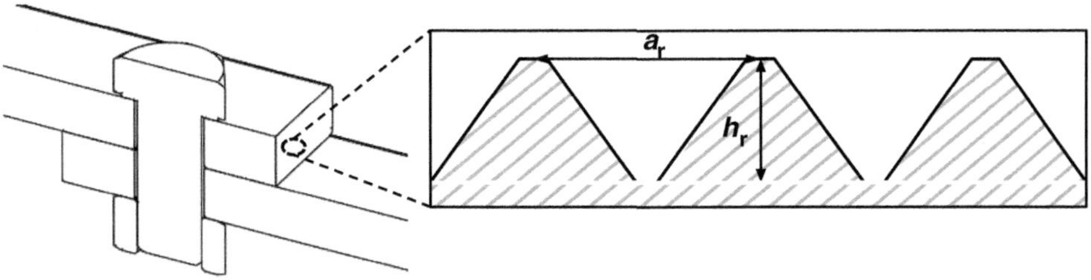

Fig. 4.60 Structuring of the contact area by grooves [6, 8, 63]

In the case of dynamic mechanical tension on the connection, the minimum average surface pressure required for stable contact behavior may be higher. A corresponding safety factor may then need to be determined [7].

With a rough surface of the contact partners, films can be broken up, large-scale micro-contacts can be created, and thus low contact resistances can be achieved during joining through elasto-plastic deformation of the micro-peaks. By pre-treating bare, uncoated conductor materials, such as copper or aluminum with, for example, a wire brush, not only are oxide layers broken up, but the surfaces are also micro-structured and thus the roughness is increased. Both lead to significantly better contact, especially in connections with contact partners made of aluminum. The contact behavior can be further improved by defined structuring of the contact areas (Fig. 4.60) [61]. For grooves, an optimal groove distance a_r of a value between (400 … 600) μm was determined (Fig. 4.61).

For this purpose, studies were conducted on screwed connections with contact partners made of aluminum[5] and grooved contact areas (Fig. 4.61). The optimal groove height h_r is (150 … 230) μm, but has less importance at the same groove distance. It is crucial that the groove height is uniform in order to create as many contact areas as possible during joining. For the design of the contact areas, a cutting manufacturing process with low cutting speed and good cooling is recommended to avoid hardening of the surface and excessive heating of the conductor material [63].

This idea is already being applied constructively in various current-carrying connections, such as screw terminations for cable accessories, parallel groove clamp for overhead lines, or insulation-penetrating connectors in low-voltage technology.

The effect of contact elements is based on the same principle. The geometry is chosen so that defined contact areas in the form of point or line contacts are created. Examples can include contact discs, metal foams, contact elements with lamellas, or even Cupal sheets (Fig. 4.62, Table 4.19).

[5] Tests on busbars made of Al 99.5 with dimensions (50 × 10) mm and a connection length s of 50 mm (apparent contact area $A_s = 2246\ mm^2$).

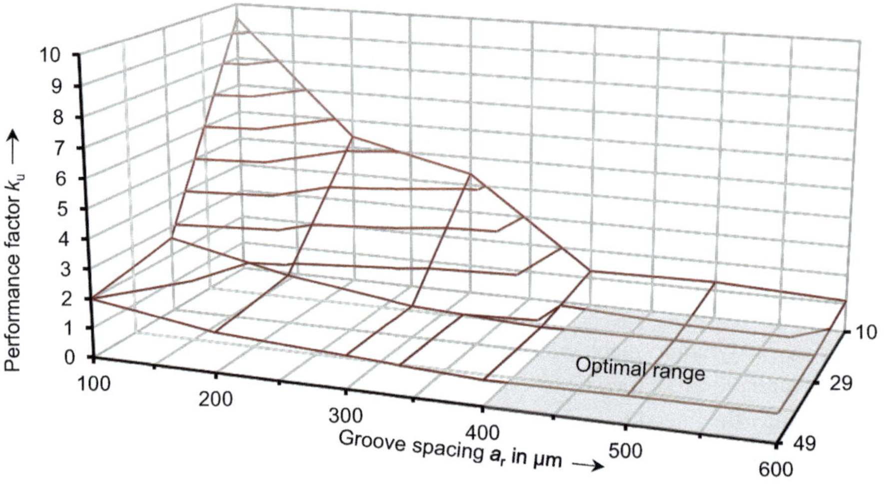

Fig. 4.61 Behavior of the performance factor k_{u0} depending on the groove distance a_r on the contact area and the contact force F_{K0} [63]

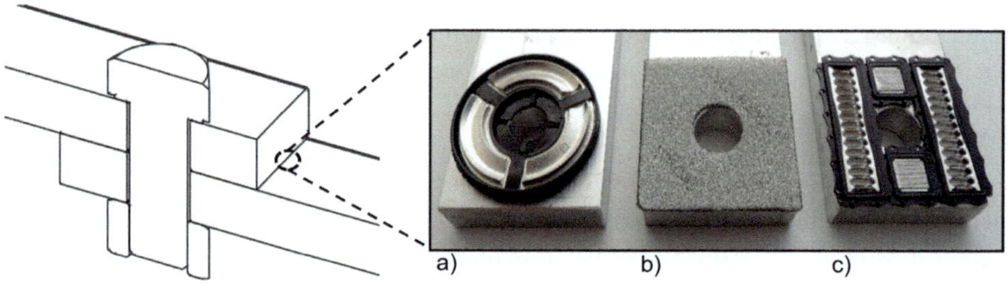

Fig. 4.62 Contact elements for plane contacts **a**) Contact disc, **b**) Metal foams from e.g. Nickel and **c**) Contact element with lamellas [52, 65]

The bridges of the lamellae have the function of pushing through the oxide layers of the two conductors to make electrical contact with them. The same applies to the grooves on the contact disc. Metal foams also press into the contact area of the conductors during the joining of the connection, creating parallel point contacts. All contact elements increase the resistance of the connection more or less strongly due to their own material resistance. However, this can, for example, stably connect different conductor materials over the long term and reduce the effort of pretreating the contact area before joining. For plane contacts, this results in different possibilities to improve the contact behavior compared to the uncoated conductor made of copper and aluminum and other material combinations (Table 4.20).

Table 4.19 Effect and application possibilities of contact elements

Contact element	Details and effect	Applicable to material pairings
Cupal Sheet	– Composite material made of aluminum and copper – Gas-tight due to material connection at the material transition aluminum—copper – Use in screw-nut and clamp connections	Al-Cu
Contact disc or contact elements with lamellas	– Penetration of films – Penetration into the contact partners – Reducing the necessary contact force when joining	Al-Cu Al-Al St-Cu St-Al
Metal foam from for example nickel or copper[1]	– Formation of many parallel contact points – Breaking up of films due to high hardness – Reducing the necessary contact force when joining	Al-Al Cu-St Al-Cu

[1]Pre-treatment of the contact areas made of aluminum still necessary/recommended

In multi-screwed connections with conductors from a width of 120 mm, the busbars should be slotted in the contact area (Fig. 4.63) [66]. In contrast, elongated holes should be avoided as they reduce the effective contact area and increase the contact resistance compared to screwed connections with circular holes. Slitting can reduce the connection resistance by (30 … 40)% compared to a connection with non-slotted busbars [65]. The reason for this is that an approximately equal mechanical stress distribution can develop in the pressure cones of the screws in the partial conductors, which increases the effective contact area by (1.5 … 1.7) times and thus improves the contact behavior with each screw through the electrical parallel connection [65].

The electrical contact resistance and the load current are crucial for the power heat losses in the connection. Both quantities thus significantly determine the temperature of the screwed connection. In the thermally effective contact resistance, the heat flows over small air gaps and foreign layers are also taken into account (Sect. 4.4, Fig. 4.64).

The thermal behavior of screwed connections can be represented using the thermal network method [60], [61]. For this, the screwed connection and the busbars are divided into discrete elements that are coupled to each other and the environment via thermal resistances and capacities (Fig. 4.65).

For stationary heating, no thermal capacities are taken into account, but only the thermal resistances for heat conduction, radiation and convection. The thermal network method is an established procedure that is currently used, for example, to model

Table 4.20 Options for improving the contact behavior of screwed connections

Contact areas Effect on	Unprocessed, bare	Defined structuring/contact element	Galvanic coating	Cold gas spraying
Effort of pretreatment	↑ Al → Cu	→	↓↓ Sn, Ag → Ni, Cu[1]	↓↓ Ag ↓ Cu (rough structure)
Necessary average mechanical tension during assembly	↑ Al → Cu	→ Al ↓ Cu	↓↓ Ag ↓ Sn ↑ Ni	↓ Cu ↓↓ Ag
Advantage	No additional manufacturing costs	– Defined arrangement of microcontacts improves contact behavior – Reducing the effort for pretreatment	– Improving contact and long-term behavior – Minimizing the effort for pretreatment	– Breaking up oxide layers during manufacturing – Material connection between conductor and coating material – Improving contact behavior through a rough contact area
Disadvantage	Pretreatment of contact areas – Specific pretreatment – Well-trained assembly personnel – Safe adherence to assembly parameters	Additional costs for processing the contact area or the contact elements	– Quality of the coating system – Additional costs	Additional costs

[1] Aluminum can be galvanically copper-plated to achieve the properties of copper in contact behavior

current-carrying components in the electrical system of vehicles [67], [68]. With the thermal network model, the temperature equivalent performance factor $k_{uT\text{-}20}$ can be calculated for screwed connections, in which the connection has the same temperature as the connected conductors under stationary current load (Sect. 4.1.2, Table 4.21). In practical applications, the cooling conditions in installation situations, the thermal coupling to neighboring components or also the thermal behavior with time-varying currents can be represented.

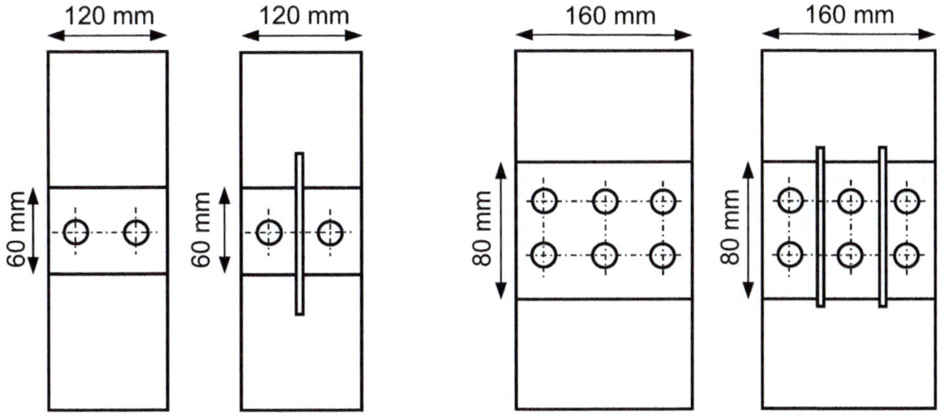

Fig. 4.63 Screwed connection with unslotted and slotted busbar in the contact area as well as **a**) two and **b**) six screws according to [65, 66]

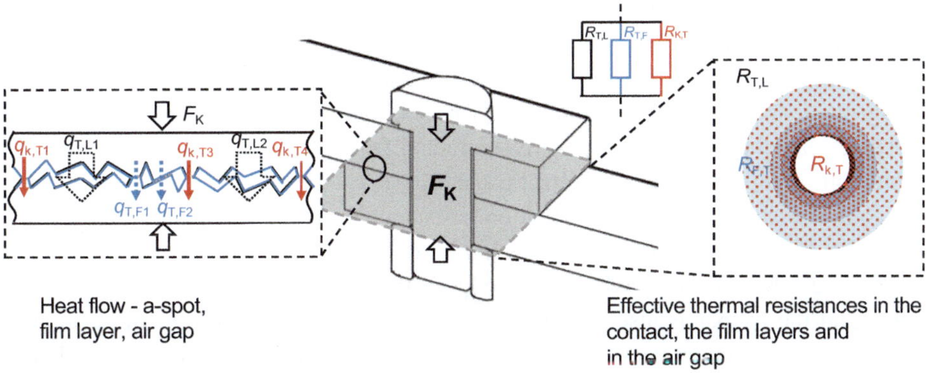

Fig. 4.64 Heat flow in plane contacts over metallic, quasi-metallic and insulating contact areas as well as thin air gaps using the example of a screwed connection with busbars (schematic)

Screw-Nut Connection with Insulated Clamping Parts/Conductors

If the contact area between the conductors is insulated, the current can only flow through the joining elements. This is the case, for example, with surface-protected body parts in a vehicle. If these are to be contacted to carry current, there are some peculiarities that must be taken into account. Thus, depending on the materials used, their surface condition, the relative movement of the individual components to each other during assembly and the resulting number and size of micro-contacts at each contact point, a contact resistance R_{Kn} results (Fig. 4.66). The sum of these contact resistances, added to the material resistances of the flowed-through areas of the joining elements and the conductor sections, then results in the connection resistance R_V (Fig. 4.66).

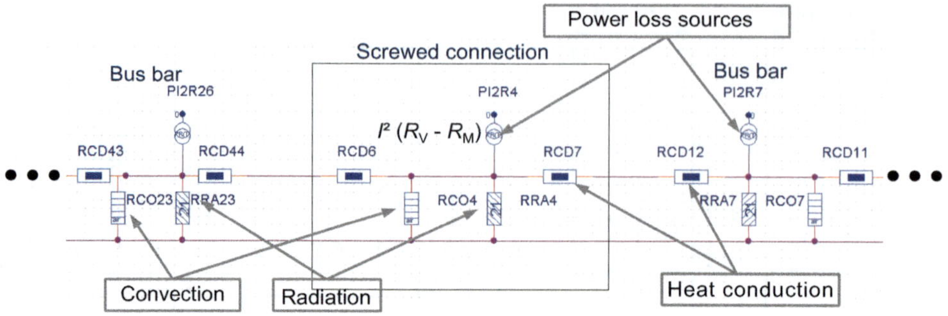

Fig. 4.65 Exemplary excerpt of a thermal network for calculating the stationary heating of a screwed connection with busbars—sources for the power heat losses in the screwed connection and the busbars, thermal resistances for heat conduction, radiation and convection are shown

Screw-nut connections were examined, in which the partial resistances over individual sections of the connection were determined. In order to estimate the contact resistances between the individual components, a FEM model was set up and the proportions of the material resistance between the potential measuring points were calculated. The contact resistance between the components was assumed to be $R_K = 0\ \Omega$. The material resistances calculated in this way were subtracted from the measured partial resistances to obtain the components of the contact resistance (Fig. 4.67).

It could be shown that the connection resistance of screw-nut connections with an insulated surface of the busbars depends in particular on the conductivity of the material used for the screw. In the case of screw-side assembly, the contact resistance between the nut and the busbar is also dominant, as there is no relative movement between the components. The contact resistance between the screw and the nut is very low and hardly noticeable in the illustration (Fig. 4.67). The relative movement between the nut and the screw and the high surface pressures, especially in the first thread turns, obviously result in good contact behavior. This is also confirmed by practical experience with tensioned nuts against threaded bolts made of copper, for example in transformer connections (Table 4.15).

The achieved performance factor of a screw-nut connection with conductors made of aluminum and joining elements made of steel is, in the example shown, about 8.5 times higher than the temperature-equivalent performance factor of $k_{uT\text{-}20} \approx 1.95$ determined for this connection. The already high contact resistance after assembly between the nut and the busbar increased further with thermal load at a temperature of 90 °C depending on the time [69].

The findings show that the current-carrying capacity of screw-nut connections with insulated conductors via the joining elements is too low to permanently carry the rated current of the connected conductor. The connection acts as a hot spot and thus limits the maximum permissible current. By using screws with different underhead geometries, the

Table 4.21 Temperature equivalent performance factor $k_{uT\text{-}20}$ for screwed connections [6]

A in mm^2	l_V in mm	Number × size of the screw	$k_{uT\text{-}20}$
12 × 5	12	1 × M5	1.95
12 × 10	12	1 × M5	2
20 × 5	20	1 × M8	1.84
20 × 10	20	1 × M8	1.96
25 × 5	25	1 × M10	1.81
25 × 5	55	2 × M10	1.72
30 × 5	30	1 × M10	1.65
30 × 5	60	2 × M10	1.62
30 × 10	30	1 × M10	1.78
30 × 10	60	2 × M10	1.69
40 × 5	40	1 × M12	1.54
40 × 5	80	2 × M12	1.54
40 × 10	40	1 × M12	1.64
40 × 10	80	2 × M12	1.58
50 × 5	50	1 × M12	1.41
50 × 5	80	2 × M12	1.46
50 × 10	50	1 × M12	1.51
50 × 10	80	2 × M12	1.52
60 × 5	60	1 × M12	1.48
60 × 5	80	2 × M12	1.43
60 × 10	60	1 × M12	1.53
60 × 5	80	2 × M12	1.47
80 × 5	80	4 × M12	1.5
80 × 10	80	4 × M12	1.54
100 × 5	80	4 × M12	1.45
100 × 10	80	4 × M12	1.49
100 × 10	100	4 × M12	1.46
100 × 15	80	4 × M12	1.56
120 × 10	80	4 × M12	1.5
120 × 15	80	4 × M12	1.56
160 × 10	80	8 × M12	1.54
160 × 15	80	8 × M12	1.59
200 × 10	80	8 × M12	1.46
200 × 15	80	8 × M12	1.52

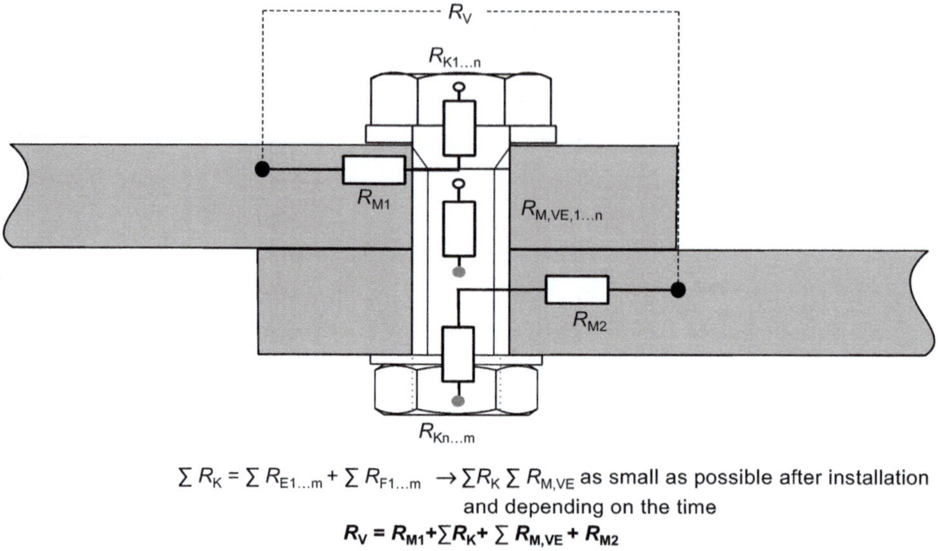

Fig. 4.66 Components of the connection resistance R_V with insulated contact area between the conductors and current flow over the joining elements

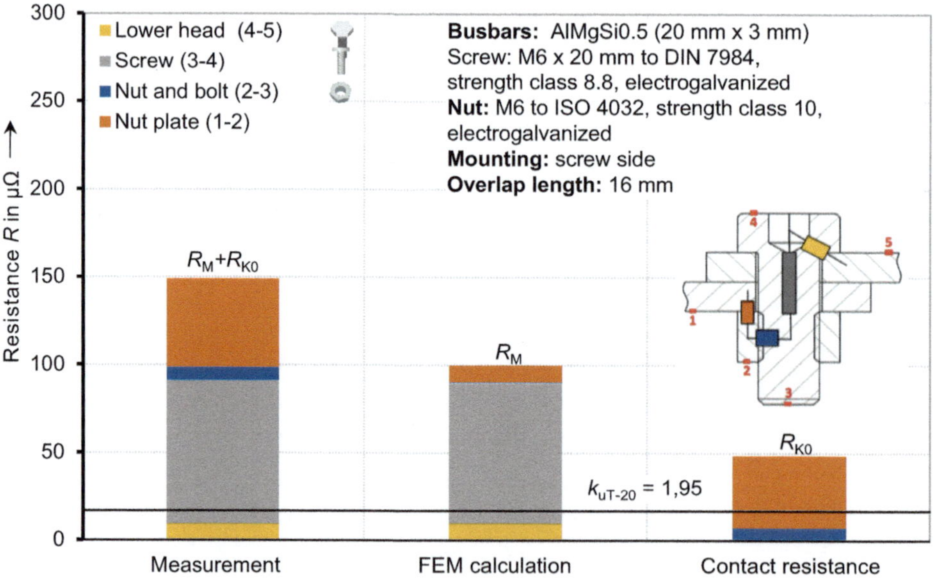

Fig. 4.67 Area-related resistances of a screwed connection with insulated contact area between two busbars [69]

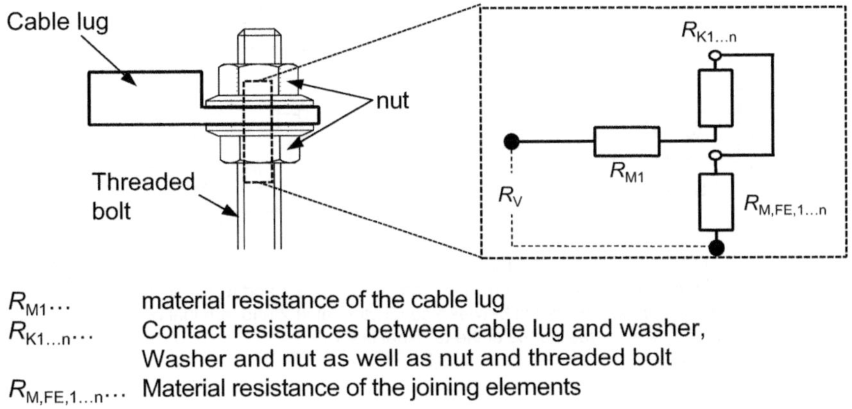

Fig. 4.68 Structure of a nut connection with clamping part and threaded bolt

contact resistance between the screw head and the conductor can be improved in the case of uncoated conductors. The use of screws made of, for example, aluminum or brass also significantly reduces the proportion of material resistance. Despite these measures, the contact behavior of the connection, which is achieved through direct electrical contact between the conductors, cannot be achieved due to the additional contact points with the sometimes high contact resistances [69].

Nut Connection with a Clamping Part and a Threaded Bolt

The nut connection with clamping part and threaded bolt is used, for example, as a transformer connection. Often, cable lugs are clamped between two nuts (Fig. 4.68).

In this case, the current commutes from the cable lug R_{M1} via possibly the joining elements disc and nut to the threaded bolt $R_{K1 \,...\, n}$. The joining elements are thus current-carrying and must have as low an electrical resistance $R_{M,FE,1 \,...\, n}$ as possible. Therefore, brass or technically pure copper such as Cu-ETP is often used in practice. At the same time, the clamping and thus the contact force must be maintained over the required operating period at the maximum permissible temperature. The electrical contact between the disc and the cable lug as well as the disc and the nut depends on the condition of the contact areas and the effective mechanical tension (see section on screw-nut connection). A low contact resistance over the thread turns must be achieved between the nut and the threaded bolt. Parts of the foreign layers are broken up by the relative movement during joining, but pre-treatment is still recommended to remove dirt and thicker oxide layers. That very low contact resistances are possible was shown in tests on screw-nut connections with insulated clamping parts (Fig. 4.67). Furthermore, it should be noted in this construction that the free elongation length of the threaded part between the screw and the nut is often short due to the low height of the clamping part/conductor. The resulting

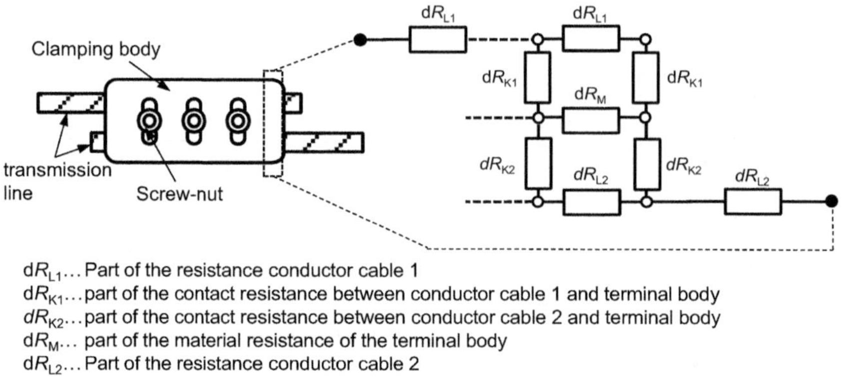

Fig. 4.69 Structure of a clamped connection using the example of a parallel groove clamp

hard spring characteristic can lead to a strong force reduction due to settling and relaxation processes. Therefore, it is recommended to use additional spring elements such as tension washers.

Clamped Connections

Typical clamped connections used in electrical energy technology are parallel groove clamp and switchgear terminals for connecting overhead line cables, round conductors, and tubes made of aluminum or aluminum-steel. They are used in overhead line construction and in switchgear and are thus permanently exposed to the open weather. The clamping bodies are cast or pressed parts and are made of aluminum alloys. Clamped connections belong to the group of force-fitting, detachable connections and are not tensile-force resistent. The assembly is done by clamping the conductor cables, round conductors, or tubes between two clamping bodies (Fig. 4.69).

The clamping force generated by the screws leads to a mechanical tension in the clamping bodies. Depending on the stiffness and the lever ratios, a contact force is established between the conductors to be connected and the clamping body. The current does not commute directly between the elongated conductors, but from one conductor via the clamping body to the second conductor. Therefore, the clamping body must have sufficient mechanical strength with the highest possible specific electrical conductivity. As multi-wire conductors are often used, not only the contact between the clamping body and the conductor is decisive, but especially the contact between the individual partial conductors, in order to achieve a desired as uniform as possible current density distribution.

Before assembly, the clamping bodies and the surface of the conductor cable are pretreated (Sect. 6.6) to improve the contact behavior. Often, the contact area of the clamping bodies for multi-wire conductors is additionally surface-structured by grooves. This

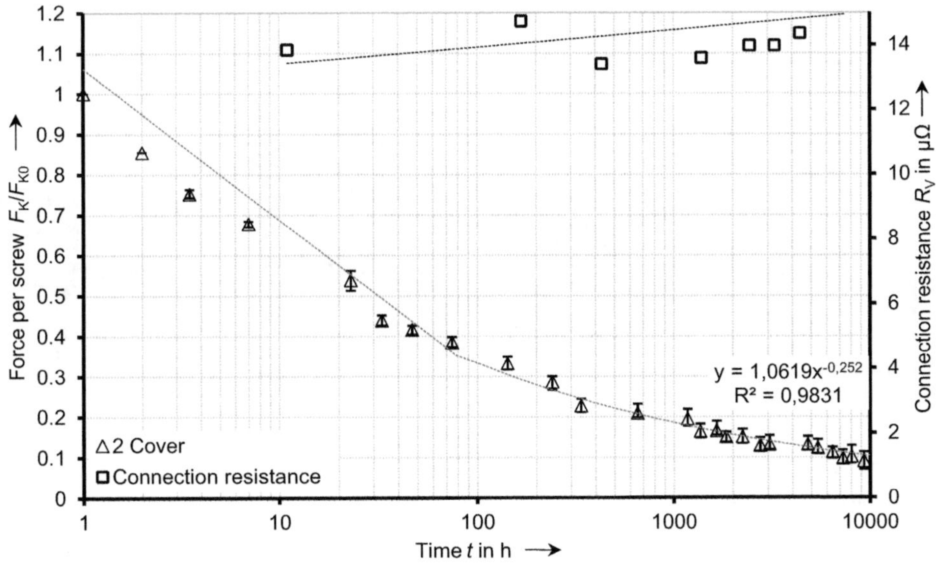

Fig. 4.70 Force curve of the screws of a 2-lid clamped connection with conductor cables of the type Curlew (525-AL1/68-ST1A) at a conductor temperature of 80 °C

creates point and line contacts between the outer layer of the conductor and the clamping body during joining. The contact force during assembly is high and is usually oriented towards the tensile strength of the screws used, usually screw size M10 or M12. This is necessary to create large-area micro-contacts through elasto-plastic deformation and to break up thin oxide layers. However, this leads to strong cold hardening locally, especially with the conductor material Al99.5. Due to the use in open-air atmosphere and the resulting influence of galvanic corrosion (Sect. 5.1.3), stainless steel screws and nuts are often used, which are tightened torque-controlled during assembly. To obtain clamping forces in a defined range, the screws are usually wetted with a special lubricant to obtain defined friction values in the thread and to prevent seizing. The contact force decreases strongly in the first hours at operating temperature when clamping multi-wire conductors, so that after 1000 h only about 20% and after 10,000 h only about 10% of the contact force compared to the initial state after assembly are present (Fig. 4.70, Sect. 5.1.1).

The pronounced hysteresis between the connection resistance and the contact force between conductors made of aluminum means that the massive force reduction does not lead to the failure of the connection (see also Fig. 5.10). However, it becomes clear that no strong external forces should act on the non-tensile connection, so that there is no relative movement between the conductor and the clamp body, which would lead to a sharp increase in the connection resistance. The form fit between the clamp body and the conductor, created by the frequently used grooves on the contact area, additionally counteracts relative movements.

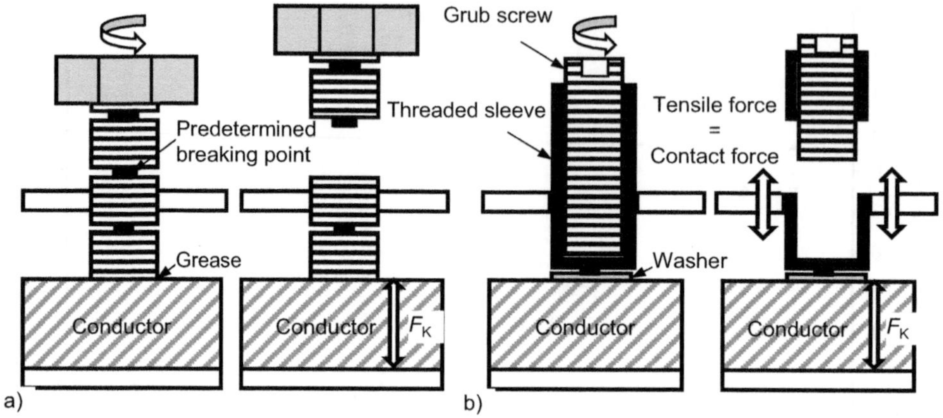

Fig. 4.71 Operating principle **a**) Shear screw with stepped predetermined breaking points, **b**) continuous shear screw (schematic)

Screw Connectors for Cable Conductors

Screw connectors are used to connect cables with cables in cable sleeves or cables with cable lugs. They belong to the group of permanent force-fit connections. With this type of connection, single-wire conductors of class 1, multi-wire conductors of class 2, and fine-wire conductors of class 5 can be connected. The advantage of screw connectors over compression connections is that a single connector can cover multiple cable cross-sections. This cannot be achieved with pressed connections, as the conductor, the sleeve cross-section, and the press inserts are coordinated.

In the case of screw connectors, non-shearing screws or shear screws are used. Non-shearing screws are screwed with a predetermined torque. Shear screws do not require a torque wrench for assembly, but shear off at a defined torque. Multi-stage shear screws are often used (Fig. 4.71).

These have predetermined breaking points in the thread, which shear off at a certain torque. The contact force between the sleeve and the conductor is built up via the torque. The friction between the screw and the conductor is set by a grease and in the thread by the material combination of the screw and the connector body (Fig. 4.50a). Due to the multiple predetermined breaking points, conductors of different heights/cross-sections can be joined. Another design variant is screw connectors with shear screws, which consist of a threaded pin with a threaded sleeve and a pressure disc at the end of the screw (Fig. 4.50b). During joining, the threaded pin is turned into the threaded sleeve until it reaches the bottom of the sleeve. Then the sleeve and threaded pin continue to turn. The disc at the bottom of the sleeve detaches and lies on the conductor surface. Thus, a relative movement only occurs between the sleeve and the disc, the friction value can be defined as with the grease in the multi-stage shear screw. A tensile force builds up in the threaded pin and the threaded sleeve, which corresponds to the contact force. If a resulting permissible tensile stress in the threaded pin and the sleeve is exceeded, they shear off.

In practice, both operating principles are used (Fig. 4.50). The advantage of both variants is that assembly can be carried out by fitters without special knowledge and no special tool such as a torque wrench is required.

As with the other types of connections, the aim is to set a defined contact force and thus a contact resistance with a low value during assembly. In addition, the conductor used must not be damaged during the joining process. The grease used or the disc should also prevent individual partial conductors from being damaged in multi-wire conductors during joining. Good electrical contacts are only created in the area of the screws, where the conductor is clamped against the connector body and elasto-plastically deformed. Several screws are usually used. This creates parallel current paths within the connection (Fig. 4.72). The electrical equivalent circuit diagram of two linear conductors can be approximated (Fig. 4.29, p. 64). To improve the contact behavior, the connector bodies are grooved inside transversely or longitudinally, so that line or point contacts are created between the conductor and the sleeve. The surfaces of the connector body are galvanically tinned and in some applications galvanically silver-plated to further improve the contact behavior. In addition, grease is introduced into the sleeve to reduce the influence of oxidation after joining in operation, by aiming for a certain gas tightness. In tests, it was found that good contact behavior after assembly leads to performance factors of $k_{u0} \approx 0.8 \ldots 0.9$.

Screw connectors are tested electrically and mechanically according to IEC 61238-1 (Sect. 7.2). Screw connectors are also used for Milliken conductors with large cross-sections over 1200 mm^2. Special solutions for this are offered by some manufacturers [70–72].

4.5.2 Compression Connections

The basic modes of operation between mechanical and electrical contact behavior in compression connections are analogous to other current-carrying connection types with plane contacts (e.g., Fig. 4.53). So far, it has only been possible to demonstrate this behavior in compression connections indirectly. For example, in hexagonal compression connections, a correlation between a higher compression factor or higher fit factor and higher contact forces during joining as well as higher remaining contact forces in FEM calculations could be determined (Fig. 4.14) [11]. In current-carrying connections with cylindrical conductors, higher contact forces were likewise calculated at higher oversizes $ü$ [36]. In both cases, the electrical contact behavior of current-carrying connections with a high compression factor or high oversize was better (Fig. 4.73) [11, 36].

Applying high contact forces $F_{\mathrm{K\ press}}$ during joining not only leads to a larger plastic deformation of the contact partners, and thus to a larger apparent contact area, but also to larger mechanically load bearing surfaces. This also increases the metallic contact area and accordingly the number of larger microcontacts (Fig. 4.74) [11]. Furthermore,

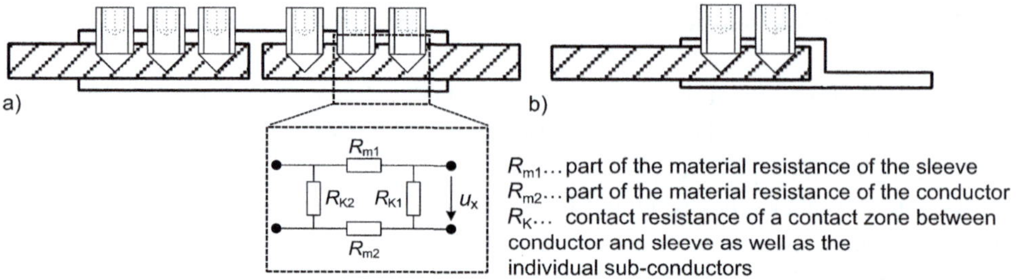

Fig. 4.72 Designs (schematic—screws should not damage the conductor) **a**) Longitudinal connector for cables and **b**) Screw cable lug as well as simplified electrical equivalent circuit diagram according to Fig. 4.29

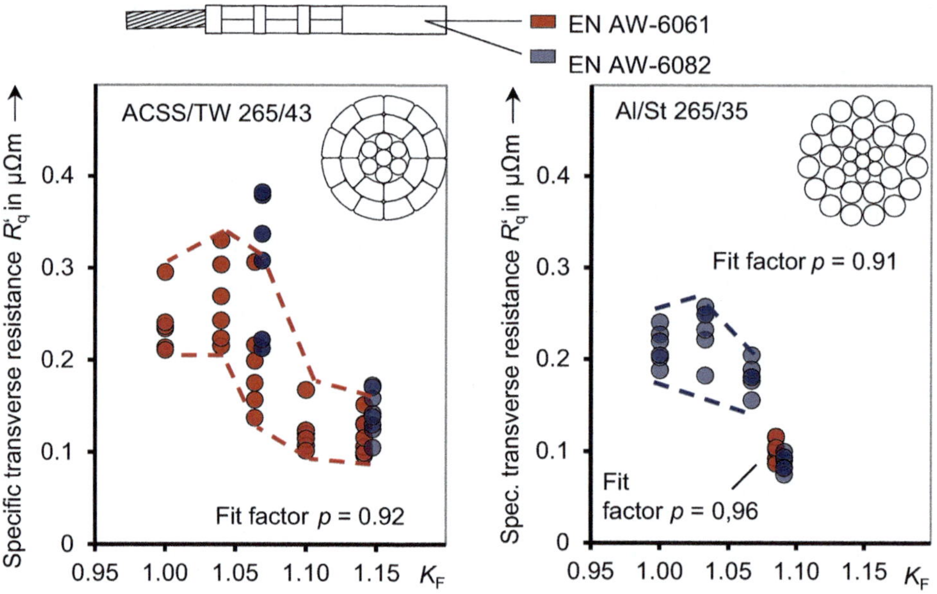

Fig. 4.73 Example of specific transverse resistance of compression connections with two conductor wires depending on the compression factor K_F and the material of the press sleeve

the electrical function of compression connections is based on the contact forces remaining on the individual equivalent areas after the pressing process. The remaining contact forces ensure the integrity of the microcontacts, especially in long-term behavior. The remaining contact force in compression connections is established by the elastic rebound of the contact partners (Sect. 4.2.2). A higher applied contact force during joining generally leads to higher remaining contact forces. For example, the cold hardening introduced into the materials of the contact partners favors the elastic rebound of the contact

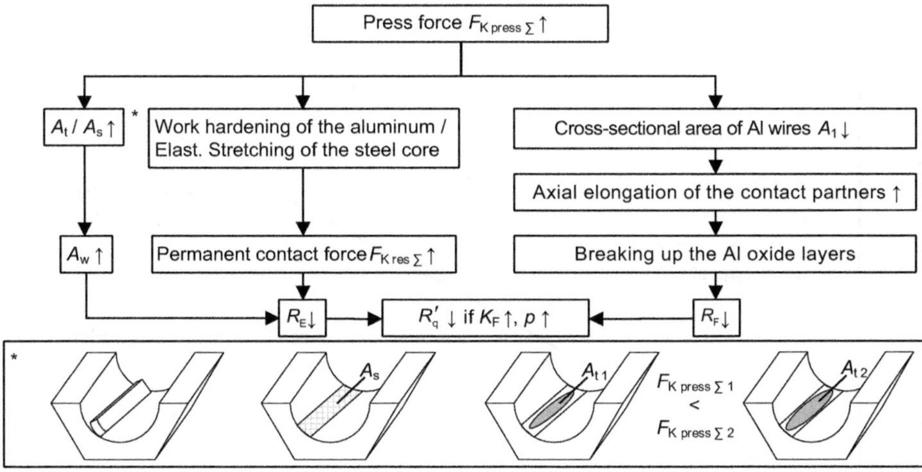

Fig. 4.74 Force effect in compression connections; correlations between mechanical and electrical contact behavior

partners. Furthermore, in the known pressing methods, the axial elongation of the contact partners is usually high. The associated shear stresses, for example, favor the tearing of the brittle aluminum oxide Al_2O_3 [78]. But they can also lead to the tearing of the coating on the contact partners [36].

The heating of compression connections is conveniently calculated using the thermal network method (see Sect. 4.5.1). The power losses occurring in these connections are determined primarily by the material resistances of the contact partners and not by the contact resistance. Furthermore, the heating of compression connections is significantly determined by the overall thermal system. For example, the surface of the connection itself as well as the type and cross-section of the conductors to be connected have a significant influence on the total heat output. A poor contact behavior can be partially compensated by an adequate thermal dimensioning of current-carrying compression connections. Often, it is crucial in practice that the connection does not heat up more than the conductor (Sect. 6.3.4). The maximum permissible temperature-equivalent performance factor $k_{uT\text{-}20}$ for a coaxial cylinder arrangement can be calculated in general (Fig. 4.75).

4.5.3 Plug-in Connections

Connectors are used in various designs (Chap. 2). Generally, two conductors are connected to each other by the connector, allowing relative movements during operation. The connecting conductors are each connected to the socket and the plug by an electrical contact. A plug-in or sliding contact is created between the socket and the plug. If

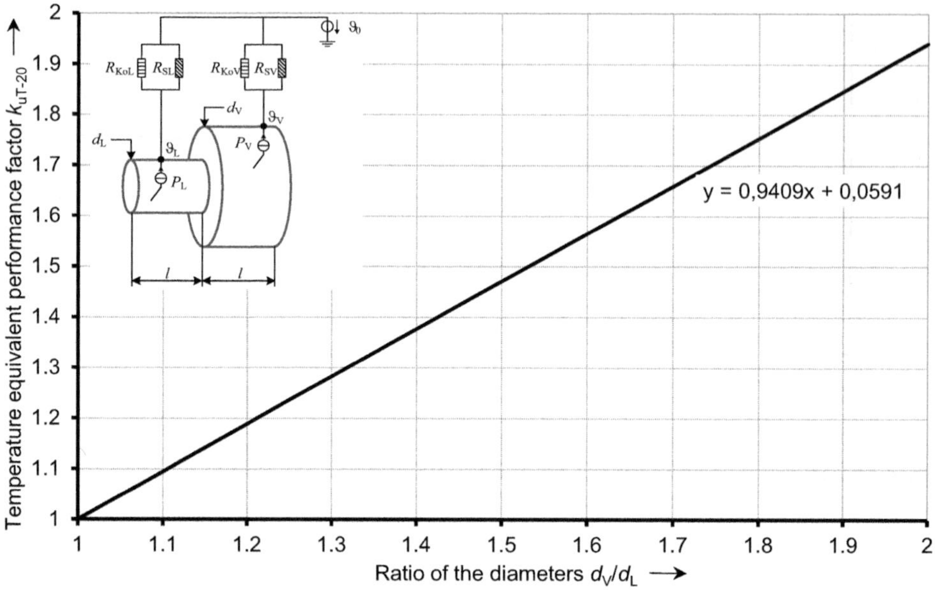

Fig. 4.75 Temperature-equivalent performance factor $k_{uT\text{-}20}$ depending on the cross-sectional ratio d_V/d_L for approximately rotationally symmetric conductors and connections according to [10]

necessary, a contact element is additionally used. As a result, several resistance components must be taken into account, which in total result in the connection resistance of the entire connector between the connecting conductors (Fig. 4.76). The following considers the plug or sliding contact, as the contact to the connecting conductors is realized by other types of connections, which are screwed, pressed, welded or material-bonded (Sects. 4.5.1 and 4.5.2).

For connectors, the spring characteristic of the used spring elements can be determined numerically, analytically or experimentally. The analytical calculation is only possible for very simple geometries, such as for an axially loaded helical spring. For radially loaded helical springs, the relationship between deflection and contact force must be calculated numerically [73]. Depending on the contact force, the coating and the contact geometry, contacts are created that can be simplified as point or line contacts.

The size of the contact area of a connector is mainly dependent on the contact or coating material used and the contact force. Assuming a plastic deformation of the contacts (Eq. (4.21)), the approximate size of the contact area can be calculated based on the contact hardness. This is shown in Fig. 4.77 for frequently used contact materials.

Since the contact hardness depends on the condition of the contact material, only a range for the expected size of the contact area can be given. Relatively soft tin forms the largest contact area. Nickel is hard, which results in a smaller contact area with the same contact force. The contact hardness of silver is lower than that of copper, resulting in larger contact areas. If the contact is free of foreign layers, a corresponding electrical

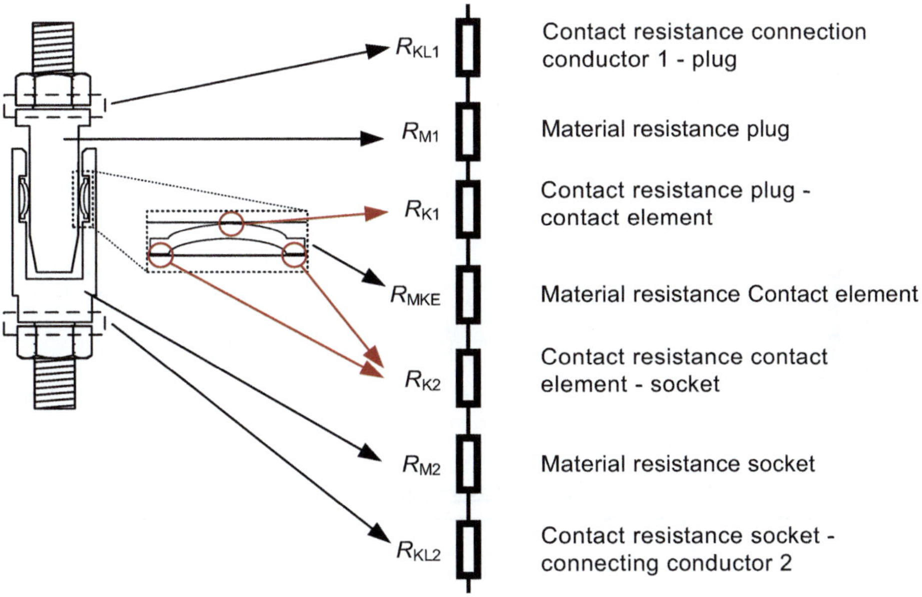

Fig. 4.76 Electrical equivalent circuit diagram of the components of a connector with contact element (exemplary)

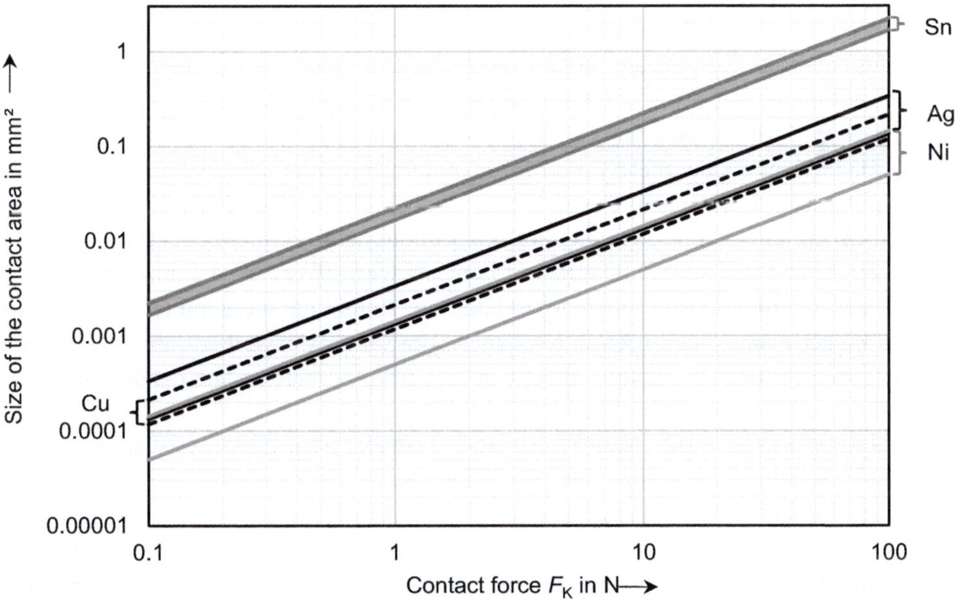

Fig. 4.77 Size of the contact area of a point contact assuming plastic deformation depending on the contact force

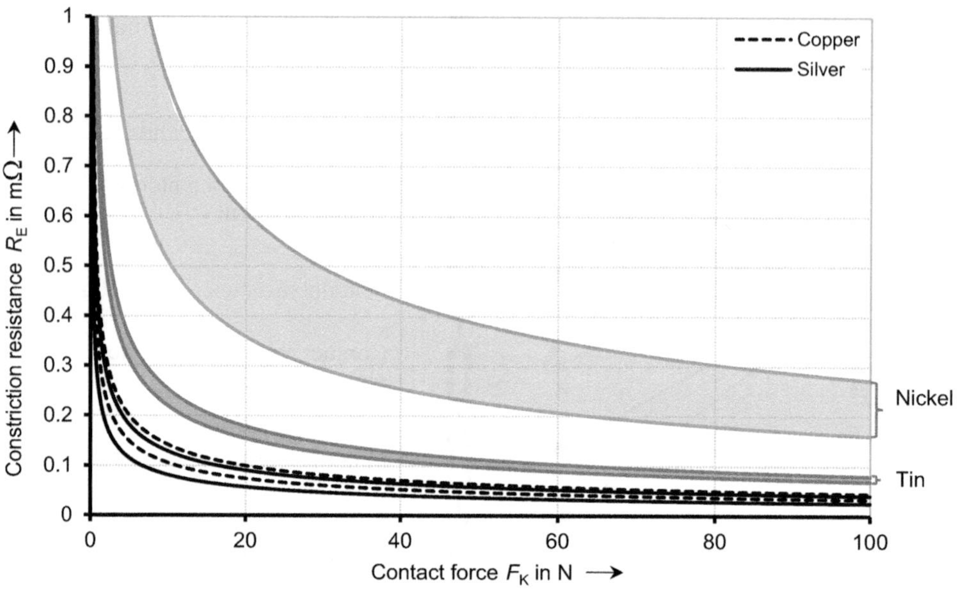

Fig. 4.78 Electrical constriction resistance of a film-free, plastically deformed point contact depending on the contact force at a temperature of 20 °C

resistance of the constriction results in relation to the size of the contact area (4.4). The larger the contact area and the higher the electrical conductivity of the contact materials, the lower the constriction resistance (Fig. 4.78). Silver creates contact points with the lowest constriction resistance, as it is comparatively soft and has the highest electrical conductivity of the metals used for coating conductors of connections. Tin forms larger contact areas compared to silver, but only has 13.4% of its electrical conductivity. The electrical constriction resistance of connections with tin-plated contact partners is therefore significantly higher than that of contacts with silver-plated contact partners. In real contacts, a film resistance occurs in addition, so that the measurable contact resistance is higher than the constriction resistance. The values shown therefore only apply to contacts free of films. Even thin films can significantly increase the contact resistance beyond the constriction resistance.

According to the Wiedemann-Franz law, good electrical conductors are also good thermal conductors. Assuming that heat transport only occurs through heat conduction, the thermal contact resistance can be calculated from the constriction resistance (Fig. 4.79). Measurements show that the thermal contact resistance is always lower than the value calculated from the constriction resistance due to additional heat transport paths through film and air-filled gaps. The larger the proportion of films, the greater the deviation (Sect. 4.4.1).

The maximum permissible load current depends on the temperature in the contact and can be calculated for different boundary conditions and assumptions depending on the

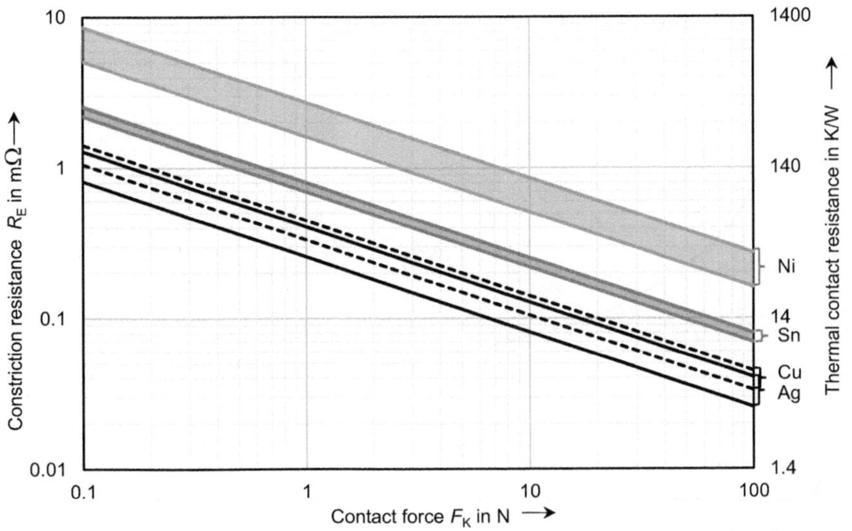

Fig. 4.79 Constriction resistance and thermal contact resistance calculated according to the Wiedemann-Franz law of a film-free, plastically deformed point contact depending on the contact force at a temperature of 20 °C (convection and radiation neglected)

load duration (Fig. 4.80). The temperature has a direct influence on the aging of the contact (see also Sect. 5.1).

In the case of continuous or alternating loads, plug-in connections age through five known aging mechanisms (Sect. 5.1), although an influence of electromigration has not yet been observed in real plug-in connections. With short, high current loads in the second range and below, the temperature increase in the contact or in the spring element can lead to thermally induced damage. For load times longer than approximately 50 ms, force degradation due to stress relaxation and bubble formation in the coating can occur. Load times shorter than 50 ms usually do not lead to a significant heating of the entire spring or contact element due to thermal inertia. Mainly the microcontacts heat up, which can weld or adhere due to the heat input. Since the microcontacts have a thermal time constant in the microsecond range, the welding threshold current increases only slightly in the range less than 50 ms to approximately 1 ms. With significantly shorter load times than 1 ms, the thermal capacity of the microcontacts can be exploited and the welding threshold current increases. The reason for this is that the load time is not sufficient to completely fill the thermal capacity of the microcontacts [75, 85].

4.5.3.1 Continuous Load with Free Convection and Forced Cooling

Just as discussed for the individual contact point in Sect. 4.4.2, an electrical flow field forms in the contact of a connector under current load. This leads to current heat losses, with the power loss density being greatest in the contact element and the contacts, as there is little current-carrying cross-section available (Fig. 4.81). Direct heat dissipation

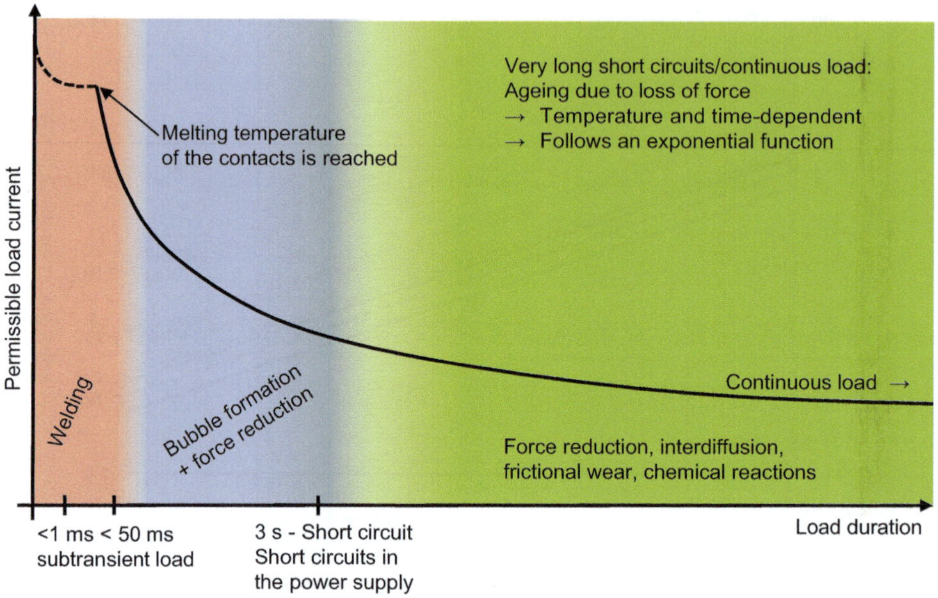

Fig. 4.80 Permissible load current depending on the load duration using the example of a silver-plated plug contact, according to [76, 77]

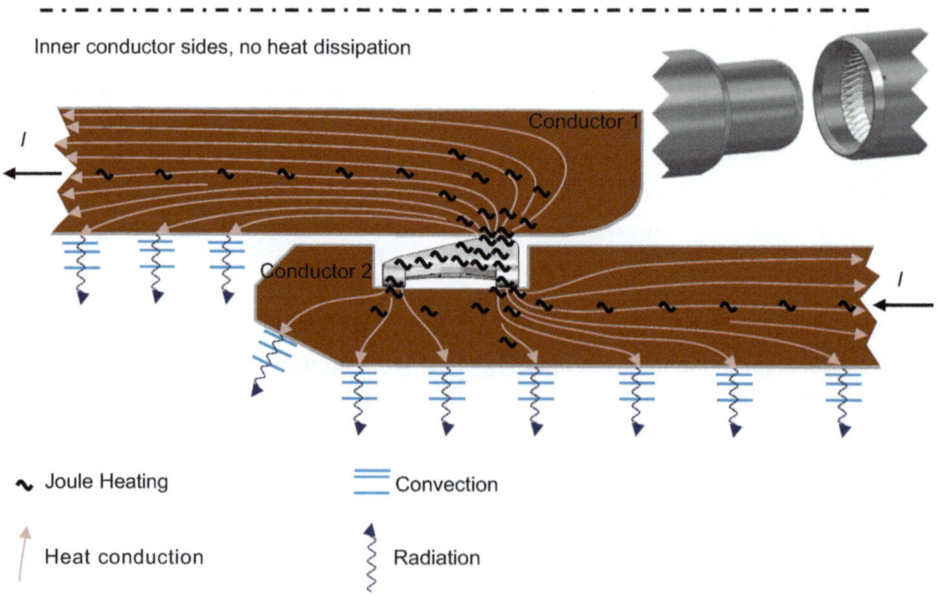

Fig. 4.81 Heat flow in a connector under continuous load

at the surface of the contact element through radiation and convection only occurs to a small extent due to the encapsulated design. Therefore, the current heat losses from the contact element are almost completely dissipated by heat conduction via the contacts into the conductors, distributed there, and emitted at the outer surfaces of the conductors by radiation and convection to the environment (see concept of the thermal local constant in [5]).

Experiments on connectors under high current load with silver-silver contacts show that the temperature-voltage relationship discussed in Sect. 4.4.2 is applicable to the "interior" of the connector including the contact element [50]. The temperature difference between the contact element and the surface of the conductor is very small for continuous operating currents and is up to 3 K with free convection. The current carrying capacity can be significantly increased with the same conductor cross-section by forced cooling of the conductors, e.g., by water [50]. The larger power loss densities and heat flows lead to temperature differences of up to 20 K. In general, due to the small temperature differences in the connector, the easily measurable conductor temperature can be used as a criterion for long-term stability in continuous operation with free convection.

4.5.3.2 Short-term Load with Direct and Alternating Current

In connectors, the lowest power loss density occurs in the connecting conductors and the highest in the contacts. The resulting heat output is stored in the large heat capacity of the connecting conductors and in the smaller heat capacities of the contact elements and contacts during short-term load. Part of the power is dissipated by heat conduction from the areas of high power loss density into the surrounding conductor (Fig. 4.82). Connectors heat up unevenly due to short-term current load. Due to the small current-carrying

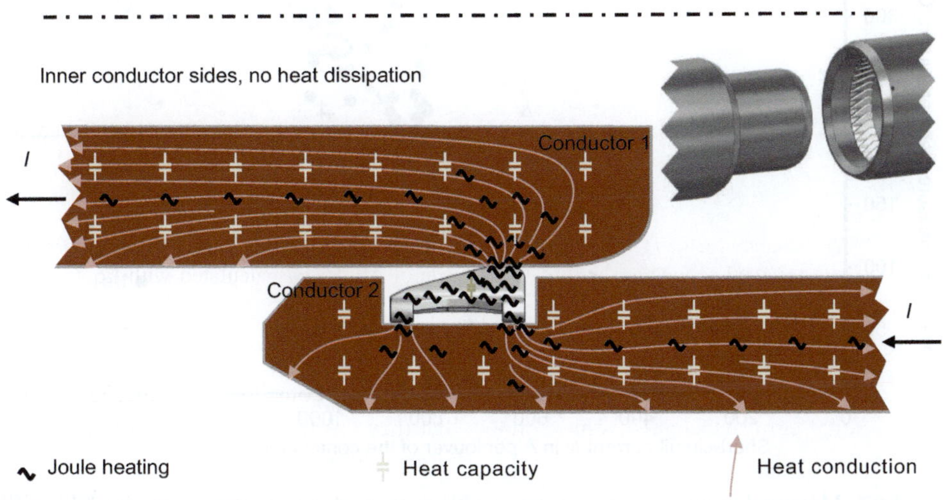

Fig. 4.82 Heat flow in a connector under short-term load

cross-section in the area of the contact elements and the resulting high current densities, short-term load, for example at high short-circuit currents, is particularly noteworthy.

The short-circuit current leads to a heating of the connector. The higher the short-circuit current and the longer the short circuit lasts, the higher is the maximum temperature in the connector (Fig. 4.83). The higher the electrical and thermal conductivities of the plug and socket, the higher is also the current carrying capacity when the connector is loaded with a short-circuit current. Therefore, manufacturers of contact elements specify reduction factors for the short-circuit current of 0.75 when using aluminum instead of copper as the base material, and 0.6 when using brass instead of copper [90].

The thermal load on the spring elements of a connector leads to a decrease in force. This correlates with the maximum temperature during the current load (Fig. 4.84).

The contact force reduces depending on the maximum temperature and the duration of the short circuit. The longer the short circuit lasts and the higher the maximum temperature, the greater the force reduction. This mechanism can lead to failure, especially in short circuits that last significantly longer than 100 ms. In addition to the force reduction, other damages can also occur. Welding of the contacts occurs when the combination of contact temperature, contact force, and exposure duration is so high that a material bond occurs. When the melting temperature of the contact materials is reached, fusion welding begins. This has been identified as the main failure mechanism for point contacts under load durations of up to several 10 ms. In addition to fusion welding, diffusion or cold welding can also occur. These processes occur at lower temperatures but longer

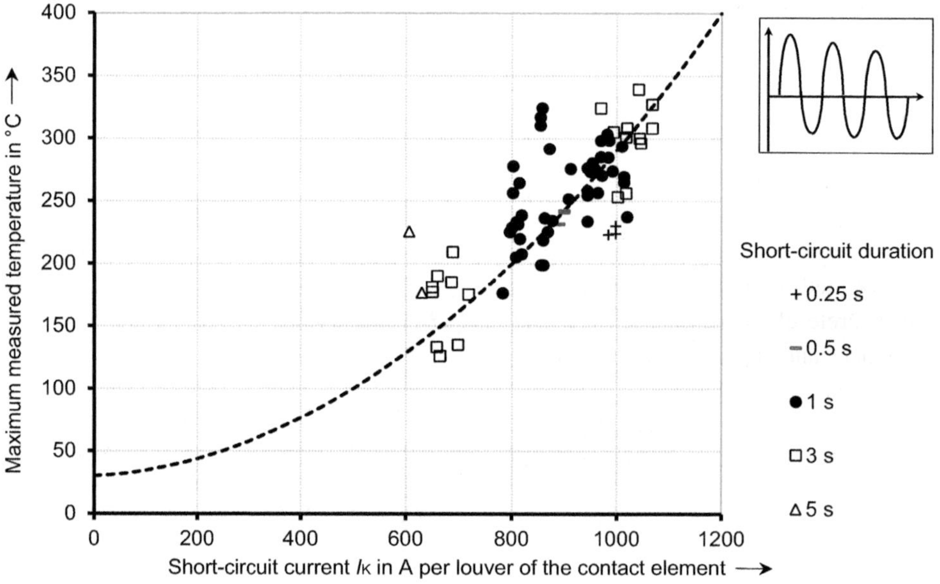

Fig. 4.83 Measured heating of a connector with contact elements depending on the current strength and the load duration

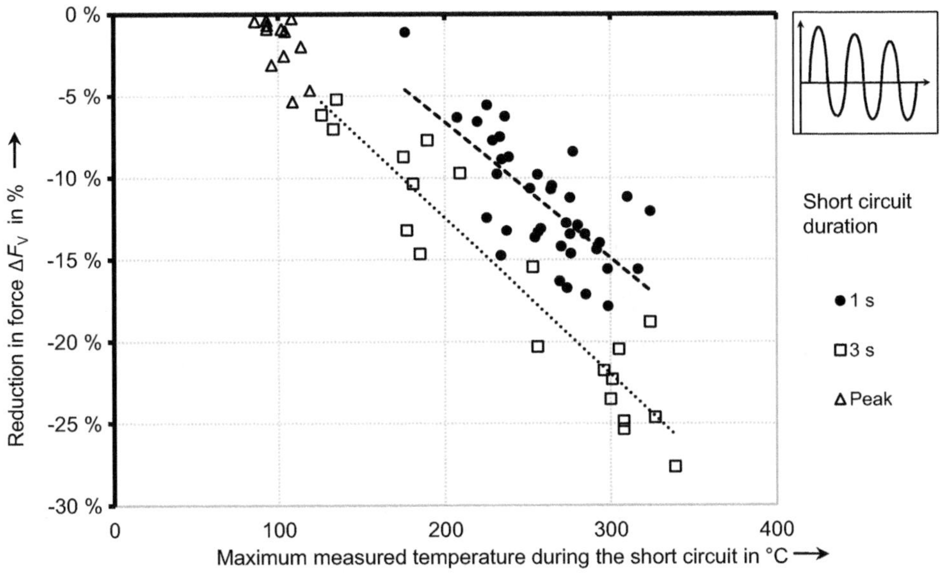

Fig. 4.84 Force reduction of a contact element made of silver-plated copper depending on short-circuit duration and maximum temperature

exposure times. Experiments suggest that these processes may already start at load times in the range of 0.3 s to several seconds and temperatures of (340 … 550) °C for galvanically silver-plated contacts [75]. The current required for welding is referred to as the welding limit current i_W.

When a connector is loaded with a short-circuit current, the highest temperature in the conductors surrounding the contacts, usually the plug and socket, is reached shortly after the end of the current load. This is because the heat generated in the contact and spring elements is partly still being dissipated into the connecting conductors by heat conduction. This can be demonstrated using the calculated progression of the temperature in Fig. 4.85.

With a calculation model, the contact behavior during the short circuit can be examined more closely. The contacts between the contact element and the plug as well as the sockets heat up almost without delay to the current load, as their thermal time constant is only a few microseconds. The contact resistance changes most strongly at a time of (0.12 … 0.35) s after the start of the short circuit (Fig. 4.86). At this time $t = (0.12 \ldots 0.35)$ s, the contact of the exemplary contact element reaches a temperature of approx. (180 … 200) °C, which is identical to the softening temperature of the contact material silver. After this process, the contact resistance continues to decrease as the temperature of the contact continues to increase and the coating material continues to soften. The apparent contact hardness decreases to 27% of the initial value and the contact area increases to 3.4 times the value. The contact resistance decreases to 54% of the initial value at the end of the short circuit. The contact resistances shown are referenced to a temperature of

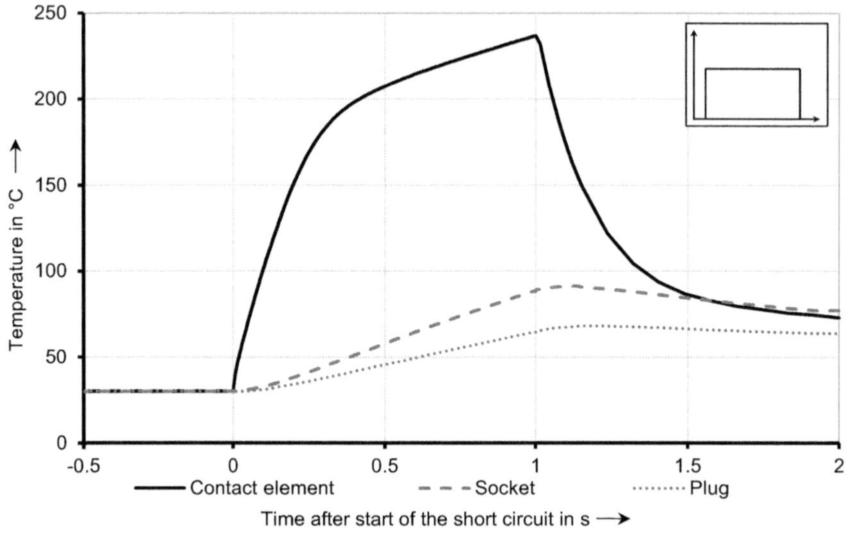

Fig. 4.85 Exemplary temperature progression of a contact element, the plug, and the socket during a 1-s short-circuit current (DC) in a round connector made of silver-plated copper (all contact areas silver-plated)

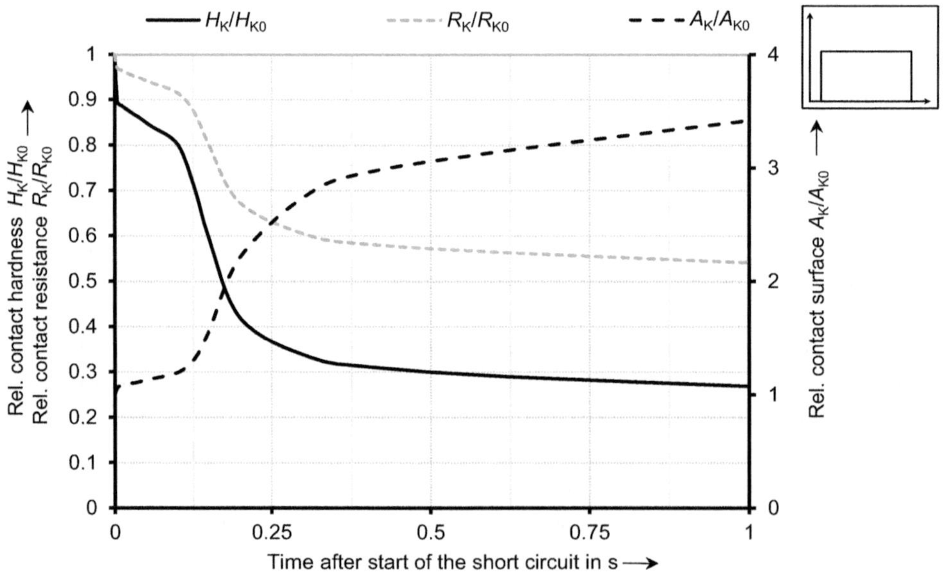

Fig. 4.86 Contact hardness, resistance, and area of a contact point for a contact element during a direct current short circuit (round connector)

20 °C for better comparison. In the calculation, the contact resistance must be related to the respective maximum temperature in the contact.

At the end of the short circuit, a voltage drop across the contact element of $U_{KS\,max} = |\varphi_{Plug} - \varphi_{Socket}| = 0.124$ V is reached in the example. Using the analytical stationary solution according to Eq. (4.54), a temperature in the contact of 257 °C would be calculated for this voltage drop. This temperature is higher than the numerically calculated maximum temperature, as $U_{KS\,max}$ includes part of the voltage drop across the plug and socket. Since the plug and socket are not in a thermally stationary state, the temperature calculated from the voltage drop for the stationary end state is higher than the numerically calculated one. For pulse-shaped short-circuit currents with a duration of $t_K < 20$ ms, the contact resistance relative to 20 °C decreases only slightly until the softening temperature of the silver is reached. Therefore, the actually effective narrow resistance differs only slightly from that calculated with Eq. 4.60, in which no resistance reduction is taken into account. The increase in contact resistance above the softening temperature is significantly lower for the numerically calculated and measured values than for the analytically determined ones, in which the temperature-dependent resistance reduction according to Eq. 4.60 is not taken into account. It turns out that the reduction of the contact resistance by increasing the contact area counteracts the resistance increase due to the decreasing electrical conductivity of the heated current narrow area. The actually effective contact resistance increases by a maximum of 21% near the softening temperature. Above the softening temperature, the contact resistance relative to the maximum temperature in the contact decreases approximately to the initial value before the short circuit R_{K0} (Fig. 4.87).

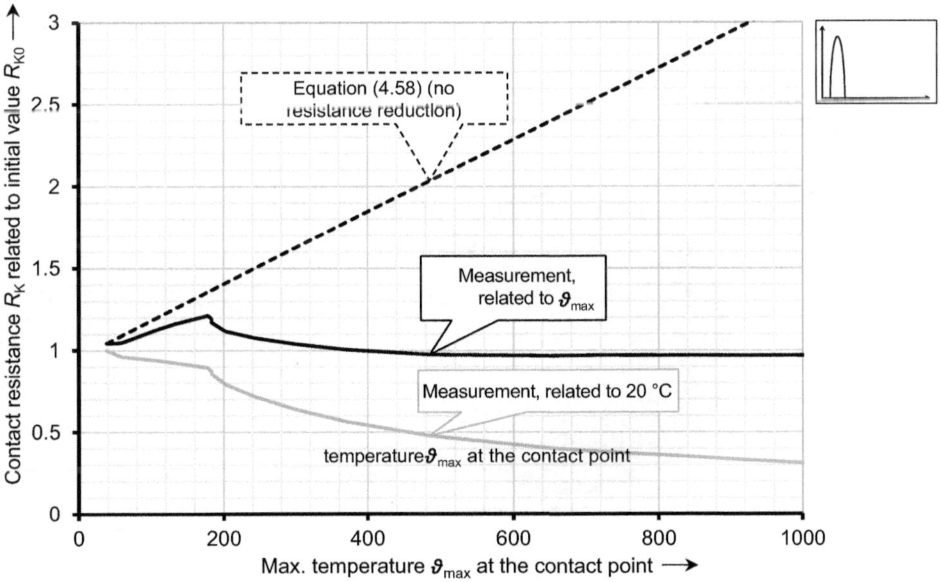

Fig. 4.87 Measured contact resistance, related to 20 °C and with Eq. (4.54) related to the maximum temperature ϑ_{max}, depending on the maximum temperature at the contact point. $T_1 = T_2 = 293.15$ K. Applies to pulse-shaped load with $t_K \leq 20$ ms

For short circuits with a duration of up to 20 ms, connectors with $F_K = (5 \ldots 10)$ N can be calculated to a first approximation with a constant, temperature-independent contact resistance. For longer short circuits, the hysteresis behavior due to the plastic deformation of the contact and depending on the design of the connection, the electromagnetic blow-off force must be taken into account (see [49, 85] and Sect. 4.2.4.2).

A load in the micro- to millisecond range is so short that essentially only the contacts heat up, whose time constants for connectors also lie in the microsecond range. In order to be able to calculate the heating of the contacts in the microsecond range, only a single point contact is considered in the following, as the surrounding contact material hardly heats up. Due to the short duration of the load, it is simplistically assumed that electromagnetic blow-off forces are not effective, as their exciting frequency components are significantly lower than the mechanical natural frequency of the system. Experiments with a current load in the microsecond range showed that the contact resistance is also reduced by this short load and that a softening of the coating at least partially sets in [75, 85]. However, the resistance reduction sometimes scatters significantly higher compared to experiments with a longer current flow duration. An enlargement of the contact areas cannot currently be numerically modeled, as this requires a coupled mechanical-electrical-thermal analysis for which no data on the material behavior in the transient state are known. The following numerical calculation is intended to illustrate the electrical-thermal processes taking place in the contact, as these cannot be detected by measurement technology.

The results of the calculation are therefore to be understood as qualitative statements, which can deviate from experiments due to the necessary simplifications. The contact area considered in the model has a size of $A_K = 0.1$ mm^2 (Fig. 4.88). The contact resistance is approximately equivalent to that of a high-current plug-in connection before

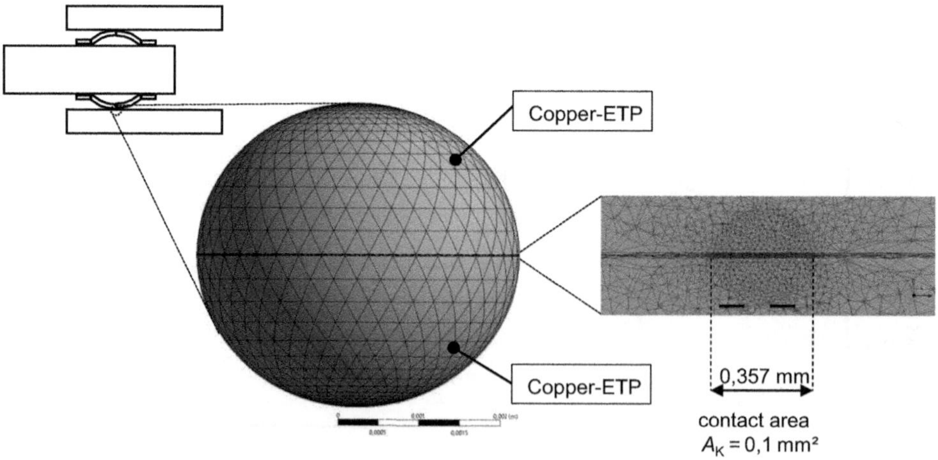

Fig. 4.88 Networked geometry for a transient electrical-thermal FE model for a point contact with a diameter of the hemispheres (left) 4 mm

a short circuit. The heating of the contact is calculated for rectangular and sinusoidal current pulses with a duration in the range of 1 μs to 10 ms. The currents are given (Fig. 4.90), at which the melting temperature of the contact material silver is reached in the contact.

The temperature distribution in the current constriction area and on the contact area depends on the duration of the load (Fig. 4.89). The maximum temperature always occurs at the edge of the contact area. With a load duration of 1 μs, the surrounding contact material is hardly heated, although the edge of the contact area has already reached the melting temperature. The longer the load lasts, the more the surrounding contact material heats up in addition to the contact area, as the generated power loss is transported out of the contact by heat conduction. The temperature distribution then increasingly resembles that of a stationary heated contact (see Sect. 4.1.3).

With a sinusoidal load current with a duration of 10 μs, the maximum temperature in the contact occurs 1.7 μs after the current maximum (Fig. 4.90). With a rectangular load current, on the other hand, the temperature maximum occurs at the end of the load. The peak value of the load current, which leads to the melting temperature of silver, is 13.6% lower with a load duration of 10 μs and rectangular current than with a sinusoidal current. The shorter the load duration, the greater the difference in the numerically calculated welding threshold currents i_w between sinusoidal and rectangular currents [75]. The effective value of the welding threshold current is computationally higher with rectangular load current than with sinusoidal load current. The reason is the thermal inertia of the contact, which can be used with a rectangular load current. With longer load durations, more heat can be conducted from the contact area into the surrounding current constriction area. This reduces the influence of the current shape on the heating. The heating of a plug-in connection differs qualitatively and quantitatively depending on the current shape and especially on the duration of the load (Table 4.22). With short load durations, for example single 50-Hz half-waves and capacitor discharges, mainly the contact points are heated. The longer the load lasts, the more the contact material of the contact element, the plug and the socket heats up in addition. Significant heating of the plug and socket typically occurs with current flow times longer than about 100 ms.

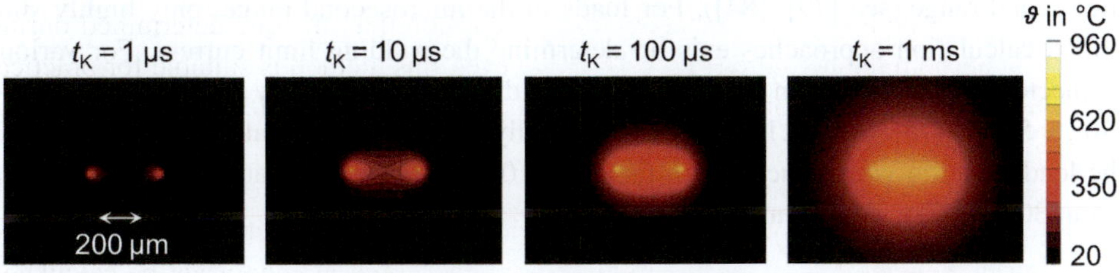

Fig. 4.89 Numerically calculated temperature distribution of a point contact at the end of a load with rectangular load current and current flow durations in the range of 1 μs to 1 ms. The current intensity is chosen so high that the melting temperature is reached at the end of the load.

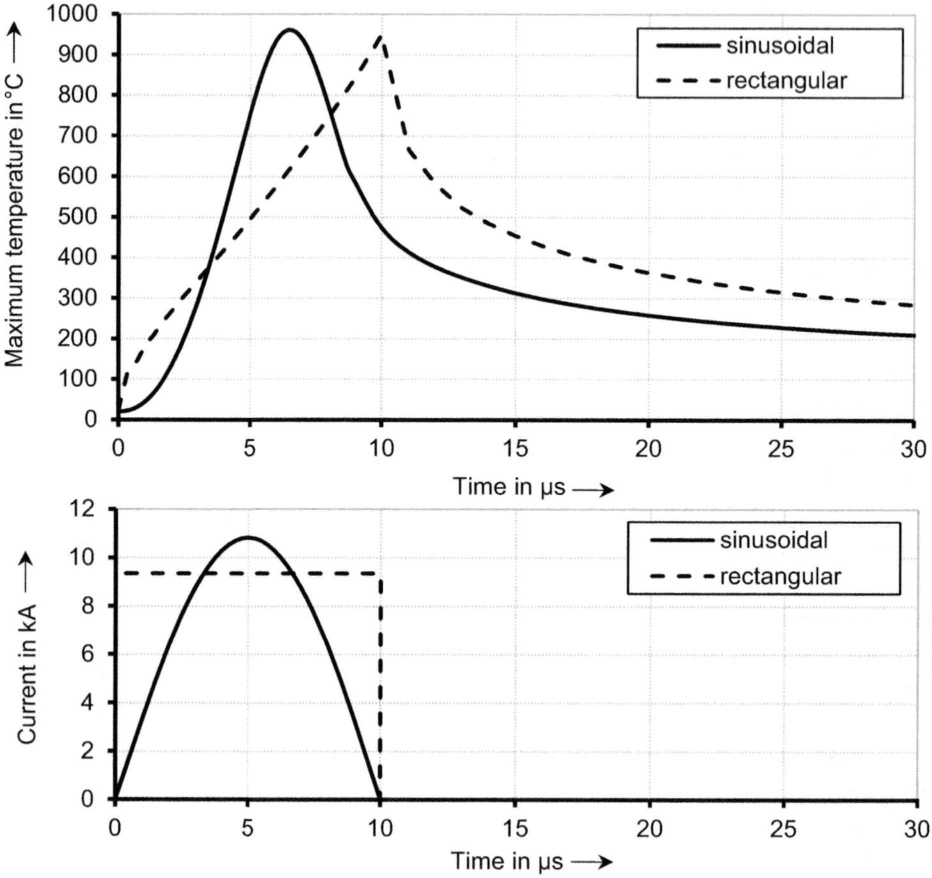

Fig. 4.90 Temperature course in the contact (above) with sinusoidal and rectangular current (below), each leading to the melting temperature of the contact material silver

If the limit temperatures for the damage mechanisms (Table 4.23) are known, a limit current can be calculated for short-circuit durations from individual 50-Hz half-waves to the second range (see [79], [84]). For loads in the microsecond range, only highly simplified calculation approaches exist to determine the welding limit current. For various connectors, for example, an increase in the welding limit current by (20 … 25)% [85] to (40 … 50)% ([75], Sect. 6.1) was experimentally and computationally determined when the load duration was reduced by a factor of 10. Short-circuit durations in the range of about 30 µs to 5 s were examined.

Table 4.22 Summary of findings on the behavior of plug-in connections with short-term currents in the range of 30 μs to 5 s

Current form	Direct current	Alternating current generator remote	Alternating current generator close	50-Hz shock alternating current	Pulse
Time course	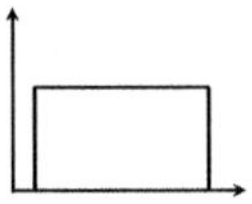	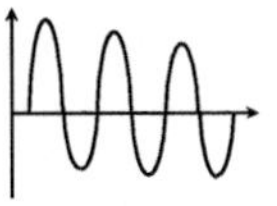	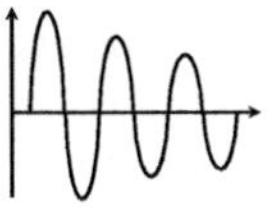	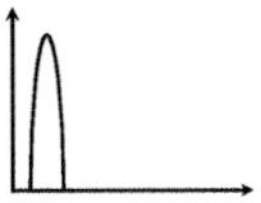	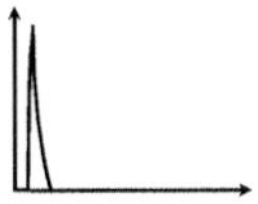
Typ. durations in s	0.1 … 4	0.1 … 4	0.1 … 4	0.01 … 0.02	<10 ms
Typ. limit currents related to the circumference [a]	≈ (0.7 … 4.4) kA/cm			≈(3.6 … 11.2) kA/cm	Mostly not spec., identical to shock current
Heating of the connection conductors	typ. $\vartheta_{max} \leq 200$ °C (see Fig. 5.37)			A few K	None
Heating contact element	A few hundred K, greater than connection conductors			Low, only near the contacts	Low
Heating of the contacts	Almost equal to the contact element	Higher than contact element at $\kappa > 1$	Up to the melting temperature of the contact material		
Limit current calculable	If limit temperatures for damage mechanisms are known			Analytically/numerically	Only order of magnitude [85]
Maximum temperature measurable with IR camera	Yes, occurs at the end of the short circuit		Probably yes, occurs during the SC	Contacts not directly measurable	Contacts not directly measurable

[a]Applies to axially loaded coil springs and lamellar contacts for switchgear

Table 4.23 Damage mechanisms of a connector under short-term current load and examples of failure limits

Damage	Observation	Mechanism	Condition for Activation	Example of Failure Limits
Electrical	Increased contact resistance after re-contacting	Welding of the contact points Electromagnetic lifting forces	Reaching the welding temperature of the coating, usually due to the thermal effect of the surge current	Increase in contact resistance by a maximum of 20%, heating run possible for resistance reduction [74]
	Reduced contact resistance after current load	Loosening of the contact material Welding of the contact points	Reaching the loosening or welding temperature of the coating	Not standardized, welding measured at contact resistances $R_K/R_{K0} = (0.26 \ldots 0.36)$, calculated minimum value $R_K/R_{K0} = (0.194 \ldots 0.324)$ [75]
Mechanical	Force reduction, plastic deformation	Stress relaxation	Heating of the mechanically stressed parts (torsion springs, leaf spring) → reaching a certain activation energy	Still sufficiently good contact behavior at (20 … 30)% force reduction [50]
		Plastic deformation	Deflection e.g. by lateral forces acting on the connector	Contact not closed if the height of the contact element is less than the gap
Optical	Bubble formation	Partial detachment of the coating	Rapid change in the temperature of the coating (temperature shock)	Not standardized, generally not permissible as contact properties can be deteriorated
	Destruction or detachment of the contact material	Welding and tearing off when separating	Reaching the welding temperature of the material, depending on the contact force and the duration of exposure	Not permissible [74], as the contact behavior can deteriorate, especially through oxidation
Chemical	Decomposition/ Evaporation of lubricating grease	Boiling or sublimation	Reaching the limit temperature of the lubricating grease, no basic investigations known	Possibly increased friction coefficient after the short circuit, not yet researched

References

1. *Vinaricky, E.*: Elektrische Kontakte, Werkstoffe und Anwendungen – Grundlagen, Technologien, Prüfverfahren. 3. Auflage, Springer Verlag, 2016.—ISBN 978-3-642-45427-1.
2. *Mballa, F.; Noël, S., et al.*: Three-level Multi-scale Modeling of Electrical Contacts—Sensitivity study and experimental validation. Konferenzbd. IEEE 61st Holm Conference on Electrical Contacts, San Diego, 11.–14.10.2015, pp. 414–422.—ISBN 978-1-4673-9341-6.
3. *Holm, R.*: Electrical Contacts: Theory and Application. 4. Aufl. Berlin, Heidelberg, New York: Springer, 1967.
4. *Erk, A.*; Schmelzle, M.: Grundlagen der Schaltgerätetechnik. Berlin: Springer, 1974.—ISBN 978-3-642-50340-5.
5. *Böhme, H.*: Mittelspannungstechnik-Schaltanlagen berechnen und entwerfen. 2. Stark bearbeitete Auflage Berlin: Verlag Technik 2005.—ISBN 3-341-01495-0.
6. *Bergmann, R.*: Zum Langzeitverhalten des Widerstands elektrischer Stromschienen-verbindungen. Dissertation, TU Dresden, Fortschritt-Berichte VDI Nr. 195. VDI-Verlag 1996.—ISBN 3-18-319521-6.
7. *Poudeu, F. S. D.; Beilner, M.; Schlegel, S.:* Einfluss von additiven thermischen Belastungen auf das Langzeitverhalten von stromführenden Schraubenverbindungen mit verzinnten und vernickelten Leitern aus Aluminium und Kupfer in elektrisch angetriebenen Fahrzeugen. Wiley Online Library, 2021 in press.
8. *Schlegel, S.*: Langzeitverhalten von Schraubenverbindungen mit Stromschienen aus Reinkupfer in der Elektroenergietechnik unter besonderer Berücksichtigung der Temperatur. Dissertation, TU Dresden, 2011.
9. *Fuhrmann, T.*: Kontakt- und Langzeitverhalten von stromführenden Schraubenverbindungen mit vernickelten und versilberten Leitern aus Aluminiumwerkstoffen. Dissertation, TU Dresden, 2019.—ISBN 978-3-86780-636-7.
10. *Fuhrmann, T., Schlegel, S., Großmann, S., Hoidis, M.:* Studies on current-carrying connections with flat surfaces—relationship between mechanical stress and electrical contact behavior. Konferenzbd. 28th International Conference on Electric Contacts, Edinburgh, 06.–09.06.2016, pp. 311–316.
11. *Hildmann, C.*: Zum elektrischen Kontakt- und Langzeitverhalten von Pressverbindungen mit konventionellen und Hochtemperatur-Leiterseilen mit geringem Durchhang. Dissertation, TU Dresden, 2016.—ISBN 978-3-00-056211-2.
12. *Babikow, M. A.*: Wichtige Bauteile elektrischer Apparate—Band 1 Theoretische Einführung. VEB Verlag Technik Berlin, 1954.
13. *Nicolaus, G.*: Lebensdauerverhalten ruhender Kontakte unter den Einsatzbedingungen der Nachrichtenvermittlungstechnik. Dissertation, Technische Hochschule Karl-Marx-Stadt, 1974.
14. *Höft, H.*: Elektrische Kontakte – Physikalische Grundlagen. Berlin, VEB Verlag Technik, 1977.
15. *Slade, P. G.*: Electrical Contacts—Principles and Applications. 2. Auflage, CRC Press Taylor & Francis Group, 2014.—ISBN 978-1-4398-8130-9.
16. *Leidner, M., Myers, M., Schmidt, H., et al.*: Enhancing the contact interface by matching the surface pressure and current density distribution. Konferenzbd. 27th International Conference on Electrical Contacts, 22.–26.06.2014, Dresden, pp. 23–28.—ISBN 978-3-8007-3624-9.
17. *Leidner, M., Schmidt, H.*: Stromdichteverteilung innerhalb realer elektrischer Kontakte in Abhängigkeit von der Kontakttopographie und Normalkraft. Konferenzbd. 21. Albert-Keil-Kontaktseminar, Karlsruhe, 28.–30.09.2011, pp. 187–194.
18. Liefervorschrift LV 215-1 (2013-02): Electrical/Electronic Requirements of Highvolt Connectors Elektrik/Elektronik Anforderungen an Hochvolt Kontaktierungen.

19. DIN EN 61851-23 (Entwurf 2018-03): Konduktive Ladesysteme für Elektrofahrzeuge – Teil 23: Gleichstromversorgungseinrichtungen für Elektrofahrzeuge.
20. *Timsit, S.*: Electrical contact resistance properties of stationary interfaces. Konferenzbd. 44th IEEE Holm Conference on Electrical Contacts, Arlington, USA, 26.–28.10.1998, pp. 1–19, https://doi.org/10.1109/HOLM.1998.722422.
21. Stäubli Elektrical Connectors AG (Hrsg.): Das MC Kontaktlamellen Prinzip.
22. *Timist, R. S.*: Contact Properties of Tubular Crimp Connections: Elementary Considerations. Konferenzbd. 54th IEEE Holm Conference on Electrical Contacts, Orlando, USA, 27.–29.10.2008, pp. 161–167.
23. *Brand, R., Kochjohann, T.*: Verschleißfreies Kontaktsystem für Ladestecker in der E-Mobility, Konferenzbd. 4. Symposium Connectors: Elektrische und Optische Verbindungstechnik, Lemgo, 06.–07.03.2013, pp. 167–176.
24. *Holm, R.*: Electrical Contacts: Theory and Application, 4. Aufl., Berlin, Heidelberg, New York, Springer, 1967.
25. *Papailiou, K. O.* (Hrsg.): Springer Handbook of Power Systems—Chapter 4 4. High Currents and Contact Technology. Singapore, Springer, 2021.—ISBN 978-981-329-937-5.
26. *Kloos, K.-H., Thomala, W.*: Schraubenverbindungen – Grundlagen, Berechnung, Eigenschaften, Handhabung. 5. Aufl., Berlin, Heidelberg. 2007.—ISBN: 978-3-540-21282-9.
27. VDI-Richtlinie 2230 (2003): Systematische Berechnung hochbeanspruchter Schraubenverbindungen – Zylindrische Einschraubenverbindungen.
28. *Pullen, J.; Williamson, J. B. P.*: On the Plastic Contact of Rough Surfaces. Proceedings of the Royal Society A: Mathematical, Physical and Engineering Sciences, vol. 327, no. 1569, pp. 159–173, 1972.
29. *Uppal, A. H.; Probert, S. D.*: The plastic contact between a rough and a flat surface. Wear, Volume 23, Issue 2, Pages 173–184, Februar 1973.
30. *Poudeu, F.; Beilner, M.; Schlegel, S.*: Einfluss der Montageparameter auf das Kontakt- und Langzeitverhalten stromführender Schraubenverbindungen. Konferenzbd. 2. VDI-Fachkonferenz, Versagensmechanismen von Schraubenverbindungen, VDI Wissensforum, 03.–04.11.2020.
31. RIBE Blauheft Nr. 30: Technische Tabellen. 3. Auflage 1990.
32. *Poudeu, F.; Beilner, M.; Schlegel, S.*: Impact of additive thermal loads on the long-term behaviour of current-carrying bolted joints made of tinned and nickel-plated aluminum and copper conductor materials in electrical vehicle. Material Science and Engineering Technology, Manuscript ID mawe.202100159, 28.06.2021.
33. *Poudeu, F*: Langzeitverhalten stromführender Schraubenverbindungen mit Stromschienen kleiner Dicke bei wechselnden Umgebungsbedingungen. Dissertation, TU Dresden 2022 (in press).
34. DIN EN 50343 (Ausg. 2014-09): Bahnanwendungen – Fahrzeuge – Regeln für die Installation von elektrischen Leitungen.
35. DIN 43673-1 (Ausg. 1982-02): Stromschienen-Bohrungen und -Verschraubungen; Stromschienen mit Rechteck-Querschnitt.
36. *Ramonat, A.*: Untersuchungen zum elektrischen Kontakt- und Langzeitverhalten von Fügetechnologien mit zylindrischen Leitern aus Aluminium. Dissertation, TU Dresden, 2018.
37. *Beyer, W.; Iancu, P.; Merkel, M.*: Verbindungen und Anschlüsse in der Elektrotechnik. Berlin, Verl. Technik, 1992.
38. *Hildmann, C.; Grossmann, S.; Dockhorn, T.*: The initial contact stress in aluminum compression connections with high temperature low sag conductors, Konferenzbd. 27th International Conference on Electrical Contacts (ICEC), Dresden, 2014. pp. 557–562.

39. *Salewski, A.*: Grundlagenuntersuchungen zur Herstellung und zu dem Betriebsverhalten elektrisch leitender Umfangpreßverbindungen. Dissertation, Technische Hochschule Leipzig, 1985.
40. *Williams, D. M.*; et al.: Forensic analysis of thermally stressed crimp connections. Konferenzbd. 61st IEEE Holm Conference on Electrical Contacts, San Diego, 11.–14.10.2015, pp. 331–337.
41. *Babiel, G.*: Bordnetze und Powermanagement -- Thermische Modellbildung für elektrische und elektronische Bauelemente. Springer Vieweg, Wiesbaden, 2013.
42. *Prin, G. R.; Courtin, T.; Boyer, L.*: A new method to investigate electrical conduction in crimp joints—Influence of the compaction ration and electrical model. Konferenzbd. 48th IEEE Holm Conference on Electrical Contacts, Orlando, USA, 2002, pp. 246–251.
43. Cigré TB 758 (Ausg.: 2020-10). Test regimes for HV and EHV cable connectors. https://e-cigre.org/publication/758-test-regimes-for-hv-and-ehv-cable-connectors.
44. DIN EN 61284 (Ausg.: 1998-05): Freileitungen – Anforderungen und Prüfungen für Armaturen.
45. *Yovanovich, M.*: Four Decades of Research on Thermal Contact, Gap, and Joint Resistance in Microelectronics. IEEE Transaction on Components and Packaging Technologies, Vol 28, 2005, pp 182–206.
46. *Kuckuck, C.; Blauth, M.; Song, J.*: Differentielles Widerstandsmodell zur Beschreibung des elektromechanischen Kontaktverhaltens von Crimpverbindungen. Tagungsbd. 25. Albert-Keil-Kontaktseminar, Karlsruhe, 09.–11.10.2019, pp. 64–73.
47. *Möcks, L.*: Die Stromverteilung in der Starkstromklemme. Elektrie, 1995, vol. 49, no. 8/9, pp. 299–303.
48. *Holm, E.; Holm, R.*: Theory of Hardness and Measurements Applicable to Contact Problems. Journal of Applied Physics 20, April 1949, pp. 319–327.
49. *Israel, T., Schlegel, S., Großmann, S., Kufner, T., Freudiger, G.*: Modelling of Transient Heating and Softening Behaviour of Contact Points during Current Pulses and Short Circuits. Konferenzbd. 65th IEEE Holm Conference on Electrical Contacts, Milwaukee, USA, 14.–18.09.2019, pp. 9–18.
50. *Gatzsche, M.*: Elektrisch-thermisches Betriebs- und Langzeitverhalten hochstromtragfähiger Kontaktelemente. Dissertation TU Dresden, Shaker Verlag, 2016.—ISBN 978-3-84404-894-0.
51. DIN 40501-2 (Ausg. 2005-03): Aluminium für die Elektrotechnik – Teil 2. Rohre, Stangen und Profile, Technische Lieferbedingungen.
52. *Schlegel, S.*: Stromführende Verbindungen und Leiterwerkstoffe der Elektroenergietechnik – Theorie zum Kontakt- und Langzeitverhalten von Schraubenverbindungen mit Flächenkontakten. Habilitation, TU Dresden, 2019.—ISBN 978-3-86780-614-5.
53. *Poudeu F., Schlegel, S.*: Bestimmung der Mindestflächenpressung von stromführenden-Schraubenverbindungen mit kleinen und großen Leiterquerschnitten unter Berücksichtigung der Umgebungsbelastungen. Konferenzbd. 26. Albert-Keil-Kontaktseminar, Karlsruhe, 30.03–01.04.2022, pp. 82–89.
54. *Kohlrausch, F.*: Ueber den stationären Temperaturzustand eines elektrisch geheizten Leiters. *Ann. Phys.*, Bd. 306, Nr. 1, 1900, pp. 132–158. https://doi.org/10.1002/andp.19003060107.
55. *Diesselhorst, H.*: Ueber das Problem eines elektrisch erwärmten Leiters. *Ann. Phys.*, Bd. 306, Nr. 2, 1900, pp. 312–325. https://doi.org/10.1002/andp.19003060211.
56. *Greenwood, J. A.; Williamson, J. B. P.*: Electrical Conduction in Solids. II. Theory of Temperature-Dependent Conductors. *Proc. Royal Soc. Math. Phys. Eng. Sci.*, Bd. 246, Nr. 1244, Juli 1958, pp. 13–31. https://doi.org/10.1098/rspa.1958.0103.
57. *Kharin, S. N.*: Mathematical Models of Phenomena in Electrical Contacts. Novosibirsk: A.P. Ershov Institute of Informatics system, Siberian Branch of RAS, 2017.

58. *Hilgarth, G.*: Zusammenhang zwischen Kontaktspannung und Kontaktstellentemperatur bei kurzzeitig belasteten Starkstromkontakten. Elektrotechnische Zeitschrift, Bd. 79, Ausgabe A, 1958, pp. 464–468.
59. *Küchler, A.*: Hochspannungstechnik, Grundlagen – Technologie – Anwendungen. 3. Aufl., Kapitel 4.2.2.2, Springer, 2009.—ISBN 978-3-540-78413-5.
60. *Böhme, H.; Löbl, H.*: Die Resistanz von Schraubverbindungen in Geräten der Elektroenergieübertragung und ihr Einfluß auf deren Strombelastbarkeit. Elektrie, Bd. 38 (1984), Nr. 5, pp. 181–183.
61. *Löbl, H.*: Zur Dauerstrombelastbarkeit und Lebensdauer der Geräte der Elektroenergieübertragung. Dissertation B, TU Dresden, 1985.
62. *Büttner, L.*: „Erwärmungsverhalten von NH Sicherungseinsätzen in Einbausituation“ Diplomarbeit, TU Dresden, 2019.
63. *Mücksch, D.*: Kontaktwiderstand von Schraubverbindungen bei Stromschienen. Diplomarbeit, TU Dresden, 1981.
64. *Braunovic, M.*: Effect of Connection Design on the Contact Resistance of High Power Overlapping Bolted Joints. IEEE Transactions on Components and Packaging Technologies, vol. 25, No. 4, Dezember 2002, pp. 642–650.
65. *Braunovic, M.*: Electrical Contacts-Fundamentals, Applications and Technology. Taylor & Francis Group, 2006.—ISBN 1-57444-727-0.
66. DIN 43673-1 (Ausg. 1982-02): Stromschienen-Bohrungen und -Verschraubungen; Stromschienen mit Rechteck-Querschnitt.
67. *Hübner, F.*: Thermische Modellierung von Komponenten im automobilen Bordnetz. Konferenzbd. 25. Albert Keil Kontaktseminar, Karlsruhe, 09.–11.10.2019, pp. 160–167.—ISBN 978-3-8007-5083-2.
68. *Füssel, U.; Johne, V.; Lamm, A.; Großmann, S.; Schlegel, S.; Oberst, M.*: Elektrisches Eigenschaftsprofil von Schraubenverbindungen im Fahrzeug- und Maschinenbau. Abschlussbericht AiF-Projekt 18231 BR, 2019.
69. *Pfisterer Holding AG*: FrontCon– Anschluss- und Verbindungstechnik für einzeldrahtisolierte Starkstromkabel. M2018-013 DE 02 11/2021, (29.11.2021) https://www.pfisterer.com/de/produkte/unsere-produkte/frontcon.
70. Patentschrift EP2226899A1: Vorrichtung zum Verbinden von zwei elektrischen Leitern. 2009. (29.11.2021) https://patents.google.com/patent/EP2226899A1/de.
71. Fa. Relibond (29.1.2021): http://www.relibond.com/cold-jointing-technology/.
72. *Lücke, N.*: Zum Langzeitverhalten elektrischer Steckverbindungen mit Kupfer-Beryllium-Kontaktlamellen. Dissertation, TU Dresden, 2014.
73. DIN EN 62271-1 (Ausg.: 2009-08): Hochspannungs-Schaltgeräte und -Schaltanlagen.
74. *Israel, T.*: Verhalten von Hochstrom-Steckverbindungen bei kurzer Strombelastung. Dissertation, TU Dresden, 2020.—ISBN 978-3-86780-653-4.
75. *Israel, T.; Schlegel, S.; Großmann, S.; Kufner, T.; Freudiger, G.*: The Influence of Peak Current and ECR on the Transmission Performance of High Power Connectors during Faults. Konferenzbd. IEEE 64th Holm Conference on Electrical Contacts, Albuquerque, USA, 14.–18.10.2018, pp. 254–261. https://doi.org/10.1109/HOLM.2018.8611641.
76. *Israel, T.; Schlegel, S.; Großmann, S.; Kufner, T.; Freudiger, G.*: Recommendations for Testing the Short Circuit Behaviour of Power Plug-In Connectors. Konferenzbd. 30th International Conference on Electrical Contacts, online, 07.11.06.2021, pp. 404–411.—ISBN 978-3-907255-11-7.
77. *Bond, N. T., Robinson, D. L., Mohajery, M.*: Fundamental consideration of aluminum electrical contact interfaces. Tagungsbd. 23rd Holm Conference on Electrical Contacts, Chicago, 01–03.11.1977, pp. 213–222.

78. *Slade, P. G.*: The Current Level to Weld Closed Contacts. Konferenzbd. IEEE 59th Holm Conference on Electrical Contacts, Newport, USA, 22.–25.09.2013, pp. 123–128. https://doi.org/10.1109/HOLM.2013.6651431.
79. *Zanetti, A.; Mauroux J.-C.*: Current Carrying Capacity of an Earthing Switch for a Generator Circuit Breaker. Konferenzbd. IEEE 64th Holm Conference on Electrical Contacts, Albuquerque, USA, 14.–18.10.2018, pp. 113–118.
80. *Faltin, C.*: Electrodynamic Repulsion Between Electric Contacts With Arbitrary Current-Density Distribution. IEEE Trans. on Compon. Hybrids Manuf. Technol., Bd. 9, Nr. 2, pp. 188–189, Juni 1986. https://doi.org/10.1109/TCHMT.1986.1136633.
81. *Zhou, L.; Man, S.; Wang, Z.; Xue, S.; Ren, W.*: „On the Relationship Between Contact a-Spots Features and Electrodynamic Repulsion Force for Electrical Apparatus", IEEE Trans. On Compon. Packag. Manuf. Technol., Bd. 8, Nr. 11, pp. 1888–1895, Nov. 2018. https://doi.org/10.1109/TCPMT.2018.2869993.
82. *Slade, P. G.; Taylor, E. D.*: The repulsion or blow-off force between closed contacts carrying current. Konferenzbd. IEEE 62nd Holm Conference on Electrical Contacts, Clearwater, USA, 09.–12.10.2016, pp. 159–162. https://doi.org/10.1109/HOLM.2016.7780025.
83. *Slade, P. G.; Taylor, E. D.; Lawall, A.*: The effect of the axial magnetic field structure on the threshold welding current for closed axial magnetic field vacuum interrupter contacts. Konferenzbd. IEEE 63rd Holm Conference on Electrical Contacts, Denver, USA, 10.–13.09.2017, pp. 253–257. https://doi.org/10.1109/HOLM.2017.8088096.
84. *Israel, T.; Müller, R.; Schlegel, S.; Schmidt, H.; Ludwig, M.*: „Untersuchung von Steckverbindungen bei Strombelastung im Milli- bis Mikrosekunden-Bereich" Elektrische und Optische Verbindungstechnik 2021, VDE-Verlag. Gehalten auf dem Symposium Connectors 2021, Lemgo.
85. *Blumenroth, F.*: Zum Langzeitverhalten von Steckverbindungen mit Schraubenfedern in Anlagen der Elektroenergietechnik. Dissertation, TU Dresden, Shaker Verlag, Aachen, 2010.—ISBN 978-3-8322-9506-6.
86. *Greiner, F. et al.*: Hochbeständige, physikalisch abgeschiedene Kontaktoberfläche auf Silberbasis für die Anwendung in Buchsenkontakten. Konferenzbd. Elektrische und Optische Verbindungstechnik 2021 zum 8. Symposium Connectors, Lemgo, 23.–24.3.2021.
87. *Oberst, M.; Israel, T.; Hildmann, C.; Schlegel, S.; Egetenmeyer, A.-K.; Kappl, H.; Willing, H.*: Elektrische und tribologische Eigenschaften von Silber-Dispersionsschichten für die Anwend ung in Steckverbindungen. Konferenzbd. Elektrische und Optische Verbindungstechnik 2021 zum 8. Symposium Connectors, Lemgo, 23.–24.3.2021.
88. *DODUCO GmbH (Hrsg.):* Datenbuch der elektrischen Kontakte. überarbeitete 3. Auflage, Stieglitz Verlag, 2012.
89. Icore International Ltd. (2009): „Datenblatt Kontakt-Band 8" und „Datenblatt Kontakt-Band 12".

5 Long-Term Behavior of Current-Carrying Connections

Current-carrying connections age depending on operating and environmental conditions. Aging increases the contact and connection resistance, which in turn increases the power loss and the temperature of the connection. Currently, five mechanisms are known that determine aging depending on the type of connection and the joining elements used (Fig. 5.1). For current carrying connections with contact partners made of the same conductor materials, force reduction, chemical reactions, and, if the current density is sufficiently high, electromigration dominate the long-term behavior. For bimetallic connections, interdiffusion must also be taken into account. If relative movements, caused by thermal expansions or during switching and plugging processes between the contact partners are possible , friction wear must also considered.

The speed of aging is strongly dependent on the thermodynamic temperature T, as almost all aging mechanisms can be described with the Arrhenius law (Eq. (5.1)).

$$v_k = D_0 \cdot e^{-\frac{E_A}{R \cdot T}} \tag{5.1}$$

For this, the activation energy E_A and the frequency factor D_0 of the acting physical process must be known, and the universal gas constant R (8.314 $J \cdot K^{-1} \cdot mol^{-1}$) must be taken into account. The course of the connection resistance R_V (t) or performance factor k_u (t) strongly depends on the type and construction of the connection, the conductor and coating material, the mechanical and electrical-thermal load, and the environmental conditions. In principle, the formation phase, the relative stability phase, the accelerated aging phase, and the intermittent phase are distinguished in the lifetime characteristic curve (Fig. 5.2) [3, 4].

S. Schlegel et al., *Contact and Long-Term Behavior of Current-Carrying Connections in Electrical Power Engineering*, https://doi.org/10.1007/978-3-662-69644-6_5

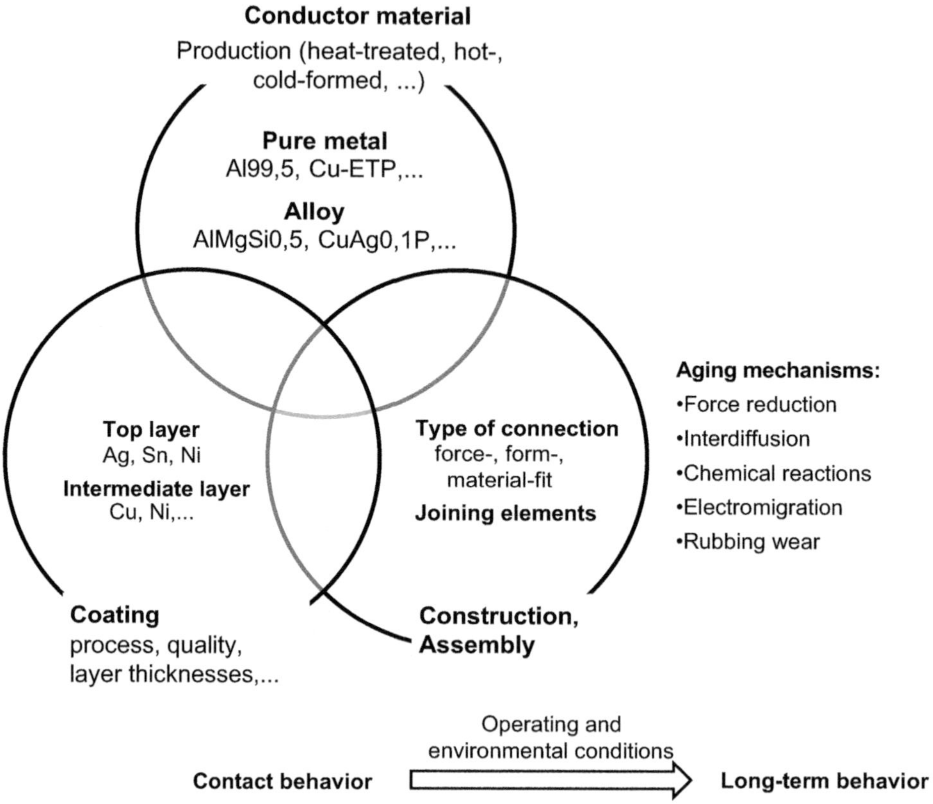

Fig. 5.1 Contact and long-term behavior of current carrying connections according to [2]

There exist various models that describe the course of aging of current carrying connections. An existing model [3, 4], which phenomenologically describes the course of the performance factor for a constant thermal load based on experimentally determined parameters from long-term tests and the Arrhenius law, will be explained more in detail below. This model assumes that the aging of current carrying connections depends particularly on the chemical reactions at the a-spots (Eq. (5.2)—1st term). Furthermore, it is assumed that the aging speed decreases with increasing thickness of film layers, as the reaction partners must diffuse through the existing film layer (Eq. (5.2)—2nd term). The failure of individual a-spots, due to insulating film layers, also changes the electrical current field, which also influences the aging speed (Eq. (5.2)—3rd term). The following approach results from these three parts [3].

$$\dot{R}_{\mathrm{V}} = d e^{-\left(\frac{b}{RT_{\mathrm{E}}}+\frac{1}{2}\right)} \cdot \frac{1}{t_0\left(\frac{t}{t_0}\right)^{\mathrm{m}}} \cdot \left(R_{\mathrm{V}}(t) - R_{\mathrm{VI}}\right) \text{ or} \tag{5.2}$$

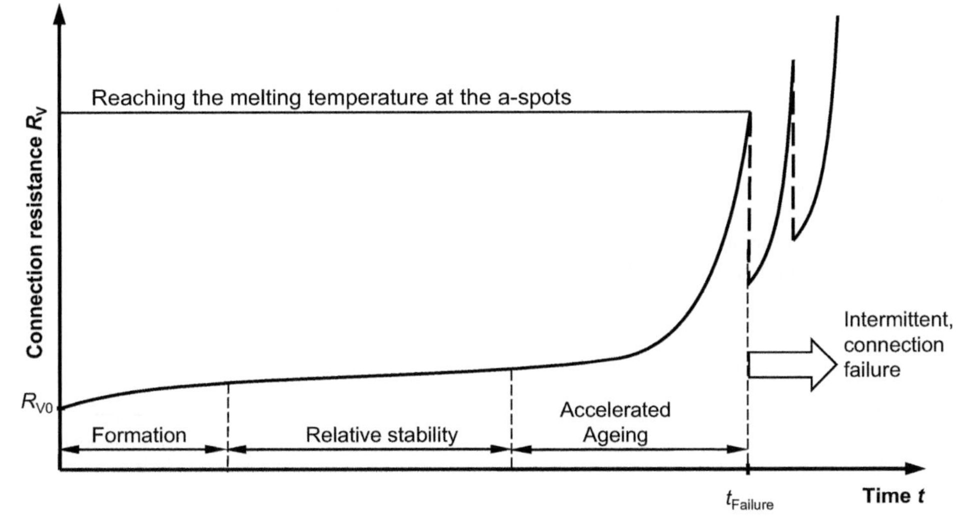

Fig. 5.2 Course of the connection resistance R_V (schematic) [4]

$$\dot{k}_u = de^{-\left(\frac{b}{RT_E}+\frac{1}{2}\right)} \cdot \frac{1}{t_0\left(\frac{t}{t_0}\right)^m} \cdot (k_u(t) - k_{uI}) \tag{5.3}$$

The aging speed depends on the experimentally determined factors m, d, b, the time t, the reference time t_o, from which extrapolation is made, the ideal performance factor k_{uI}, the performance factor k_u (t) at time t, the universal gas constant R, and the thermodynamic temperature T_E at the constriction point. The differential equation (Eq. (5.2)) is not solvable in closed form, but can be transformed into a difference equation (Eq. (5.4)), with which the performance factor k_u at time $t + \Delta t$ can be calculated from the previous value $k_u(t)$.

$$R_V(t+\Delta t) = R_V(t) + \frac{(R_V(t) - R_{VI}) \cdot d \cdot e^{-\left(\frac{b}{R \cdot T_E}+\frac{1}{2}\right)}}{t_0 \cdot \left(\frac{t}{t_0}\right)^m} \cdot \Delta t \quad \text{Order} \tag{5.4}$$

$$k_u(t+\Delta t) = k_u(t) + \frac{(k_u(t) - k_{uI}) \cdot d \cdot e^{-\left(\frac{b}{R \cdot T_E}+\frac{1}{2}\right)}}{t_0 \cdot \left(\frac{t}{t_0}\right)^m} \cdot \Delta t \tag{5.5}$$

Since the parameters m, b, d must be determined from long-term experiments, the effect of aging mechanisms is described as an integral quantity. The effect of individual physical processes is not considered in this model. Nevertheless, the lifetime of the connections can be extrapolated with known parameter triples (Table 5.1) [3]. The physics of

Table 5.1 Parameter triples for screw-nut connections with busbars [3]

Conductor material	Contact partner/pre-treatment	m	b in J/mol	d
Al99.5 F7[1)]	Spirits + wire brush	0.487	9709	0.1
Cu-ETP	No description possible, as no significant aging was detected during long-term experiments over thousands of hours			
Al99.5 F7-Cu-ETP[2)]	Spirits + wire brush	0.616	14,432	5

[1)] applies in the temperature range of (90 … 150) °C
[2)] applies for a temperature of 90 °C

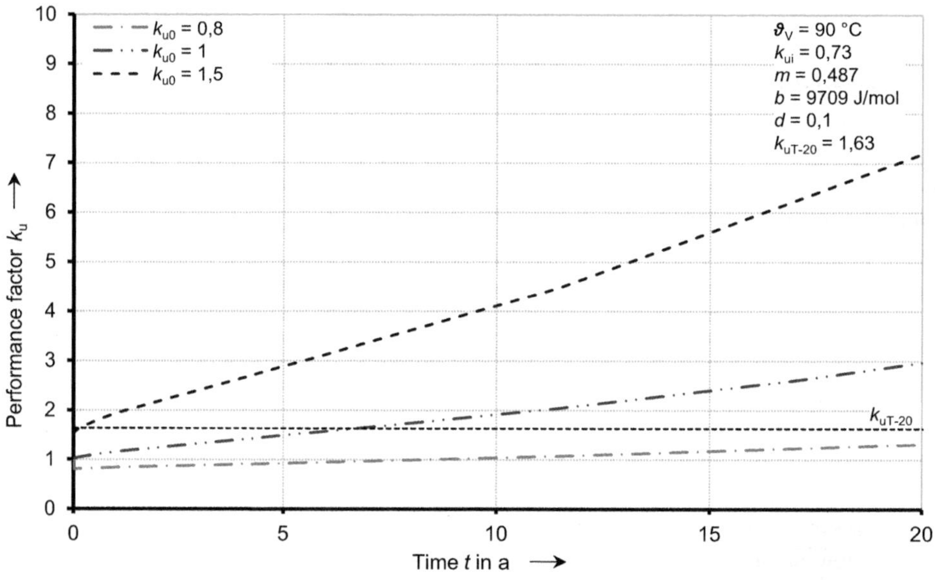

Fig. 5.3 Performance factor k_u depending on the time t and the initial value of the performance factor k_{u0} of a screwed connection with busbars made of Al99.5 F7 calculated with Eq. (5.5) and Table 5.1

aging must not change when different temperatures are considered. Numerous studies on individual aging mechanisms have shown that they work independently of each other [8, 9, 11, 12] (Fig. 5.3).

Depending on the current load of the connection, they jointly determine its long-term behavior, but to varying degrees. With the knowledge gained so far, it is therefore not yet possible to provide a closed analytical or numerical aging model that describes the complex processes of the individual aging mechanisms and their interaction. Therefore, it must be estimated for the individual types of connections and their operating conditions which mechanisms significantly determine the course of the lifetime characteristic curve.

Models and approaches have been developed for the effect and influence of individual mechanisms on, for example, the connection resistance or the performance factor, which will be examined in more detail below.

5.1 Aging Mechanisms

5.1.1 Force Reduction

The force-fit is necessary for good contact and long-term behavior in all non-material closure connections and depends in particular on the design and the load during operation. The contact force significantly influences the contact resistance (Sect. 4.1). When assembling a current-carrying connection, a sufficiently high contact force or mechanical tension must first be generated in the contact area so that large-scale a-spots can form. During operation, the contact force must not fall below a minimum value, otherwise the a-spots could break up. The contact and thus the connection resistance would increase abruptly. The minimum contact force $F_{K'min}$ depends on the mechanical and thermal loads of the connection. In normal operation, for example, forces due to thermal expansions and vibrations of the conductors must be taken into account. In the event of a fault, for example due to high short-circuit currents, additional electromagnetic forces acting between the conductors can significantly influence the contact resistance (Sect. 4.2.4). The force reduction in a connection strongly depends on the materials used as well as the geometry of the conductors, the additional separating joints due to joining and contact elements, the maximum mechanical tension and its distribution as well as the temperature. Depending on the type of connection, relaxation or relaxation with a creep component leads to force reduction (Fig. 5.4) [1].

The force reduction can be divided into settling, primary and secondary areas (Fig. 5.5).

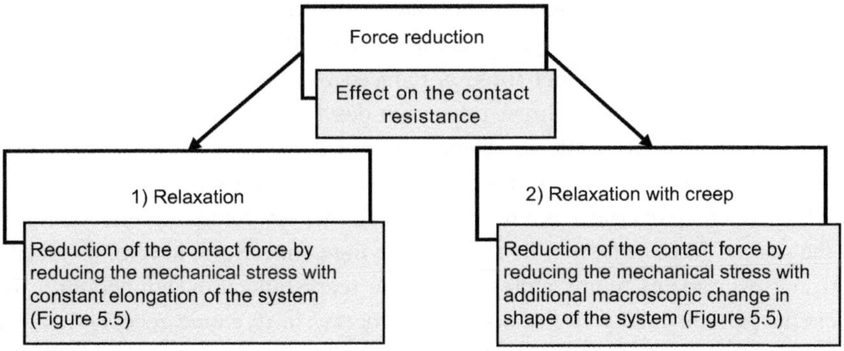

Fig. 5.4 Material physical processes during force reduction and their effect on current-carrying connections

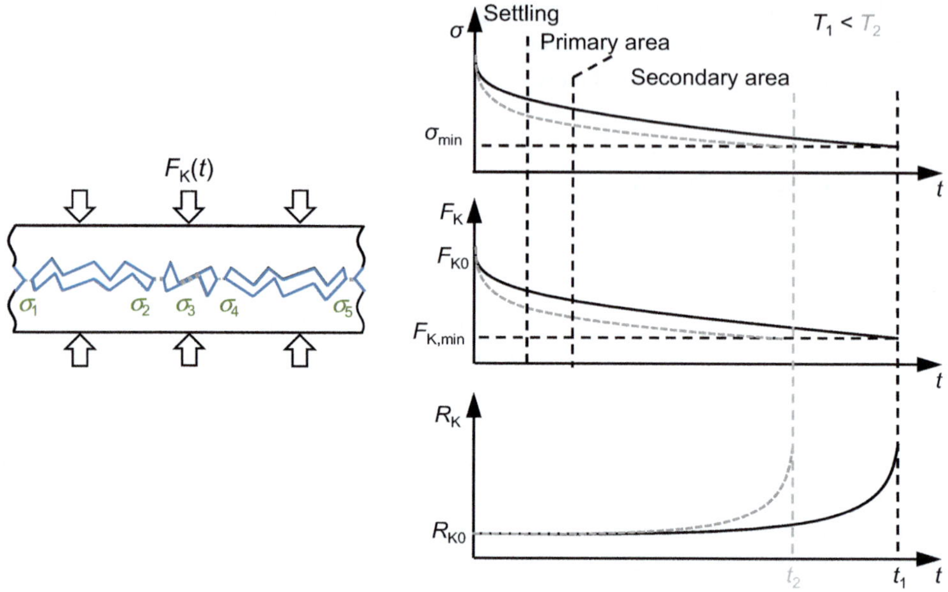

Fig. 5.5 Principal effect of force reduction on the course of contact resistance R_K [11]

Settling describes the leveling of surface roughness in the load bearing contact surfaces. Settling processes reduce the contact force immediately after assembly. The amount of settling depends in particular on the mechanical properties of the materials, the type of stress, the number of separating interfaces and the roughness of the surfaces. In addition, the amount of settling is influenced by different thermal expansion coefficients and by temperature-dependent Young's modulus of the clamped parts [2].

In the primary area of force reduction, compensation processes occur in the material, especially in areas of highest mechanical stress, until a state of equilibrium of the processes described below is reached in the secondary area. To understand the effect of force reduction, a distinction must be made between the processes and the physical mechanisms. Depending on the type of connection and the joining elements, the mechanical tension and thus the contact force is reduced by stress relaxation with or without a creep component (Fig. 5.4). Stress relaxation describes the reduction of an externally applied mechanical stress while the eleongation of the system remains constant. In the case of stress relaxation with a creep component, a macroscopic deformation of the material due to creep overlays as a result of an external directed mechanical stress [1].

In the case of force reduction in current-carrying connections, several physical mechanisms can overlap. At temperatures below the recrystallization temperature, dynamic recovery in particular acts as the dominant process. In dynamic recovery, mechanical stresses in the material are reduced by dislocation movements and the annihilation of

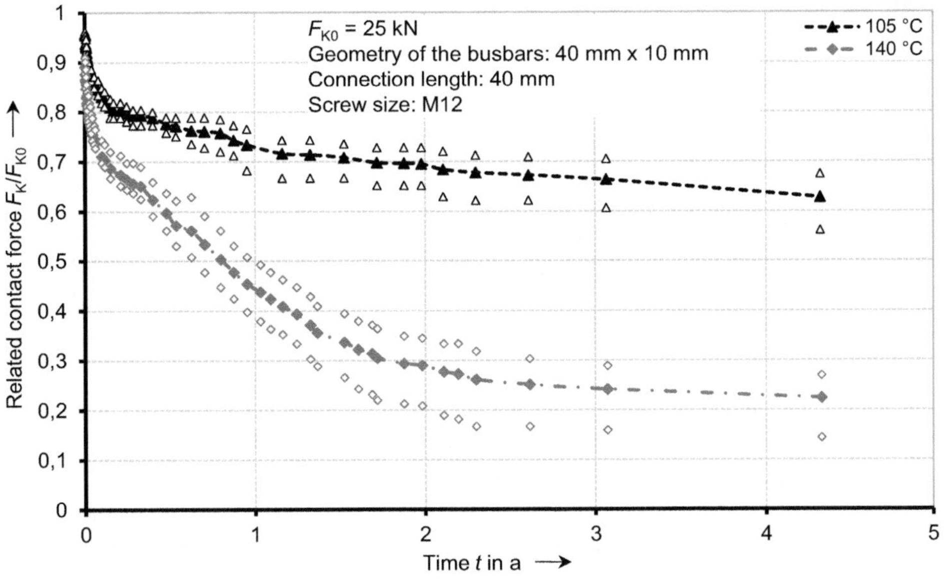

Fig. 5.6 Example—Progression of the contact force (mean value and range) of a screw-nut connection with busbars made of Cu-ETP R300 without spring joining elements [11]

dislocations. At the same time, new dislocations are created by the external mechanical stress. If the temperature is high enough, material softening due to recrystallization can overlap and thus significantly accelerate the force reduction [1].

Since recrystallization greatly reduces the mechanical strength of the conductors, this mechanism must be avoided (Fig. 5.6). The maximum operating temperature is therefore dependent on the used conductor material (Sect. 3.1) [2, 5].

The force reduction in connections can be calculated. For this, material models must be developed for the materials used, which describe the relaxation depending on the time and temperature. A very simplified approach, which can only describe the secondary area of force reduction, is the phenomenological creep equation (Eq. (5.6)).

$$F_K(\dot{\varepsilon}_s, t) \quad \text{mit } \dot{\varepsilon}_s = C \cdot \sigma^n \cdot e^{\left(-\frac{E_A}{R \cdot T}\right)} \tag{5.6}$$

As material parameters, the stress exponent n and the activation energy E_A must be determined in creep tests as well as the proportionality factor C in long-term tests with the relevant connection (Table 5.2).

The strain rate $\dot{\varepsilon}_s$ can be calculated depending on the mechanical stress σ, the thermodynamic temperature T and the universal gas constant R. This is done iteratively numerically, by calculating a mechanical stress σ for each strain ε, which in turn is the input size for the next calculation step. Eq. (5.6) may only be applied if the physical processes

Table 5.2 Creep parameters for conductor materials of contact elements and busbars

Type of connection/Source	Material	Temperature in °C	n	E_A in kJ/mol
Plug-in connection with coil spring [62]	CuCo2Be	105	8.3	46.6
	CuCr1Zr		17.5	37.6
Plug-in connection with contact lamella [63]	CuBe2	90	3.45	43.14
		115	3.37	
Screwed connection with busbars [19]	CuAg0.1P	110	7.02	27.4
		130	9.96	31.9
		150	11.31 ... 16.01	51.8 ... 78.5

that act during the creep test do not change within the time considered for the calculation. More complex material models, which additionally describe the primary area, must be developed taking into account the physical processes.

Alternatively, the force reduction can be estimated by extrapolation methods based on experiments. However, this is only possible if the course of the contact force can be measured. One method is, for example, the Larson-Miller method, which is derived from the Mont-Grant relationship and the Arrhenius law (Eq. (5.7)) [6].

$$\frac{F_K(t)}{F_{K0}} = m_K \cdot P' + n_K \text{ with } P' = T \cdot \left(C' + \lg t\right) \tag{5.7}$$

The specific parameters of the connection m_K and n_K are to be determined experimentally. The specific material constant C' can be determined for the material used from test results at the same applied mechanical stresses and different temperatures $T_{1,2}$ and times $t_{1,2}$, at which a defined mechanical stress is reached (Eq. (5.8)). For metals, $C' = 20$ can be assumed approximately [1].

$$C' = \frac{T_2 \lg t_2 - T_1 \lg t_1}{T_1 - T_2} \tag{5.8}$$

The advantage of extrapolation methods is that changing or overlapping physical processes can be taken into account within the test time (Fig. 5.7). The disadvantage is the longer test time required to make reliable statements.

The conductor structure has a significant influence on the force reduction of screwed connections with stranded conductors. In particular, due to the settling in the numerous separation interfaces and the pronounced primary area, the contact force in such connections decreases disproportionately in the first approx. 500 operating hours ($P' \approx 8000$) (Fig. 5.8) to up to approx. 20% of the assembly preload force. From this point on, the gradient of the force course is comparable to that of screwed connections with Al99.5 busbars with a low degree of cold work hardening in the H14 state.

Another extrapolation method is the one by Stennet and Campbell (Eq. (5.9)) [7].

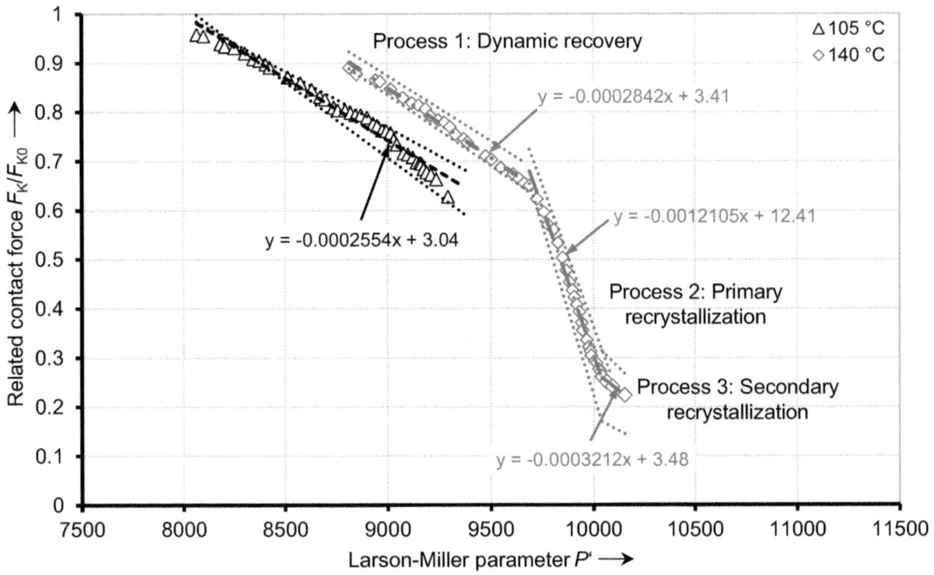

Fig. 5.7 Evaluation of the course of the contact force of a screwed connection with Cu-ETP R300 busbars without spring joining elements over the Larson-Miller parameter, acting physical mechanisms and certain values for m_K, n_K [11]

$$\frac{F_K(t)}{F_{K0}} = 100 - B' \cdot \ln\left(\frac{t}{D'} + 1\right) \text{ with } D' = e^{\left[\frac{a-100}{b}\right]} \tag{5.9}$$

The specific material characteristics B' and D' must also be determined here in long-term tests at different temperatures. The influence of the parameter D' on the logarithmic function is very small, so that regression functions (Eq. (5.10)) can be calculated from the measured force curve with good approximation (Fig. 5.9).

$$\text{Regressionsfunktion}: y = m \ln(x) + n \text{ with } m = b = B' \tag{5.10}$$

A comparison of the individual methods shows that the iterative calculation and extrapolation methods have advantages and disadvantages that must be taken into account according to the required results (Table 5.3).

The minimum contact force must be determined for each respective current-carrying connection, which must not be undercut by force degradation within the lifetime, as this would significantly increase the contact and thus the connection resistance. In the event that no additional mechanical load is applied, the minimum contact force can be determined as a first approximation using the hysteresis curve recorded during relaxation (Fig. 5.10). The minimum contact force determined in this way is very conservative, as further time- and temperature-dependent processes cannot be taken into account,

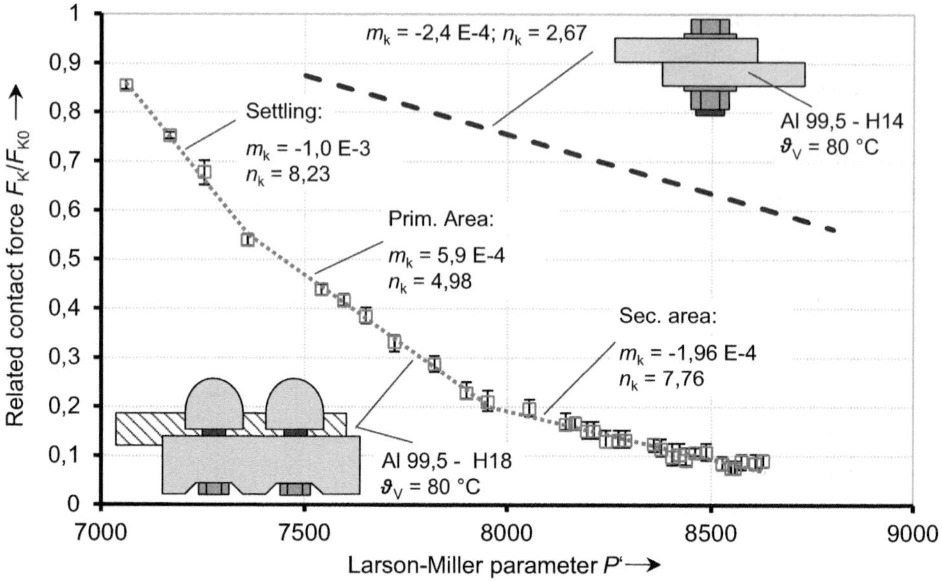

Fig. 5.8 Course of the contact force for screwed connections with Al99.5 busbars without spring joining elements according to [5] and that of a connection of a switchgear terminal with the conductor cable Curlew (525-AL1/68-ST1A)

but these can have a significant influence through, for example, contact adhesion (Sect. 5.1.2). In addition, the minimum contact force determined in this way only applies to a static mechanical load.

Settling processes at room temperature and especially at high temperature result in a significantly lower value of the minimum contact force due to the time-dependent plastic deformation of the a-spots or possible contact adhesion than the one determined during relaxation at room temperature (Fig. 5.10). However, this method is suitable for practical applications, as a higher, conservative minimum contact force is determined than the one that may be necessary.

It must be checked whether this is sufficient for high dynamic mechanical loads, such as those that occur in the event of a short circuit with small conductor distances. Due to the many influencing factors on the minimum contact force, it cannot yet be calculated conclusively. The influence of force degradation can be estimated simplistically with Eq. (5.11). It should be noted that abrupt changes in contact resistance can occur at very small contact forces, which cannot be described with Eq. (5.11).

$$R_K(F_K(t)) = \frac{k_E}{(0.1 \cdot F_K(t))^{m_E}} \tag{5.11}$$

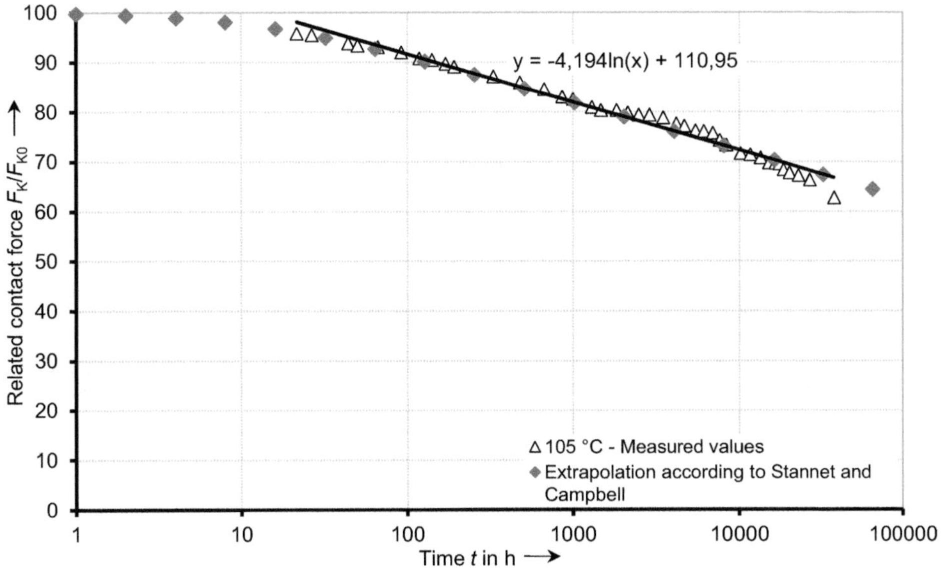

Fig. 5.9 Example—Evaluation of the contact force progression of a screwed connection with busbars made of Cu-ETP R300 without spring joining elements according to Stennet and Campbell $m = 4.194$ $a = 110.95$ [11]

Table 5.3 Comparison of the methods for determining the force reduction

Method	Calculation approaches	Extrapolation methods
Approach*	Modeling of material properties in a material model based on creep and long-term tests	Evaluating the experimental results/ Determining the required parameters from long-term tests
Advantage	Calculation of the force reduction for different initial forces and temperatures possible on the verified calculation model; system-independent material parameters of the conductor materials	Calculation of the force reduction at different temperatures and same initial forces possible
Disadvantage	Calculation not independent of experimentally determined parameters on conductor materials and connections	Only applies to the examined connections under the same boundary conditions
Application	Development accompanying	Operation accompanying

* The approaches apply as long as there is no change in the acting physical mechanisms

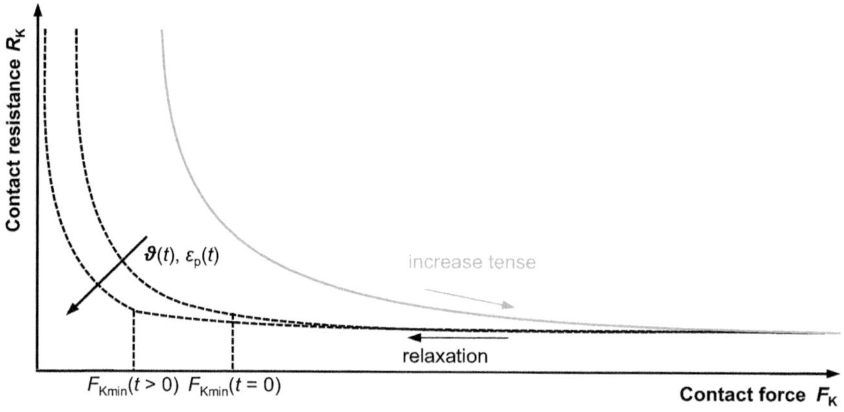

Fig. 5.10 Minimum contact force $F_{K,min}$ dependent on the temperature ϑ, the time t, the time-dependent plastic deformation ε_p and structural changes

k_E and m_E are to be determined experimentally in relaxation tests. For plug-in and screwed connections, the course of the contact resistance depending on the contact force during relaxation can be measured. This is not possible with compression connections, so only estimates can be made here using model arrangements. The material parameters for the deformed state must be known for the models.

5.1.2 Diffusion

Diffusion is a physical process that leads to the complete mixing of two or more substances. It is based on thermally activated particle movement, which takes place in a directed manner when there is a concentration difference [1]. The long-term behavior of current-carrying connections is influenced by self-diffusion, interdiffusion, and the diffusion of gases (Fig. 5.11)

The physical principles of diffusion depend on thermodynamics and kinetics and can be described with Fick's laws. The first Fick's law describes the particle flux density J as the number of particles N that flow through a cross-section A per time t (Eq. (5.12)) [1].

$$J = \frac{N}{A \cdot t} = -D\nabla c \tag{5.12}$$

The second Fick's law describes the temporal change of the location-dependent concentration c of the particles (Eq. (5.13)).

$$\frac{dc}{dt} = \nabla(D\nabla c) \tag{5.13}$$

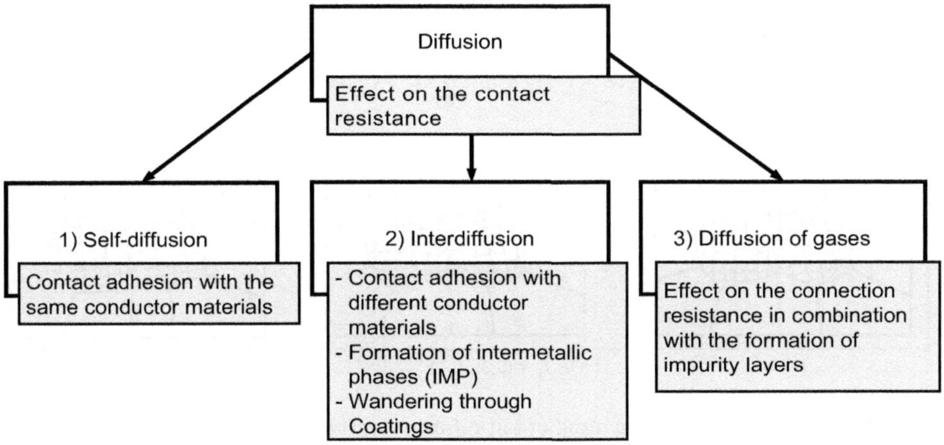

Fig. 5.11 Diffusion and its effect on current-carrying connections

Both processes are determined by the diffusion coefficient D, which is a measure of the speed of diffusion (Eq. (5.14)). The diffusion coefficient depends on the frequency factor D_0, the activation energy Q, the Boltzmann constant k_B and the temperature T [1].

$$D = D_0 \cdot e^{\left(\frac{Q}{k_B \cdot T}\right)} \tag{5.14}$$

Fick's equations, however, only apply to specific boundary conditions. For example, the diffusion coefficient must be constant. Then an analytical solution can be given for infinite and finite half-spaces. For concentration-dependent diffusion coefficients, the Boltzmann-Matano method can be used [10].

The **self-diffusion** is a crucial mechanism of recrystallization and thus of grain reformation. If two metals touch, grain reformation can cause both metal lattices to grow together. This can lead to contact adhesion occurring, especially in connections that are operated for a long time above the recrystallization temperature of the conductor or coating materials [2]. Uncoated, base contact partners are usually abrasively pretreated before assembly. This locally cold-hardens the conductor material. During joining, metallic contact points are created and adhesion forces act in the a-aspot between the contact partners. The orientation of the metal lattices of both contact partners is random and usually does not match. Grain boundaries are formed. If the temperature is raised above the local recrystallization temperature of the structure, new grains form controlled by diffusion. Partial recrystallization dissolves the grain boundary between the metal lattices in a new grain. This connects both contact partners materially and the original interface disappears (Fig. 5.12). This effect has no influence on the connection resistance, but in force-fit connections, it makes it difficult to separate the contact partners when

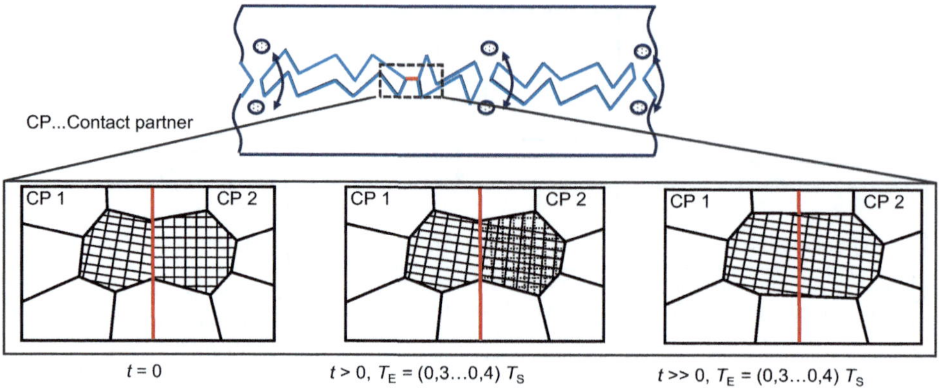

Fig. 5.12 Grain reformation as a result of self-diffusion and cause of contact adhesion

opening the connection[1] [2]. In connections with equally coated contact partners, this effect can occur through recrystallization of the coating material. This has been observed in particular in connections with silver-coated contact partners at temperatures from 100 °C. From this temperature, a softening of silver is also likely. In combination with growing intermetallic phases (IMP) between the top and the intermediate layer, the coating can additionally be torn off locally or over a large area when opening the connection, which prevents good contact behavior when rejoining [2].

Self-diffusion in the contact area can also have a positive effect on the long-term behavior in compression connections, as no solubility is required and the material connection permanently connects the current-carrying conductors. In plug-in connections, a tearing off of the coating is likely if the contacts are operated above the softening temperature for a longer period of time in the plugged state. Generally, this is a reason for failure, as the bare conductor material oxidizes in the open state and the contact properties are significantly deteriorated when plugged in again. This is intensified if the contact force has already been reduced due to force degradation.

The **interdiffusion** describes the particle transport of atoms to foreign lattice components between at least two materials due to concentration differences with the aim of balancing these. A distinction is made between interface and volume diffusion. At low temperatures, interface diffusion dominates at defects in the atomic lattice such as grain boundaries and dislocations. The higher the temperature, the stronger the volume diffusion. The activation energy for grain boundary diffusion is lower than for volume diffusion $Q_{GB} \approx (0.5 \ldots 0.6) \cdot Q_V$ [8]. Intermetallic phases (IMP) can form through interdiffusion. If these grow flat, the thickness d_{IMP} can be estimated with the parabolic time law (Eq. (5.15)).

[1] At connections where relative movements between the conductors must be compensated, this can lead to increased wear.

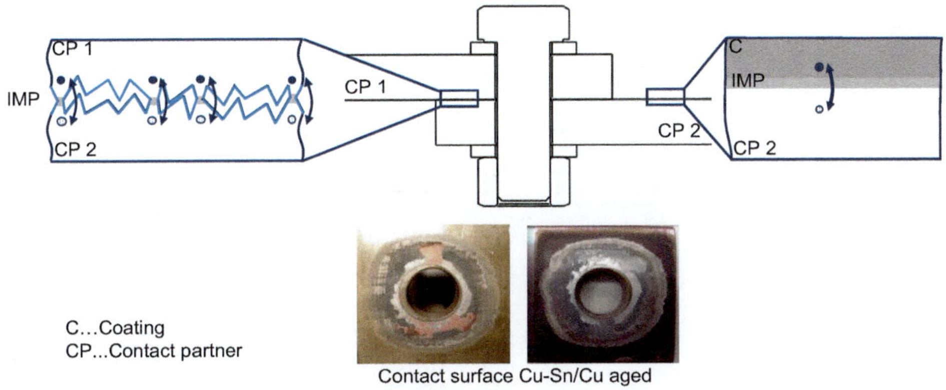

Fig. 5.13 Interdiffusion as a cause for the tearing off of coatings [11]

$$d_{\mathrm{IMP}} = \sqrt{k(T) \cdot t} \tag{5.15}$$

The growth constant k can be determined from the thickness of the intermetallic phase d_{IMP} for two different diffusion times t at the same temperature T. Intermetallic phases can form in the a-spots and at the interfaces between the conductor and the coating material due to interdiffusion (Fig. 5.13).

The IMPs usually have higher specific electrical resistances than the pure metals and are harder and more brittle than these. The properties of the IMPs such as temperature-dependent growth constant k and specific electrical resistance ρ_{IMP} were determined both from the literature and through experimental investigations (Tables 5.4, 5.5 and 5.6). To estimate the influence of the flat-growing IMPs in the a-spots, an extended approach for calculating the film layer resistance can be used. The thickness of the IMP can be calculated using the parabolic time law (Eq. (5.15)) and the resistance R_{IMP} can be approximately determined (Eq. (5.16)).

$$R_{\mathrm{IMP}}(t) = \frac{\sqrt{k(T) \cdot t} \cdot \rho_{\mathrm{IMP}}}{A_{\mathrm{w}}} \tag{5.16}$$

If several IMPs are present in a system, the resistance R_{IMP} is:

$$R_{\mathrm{IMP}}(t) = \frac{1}{A_{\mathrm{W}}} \left(\sum_{d=1}^{m} \sqrt{k_{\mathrm{d}}(T) \cdot t} \cdot \rho_{\mathrm{IMP\ W}} \right) \tag{5.17}$$

This extends Eq. (4.3) by the resistance of the intermetallic phases R_{IMP} (Eq. (5.18)).

$$R_{\mathrm{V}}(t) = R_{\mathrm{M}}(t) + R_{\mathrm{E}}(t) + R_{\mathrm{F}}(t) + \sum_{d=1}^{\mathrm{m}} R_{\mathrm{IMP\ d}}(t) \tag{5.18}$$

Table 5.4 Properties of intermetallic phases in systems with aluminum in the temperature range up to 200 °C [8, 18, 22–29]

System	Al-Cu			Al-Ag		Al-Ni					Al-Sn
Phase	Al_2Cu	AlCu	Al_4Cu_9	Ag_2Al	Ag_3Al	Al_3Ni,	Al_3Ni_2	NiAl	Al_3Ni_5	Ni_3Al	
Lattice structure	Tetragonal space-centered	Orthorhombic space-centered	bcc	Mg-Type (hP2)	Mn-Type cP20	Pnma	P3m1	Pm3m	Cmmm	P4/mmm	No IMP
Melting temperature ϑ_S in °C	550	591	1030	–	–	854	1133	1638	–	1395	
Activation energy Q in kJ/mol	138	134	147	–	–	–	–	–	–	–	
Enthalpy in ΔH_f in kJ/mol	−13.05	−19.92	−21.69	–	–	−150.6	−282.4	−118.4	–	−153.1	
Density ρ in g/cm³	4.36	2.7	6.85	8.12	8.7	3.93 … 4.1	4.9	5.88 … 6	6.7	7.5	
Young's mudulus E in kN/mm²	105	138	150	96 … 104[1)]	97 … 106[1)]	143	224	107 … 294	196	179 … 197	
Thermal expansion coefficient α in $1/K \cdot 10^{-6}$	16.1	11.9	17.6	–	–	–	–	11.9	–	12.3	
Hardness H in $HV_{0.01}$	632	794	636	291 … 203	437 … 488	–	–	270	–	325	

(continued)

Table 5.4 (continued)

System	Al-Cu			Al-Ag		Al-Ni					Al-Sn
Phase	Al_2Cu	AlCu	Al_4Cu_9	Ag_2Al	Ag_3Al	Al_3Ni,	Al_3Ni_2	NiAl	Al_3Ni_5	Ni_3Al	
Electrical conductivity κ in $m/\Omega \cdot mm^2$	11.2 … 15	7.6 … 13.3	2.8 … 5.3	2.9 … 3.1	1.7 … 1.9	–	–	6.3 … 16.7	–	–	
Temperature coefficient of electrical resistance α_{20} in $1/K \cdot 10^{-3}$	3.5 … 3.8	2.7 … 3.3	0.8 … 1.7	0.93 … 0.94	0.55 … 0.66	–	–	–	–	–	

Table 5.5 Properties of intermetallic phases in the systems with copper in the temperature range up to 200 °C [2, 17, 30, 53]

System	Cu-Sn		Cu-Ni	Cu-Ag
Phase	Cu_6Sn_5	Cu_3Sn		
Lattice structure	NiAs-Structure	Orthorhombic Cu_3Ti-Type	No IMP	No IMP
Melting temperature ϑ_s in °C	415	676		
Growth constant k in cm^2/s (Temperature ϑ in °C)	$3.5\ 10^{-11}$ (212) $3.35\ 10^{-12}$ (171) $7.4\ 10^{-14}$ (110) $3.3\ 10^{-14}$ (68)	$7.9\ 10^{-12}$ (212) $1.2\ 10^{-12}$ (171) $7.4\ 10^{-14}$ (110) $1.1\ 10^{-14}$ (68)		
Enthalpy ΔH_f in kJ/mol	−67.1	−32.8		
Density ρ in g/cm^3	8.28	8.90		
Young's modulus E in kN/mm^2	90[1)]	110[1)]		
Thermal expansion coefficient α in $1/K \cdot 10^{-6}$	16.3	19		
Hardness in $HV_{0.005}$	539	472		
Poisson's ratio ν	0.32	0.33		
Electrical conductivity κ in $m/(\Omega \cdot mm^2)$	8.2	4.9		
Temperature coefficient of electrical resistance α_{20} in $1/K \cdot 10^{-3}$	0.45	0.2		
Thermal conductivity λ in $W/(m \cdot K)$	34.1	70.4		
Specific heat capacity c_p in $J/(g \cdot K)$	0.286	0.326		

[1)] Determined with ultrasound

In surface contacts such as screwed connections with busbars and compression connections, the distribution and exact location of the a-spots is not known. In this case, the influence of the IMPs can be estimated with the ratio of the contact resistances q_K in the initial state and aged state. The contact resistance in the aged state is related to that in the inittial state (Eq. (5.19)). The influence of foreign layers and the time-dependent plastic deformation of a-spots are neglected in this case [8]. The related resistance change q_k is particularly suitable for evaluating connections that have contact partners joined at the end face, such as point contacts with cylindrical conductors and material closure connections (Fig. 5.14).

$$q_K(t) = \frac{R_K(t > 0)}{R_{K0}} = \frac{R_V(t > 0) - R_M}{R_{V0} - R_M} \tag{5.19}$$

Table 5.6 Properties of intermetallic phases in the systems with coating materials up to 200 °C [30–33]

System	Ag-Sn[1)]		Ni-Sn[2)]
Phase	Ag_3Sn	Ag_4Sn	Ni_3Sn_4
Lattice structure	Cu_3Ti-Type (oP8)	Mg-Type (hP2)	–
Melting temperature ϑ_s in °C	480	724	–
Growth constant k in cm²/s (Temperature ϑ in °C)	$6.8\ 10^{-11}$ (212) $2.2\ 10^{-12}$ (171) $3.6\ 10^{-13}$ (110) $3.2\ 10^{-14}$ (68)	–	$1.2\ 10^{-13}$ (160) $6.5\ 10^{-14}$ (125) $1.9\ 10^{-14}$ (100) $7.2\ 10^{-15}$ (75)
Enthalpy ΔH_f in kJ/mol	–	–	–
Density ρ in g/cm³	9.93	10.0	8.64
Young's modulus E in kN/mm²	11.06 … 78.9	–	119 … 152
Thermal expansion coefficient α in $1/K \cdot 10^{-6}$	–	–	–
Hardness in HV	138 (KHN)	–	540 ± 87
Poisson's ratio ν	–	–	–
Electrical conductivity κ in $m/\Omega \cdot mm^2$	8.03 ± 0.6	1.72 ± 0.25	3.51
Temperature coefficient of electrical resistance α_{20} in $1/K \cdot 10^{-3}$	0.0036	0.0006	0.0031
Thermal conductivity λ in W/mK	–	–	19.6

1) No information on electrical properties and little information on mechanical properties (New findings in the project—IGF No. 17EWBG—Title "Reliability and safety of operating resources in low-voltage distribution stations for use in the field of wind power and photovoltaic systems")
2) Two additional phases present at room temperature, which are preferably not formed in the solid state

Experiments were conducted on point contacts with conductors made of EN AW-1350 (E-Al99.5) and a conical as well as a circular, flat contact surface. For this, both contact partners were directly silver-plated, so that IMP could grow between the conductor and coating material. The results were compared with the calculated resistance changes according to Eq. (5.16) and Table 5.4 (Fig. 5.14) [8, 39].

It becomes clear that the calculated and measured relative resistance change match well and can be estimated approximately with the approach of the course (Fig. 5.14).

The results of these investigations on point contacts can be transferred to stationary plug-in connections, as the shape of the contacts is very similar. In plug-in connections, the IMP between the conductor and the coating material can cause the coating to tear off

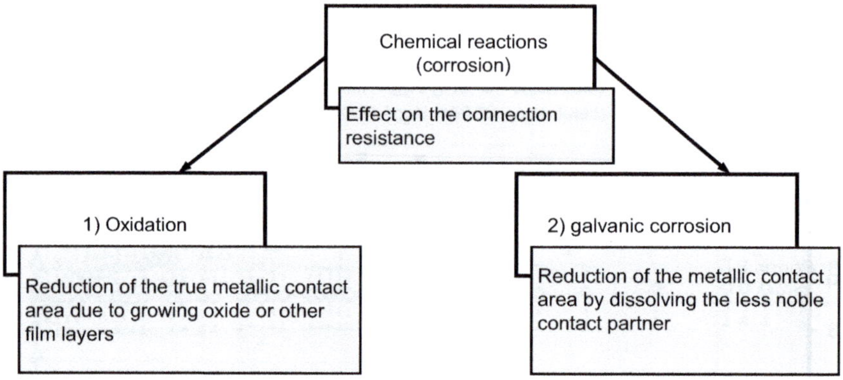

Fig. 5.16 Chemical reactions and their effect on the long-term behavior of current-carrying connections [9]

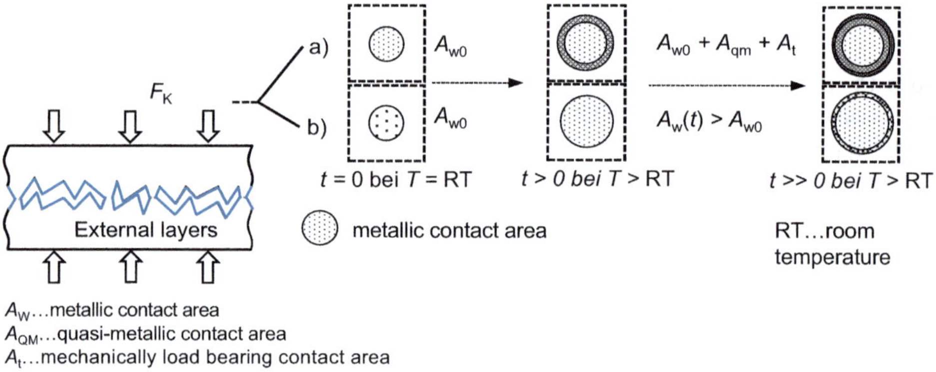

Fig. 5.17 Enlargement of the contact area by a time-dependent, plastic deformation of the micro-contacts **a**) base metals, **b**) noble metals (schematic) [9]

formed oxide layers. This behavior could be demonstrated in creep tests on point contacts with crossed cylinders made of aluminum and copper as well as silver-plated, tin-plated and nickel-plated conductors made of copper (Fig. 5.17) [12].

A-spot in connections with base conductors made of the same metals act like a boundary between grains of different or the same orientation in the structure of metals. The electrons cross the interface, are diffracted and the resistance is ideally comparable to that of a natural grain boundary in the structure of the conductor material. It is in the range of several 10 nΩ [13, 14].

If thin, tunnelable oxide layers are present in the a-spot, the skin resistance in the range of a few μΩ also acts (Sect. 4.1.1). This does not lead to thermal failure of the connection during operation. Only with thicker oxide layers can the contact resistance become so high that it leads to failure. The influence of growing oxide layers can be described for thin layers with the skin R_H and for thick layers with the volume resistance of the film layer R_F (Eq. (5.20))—Table 5.7.

$$R_H = \frac{\sigma_H}{A_w} \quad \text{or} \quad R_F(t) = \frac{\rho_F \cdot l_F(t)}{A_w} \tag{5.20}$$

The thickness of the film layer l_F strongly depends on the time t, the temperature θ and the accessibility of oxygen to the surface. *Slade* provides a general equation that can be used to calculate the specific skin resistance depending on the thickness of the oxide layer l_F [17].

$$\text{Slade}: \sigma_H = \frac{l_F \cdot h^2}{3 \cdot e^2 \sqrt{2 \cdot m \cdot \Phi}} \cdot e^{\frac{4 \cdot \pi \cdot l_F \cdot \sqrt{2 \cdot m \cdot \Phi}}{h}} \tag{5.21}$$

Here, h is Planck's constant, m is the electron mass, e is the elementary charge and Φ is the height of the potential barrier specific to the oxide layer in eV. The specific skin resistance σ_H in Ωcm^2 according to *Holm* can be estimated for an oxide layer with a thickness $l_F < 1.5$ nm using Eq. (5.22) [15].

$$\text{Holm}: \sigma_H = 0.5 \cdot 10^{-22} \cdot \frac{A^2}{1 + AB} e^{AB} \text{ with} \tag{5.22}$$

$$A = 7.32 \cdot 10^6 \cdot \left(l_F - \frac{0.72 eV}{\phi} \right) \text{ and } B = 1.265 \cdot 10^{-6} \sqrt{\phi - \frac{1\,\text{nm}}{l_F \cdot \varepsilon_r}}$$

ε_r is the relative permeability and l_F is the thickness of the tunneled oxide layer in nm. If the oxide layers are thicker than $l_F > 1.5$ nm, *Holm* recommends using the approach according to *Simmons* [16] Eq. (5.23) to use.

$$\text{Simmons}: \sigma_H = \frac{\Delta s}{3.16 \cdot 10^9 \cdot \sqrt{\phi_L}} \cdot e^{10.25 \cdot \Delta s \sqrt{\phi_L}} \text{ with } \Delta s = s_2 - s_1, s_1 = \frac{0.6 eV}{\varepsilon_r \phi},$$

$$s_2 = l_F - \frac{0.6 eV}{\varepsilon_r \phi} \text{ and } \phi_L = \phi - \frac{0.575}{\varepsilon_r \cdot (s_2 - s_1)} \cdot \ln \left(\frac{s_2 \cdot (s - s_1)}{s_1 \cdot (s - s_2)} \right) \tag{5.23}$$

Both approaches differ particularly in the mathematical approximation of the shape of the potential wall. The basic form is an exponential function, in which the prefactor and the exponent increase with increasing layer thickness.

The specific skin resistance of aluminum oxide, for example, shows that this changes over several decades even with layers only a few nm thick (Fig. 5.18). Growth laws have been derived for oxide layers, for example, on metal surfaces exposed to air [1]. The example of a point contact shows that even layers a few tenths of a nm thick have a significant influence on the contact resistance (Fig. 5.19).

Table 5.7 Properties of film layers in the temperature range up to 175 °C [45–57]

Foreign layer	Cu_2O Copper(I) oxide	Al_2O_3 (>99 %) Aluminum oxide	SnO Tin(II) oxide	α-Ag_2S Silver(I) sulfide	Ag_2O Silver(I) oxide	NiO Nickel(II) oxide
Crystal structure	cP6	hR10	tP4	Monoclinic		
Color	Transparent, yellow, brown, red, black (depending on thickness)	Transparent (thin layers), white	Blue, black	Lead gray, gray, iron black	Almost black	Light yellow, green, gray-green, black
Melting temperature ϑ_s in °C	1242	2040 … 2055	1042 (decomposes into Sn and tin dioxide)	150 … 300 (decomposes)	230 (decomposes)	1957
Density ρ in g/cm³	5.8 … 6.1	3.94	6.45	7.3	7.2	7.45
Band gap in eV	1.8 … 2.1	8.7	–	1.3	–	3.7
Relative primitivity ε_r	7 … 16.2	9 … 10	9.72	–	–	–
Dielectric strength E_d in kV/mm	–	35	–	–	–	–
Specific electrical resistance ρ_{20} in Ωm at 20 °C	$5.5 \cdot 10^4$	10^{12}	10^2	3.6	$9.7 \cdot 10^{-1}$ … $2.6 \cdot 10^{-5}$	10^{11}
Thermal conductivity λ in W/K · m	6.3 … 8	19 … 30 (30 … 100) °C	–	–	–	12.3
Specific heat capacity c_p in J/(kg · K)	–	850 … 1050 (30 … 600) °C	–	–	–	–

(continued)

Table 5.7 (continued)

Foreign layer	Cu_2O Copper(I) oxide	Al_2O_3 (>99 %) Aluminum oxide	SnO Tin(II) oxide	α-Ag_2S Silver(I) sulfide	Ag_2O Silver(I) oxide	NiO Nickel(II) oxide
Young's modulus E in kN/mm^2	–	360	–	–	–	–
Bending strength in N/mm^2	–	520	–	–	–	–
Linear expansion coeffi-cient α in $1/K \cdot 10^{-6}$	–	7 … 8 (30 … 600) °C	–	14	–	–
Hardness H in HV	109 … 189	1730 … 2060	36 … 60	36 … 60		535 … 550

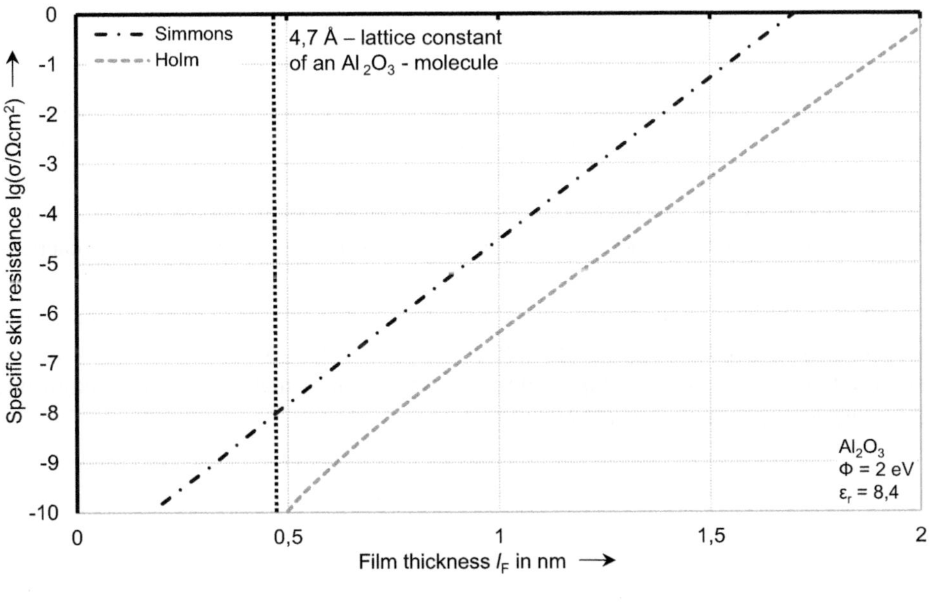

Fig. 5.18 Specific skin resistance σ_H for aluminum oxide Al_2O_3 between two aluminum conductors depending on the Film thickness l_F (material values from [38])

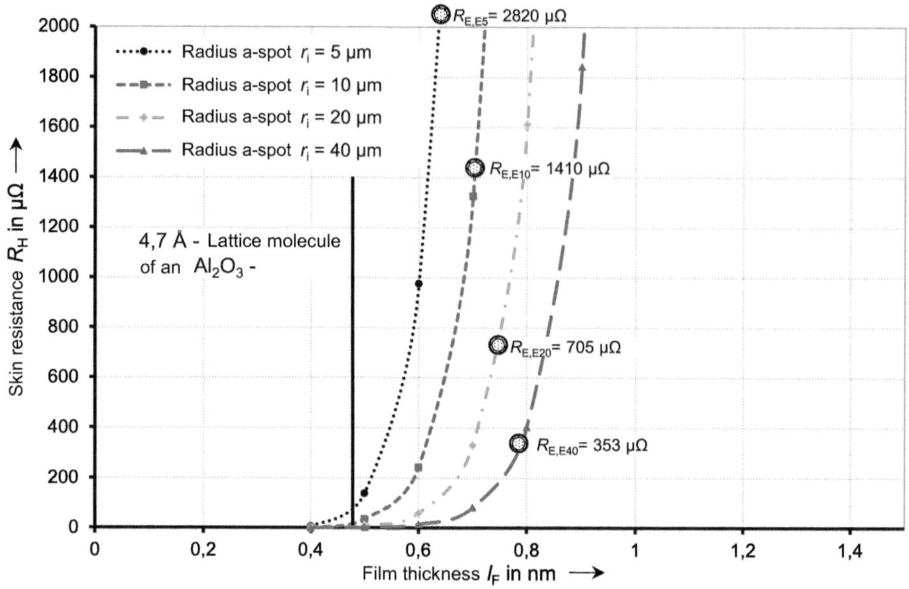

Fig. 5.19 Skin resistance R_H of aluminum oxide Al_2O_3 for different sized point contacts between two aluminum conductors depending on the thickness of the oxide layer l_F Eqs. (5.22 and 5.21) compared to the constriction resistance $R_{E,E}$ Eq. (4.4)

Oxide layers with a thickness of (0.5 … 0.7) nm already lead to skin resistances that are in the range of the constriction resistance. Oxide layers with a greater thickness significantly increase the contact resistance. This also applies to parallel-connected point contacts as they occur, for example, in screw connections with crossed groove structures on the contact surface of the busbars (Fig. 5.20). A critical layer thickness is reached at values >1 nm.

An approach with which the contact resistance of a single or several parallel a-spots, whose radii are statistically evenly distributed, can be described depending on the time and the thickness of the forming oxide layer for the quasi-metallic contact area through diffusing oxygen is as follows [37]:

$$\text{one a-spot } R_K(t) = \frac{R_{K0}}{1 - \frac{2D}{r_i^2 l_q / l_{F\,max}} t} \tag{5.24}$$

$$\text{multiple a-spots } R_K(t) = \frac{R_{K0}}{\left(1 - \frac{2D}{\overline{r_i}^2 l_q / l_{F\,max}} t\right)\left(1 - \sqrt{\frac{2D}{\overline{r_i}^2 l_q / l_{F\,max}} t}\right)} \tag{5.25}$$

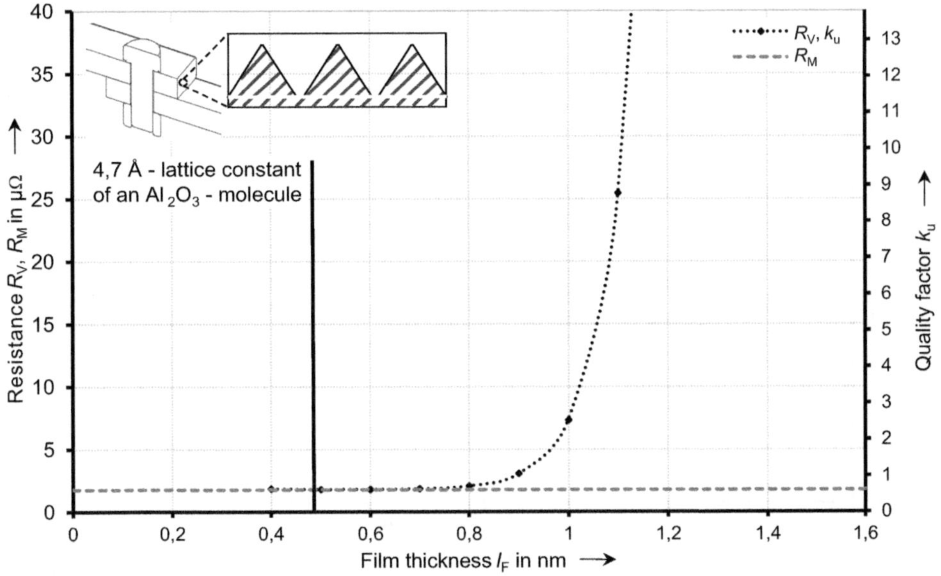

Fig. 5.20 Connection resistance and performance factor depending on the thickness of the oxide layer l_F using the example of a screwed connection with grooved busbars made of aluminum, an apparent contact area of 1477 mm² and 5869 parallel-connected a-spots with a size of 40 µm (no interaction between the a-spots)

Here, $l_{F\,max}$ is the maximum thickness of the oxide layer outside the a-spot. This depends on the type of material and the environmental conditions. l_q is the thickness of the quasi-metallic, tunnelable oxide layer in the contact and is usually unknown. R_{K0} is the contact resistance of the point contact in the initial state and D is the diffusion coefficient, which describes the speed of the incoming oxygen into the a-spot. The diffusion coefficient is almost unknown and depends on the temperature T, the activation energy Q and the frequency factor D_0 [37].

Example from [21]

A point contact between a conical cylinder made of copper and a flat cylinder made of aluminum with parallel a-spots is considered, which have an average radius of $r_i = 7\ \mu m$. The point contact is aged at a temperature of 115 °C. The measured output resistance was $R_{K0} = 248\ \mu\Omega$ and the diffusion coefficient $D_{115\,°C} = 4.2 \cdot 10^{-16}\ cm^2/s$. The maximum thickness of the Al_2O_3 oxide layer outside the contact was assumed to be $l_{F\,max} = 2.5$ nm and the tunnelable layer thickness was assumed to be $l_q = 15$ nm. The average measured and calculated contact resistance over 4000 h was compared (Fig. 5.21). The following is the calculation step for a time point after 4000 h for explanation:

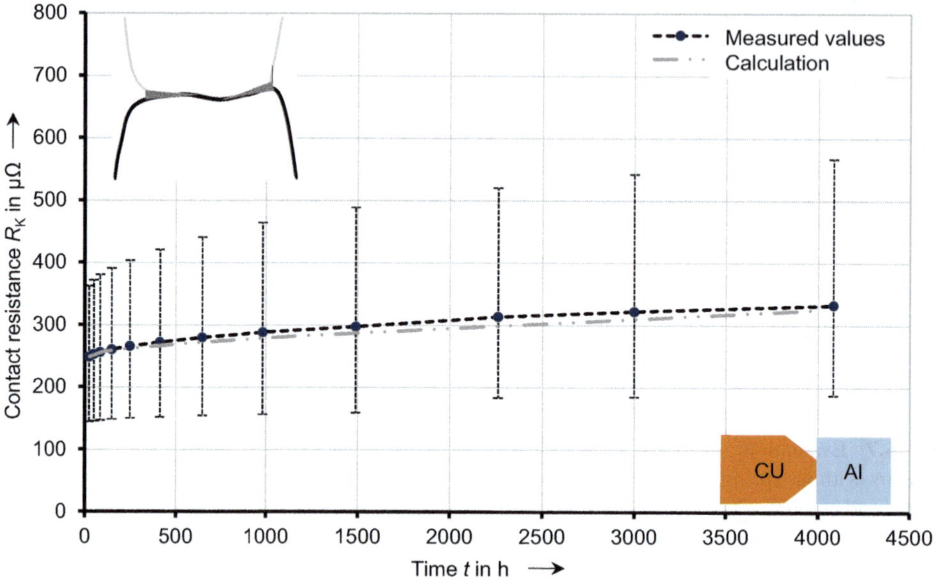

Fig. 5.21 Calculated and measured course of the contact resistance of a point contact with conductors made of aluminum and copper according to [21]

$$R_K(4000\ \text{h}) = \frac{248\ \mu\Omega}{\left(1 - \frac{2 \cdot 1{,}5 \cdot 10^{-12}\ \text{cm}^2/\text{h}}{(0{,}0007\ \text{cm})^2 \cdot 15\ \text{nm}/25\ \text{nm}} 4000\ \text{h}\right)\left(1 - \sqrt{\frac{2 \cdot 1{,}5 \cdot 10^{-12}\ \text{cm}^2/\text{h}}{(0{,}0007\ \text{cm})^2 \cdot 15\ \text{nm}/25\ \text{nm}} 4000\ \text{h}}\right)} = 325\ \mu\Omega$$

◀

During the formation phase, the contact resistance can slightly increase due to growing oxide layers in combination with settling processes. How pronounced this effect is depends on the type of connection and conductor as well as the distribution of mechanical stress in the contact plane. For example, the force reduction in connections with stranded conductors made of aluminum is particularly pronounced due to the settling in the numerous contact planes within the first hours, much more than in connections with solid conductors. In addition, the high number of contact planes allows oxygen to easily penetrate the contact surfaces and growing oxide layers to interrupt smaller a-spots (Fig. 5.22). Accordingly, a slight increase in the connection resistance in the formation phase is typical for current-carrying connections with stranded conductors.

Further oxidation processes can only take effect once the contact force has further significantly reduced and/or relative movements occur between the contact partners. The contact resistance of bimetallic connections can be influenced by the migration of oxygen along the interfaces (Fig. 5.23a) or by the conversion from one oxide to another within the first operating hours (Fig. 5.23b). The Ellingham-Richardson diagram shows

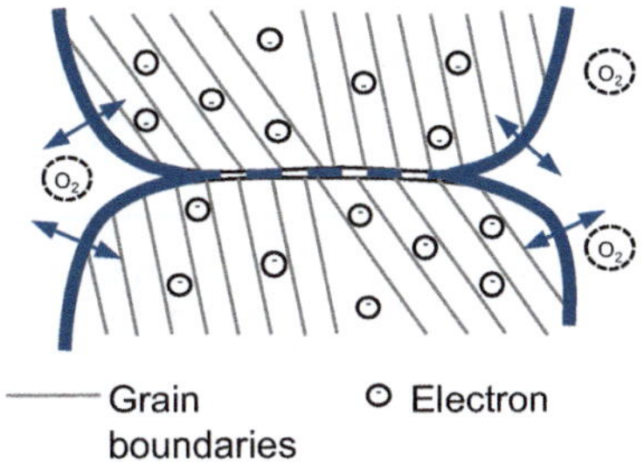

Fig. 5.22 Hypothesis on the influence of growing oxide layers on the a-spots of connections with the same, ignoble contact partners (schematic) [9]

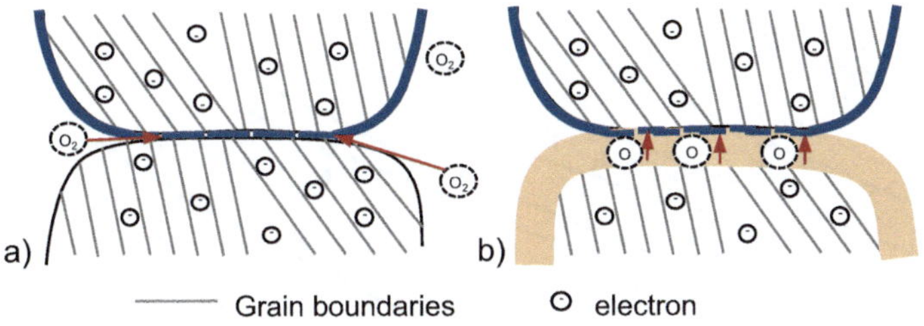

Fig. 5.23 Hypothesis on the influence of growing oxide layers on the a-spot of connections with **a**) a ignoble and noble contact partner and **b**) different ignoble contact partners (schematic) [9]

(Fig. 5.24) that aluminum oxide has the lowest energy state compared to the other oxides, providing a drive for oxygen to be released from the bond of one oxide to form another oxide [21, 39, 40].

Oxygen can oxidize the less noble contact partner in the a-spots. This is conceivable, for example, in connections with conductors made of aluminum and copper. Furthermore, oxygen from the environment can migrate along the grain boundaries to the surface of the less noble contact partner. This phenomenon could be observed in connections with an uncoated and a silver-plated contact partner made of aluminum (see also Fig. 5.40—Sect. 5.3) [21].

Silver hardly forms oxide layers on the surface in normal atmospheric air. Silver(I) oxide (Ag_2O) is very soft and only weakly bound to the surface. A negative influence on the contact behavior is not known [18, 19]. However, it has been shown that at temperatures greater than 200 °C, oxygen migrates along the grain boundaries in the silver layer [19, 20]. At higher temperatures >380 °C, volume diffusion through the crystal structure of the silver is also possible [43]. The driving force for this is the strong concentration gradient of the oxygen to the surface of the aluminum.

For aging through oxidation, it is crucial whether oxygen reaches the a-spots and thus an oxide layer can grow on the metallic contact area. Therefore, it is decisive whether a

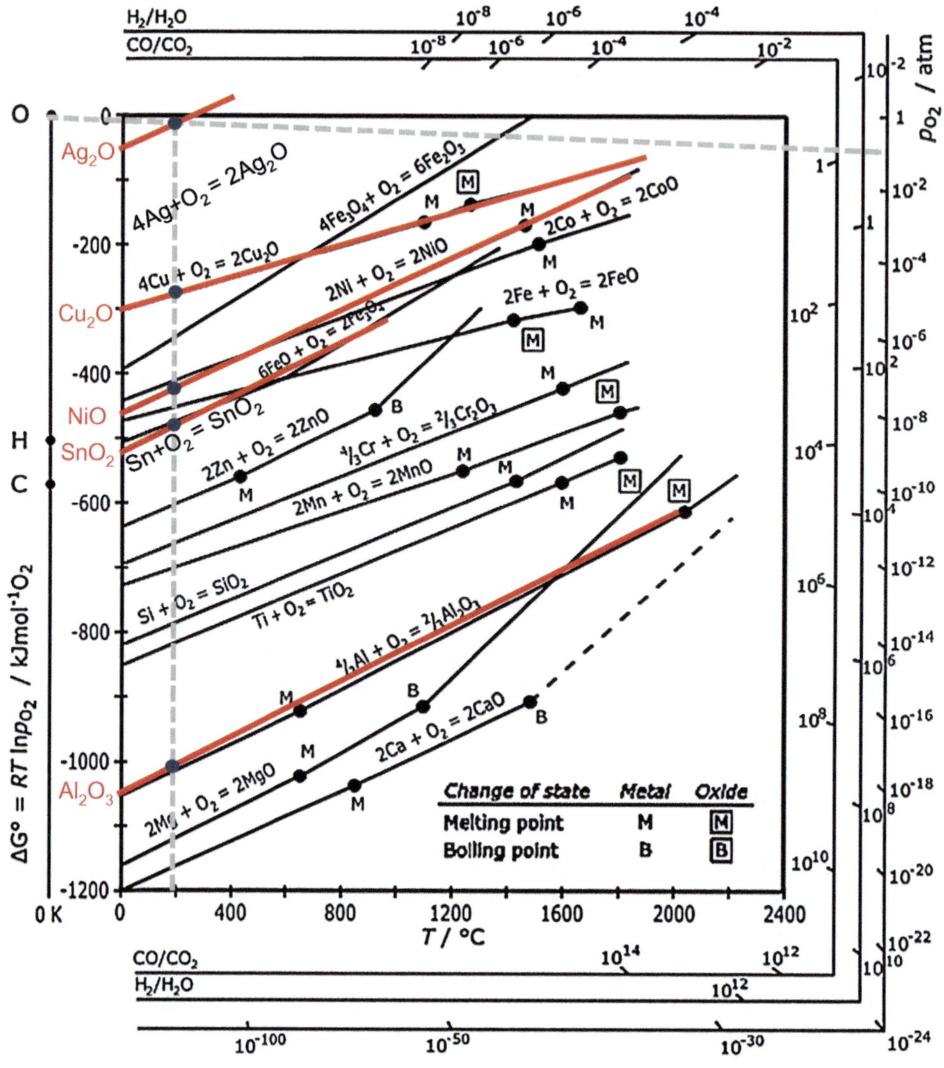

Fig. 5.24 Ellingham-Richardson diagram (oxides highlighted in red, tin oxide added)—As an example, the upper gray dashed line for $p_{O2} = 0.2$ atm (air) is drawn. Oxides whose ΔG_0 value lies below the line are thermodynamically stable in normal atmospheric air. Additionally, a gray line at a temperature of 200 °C is drawn. The intersections (blue points) show the ΔG_0 value [40]

connection is gas-tight or not. One approach to test this is the probabilistic percolation theory[2] [42]. With this, the specific percolation probability for two- or three-dimensional

[2]Percolation theory describes the probability with which a randomly occupied grid of a certain type forms connected areas.

structures of different fill factors f and the critical percolation density p_C from the product of these two can be determined. Above the critical percolation density, there are definitely an unlimited number of connected areas or clusters of contact points in a microstructure or a grid, so that, for example, the penetration of oxygen from the outside into the interior of the protected areas can be excluded. A-spots located inside the protected areas are thus protected from further oxidation. For two-dimensional structures, the critical percolation density can be determined to be (42 ... 46) % and for three-dimensional structures (14 ... 16) % [44]. In a first approximation, the two-dimensional contact plane can be distinguished into touching and non-touching surfaces. Consequently, if the proportion of the mechanically load bearing conact area A_t to the apparent contact area A_s is greater than 44%, a connection would have gas-tight areas in the contact plane. To calculate the necessary contact force or average mechanical stress, the empirical Eq. (5.26) can be derived and applied from Eq. (4.10) [42]:

$$F_K = \frac{H_K A_S A_t}{A_S - A_t} \quad \text{or} \quad \frac{F_K}{A_{Ers}} = \sigma_m = \frac{H_K A_S A_t}{A_S - A_t} \tag{5.26}$$

Here, H_K is the contact hardness of the softer contact partner. An initial evaluation of the approach has shown for real connections that the forces calculated with Eq. (5.26) are too high compared to those that led to gas tightness in experiments [21]. In Eq. (5.26), it is assumed that the force in the apparent contact area is evenly distributed, which is practically not the case in real connections. Therefore, the mechanical stress distribution must be taken into account. Contrary to the assumptions of percolation theory, the connected areas do not form randomly, but preferably in areas of high mechanical stress. Accordingly, depending on the material combination, gas tightness of the a-spots can possibly be achieved with lower contact forces. Consequently, it is crucial whether, on the one hand, local gas tightness is achieved in the contact plane of a surface contact or point contact and whether the a-spots are gas-tight. Both cases have so far been insufficiently investigated, so the applicability of percolation theory must be validated in the future to determine limit values for required contact forces depending on the material combination.

An experimentally determined criterion for a gas-tight connection in the initial state was derived from the related resistance change q_k from the contact resistance after 1 h and 24 h of operating time $R_K(1\text{ h})$ and $R_K(24\text{ h})$ [21]:

$$q_K = \frac{R_K(24\,\text{h})}{R_K(1\,\text{h})} < 1,5 \tag{5.27}$$

Phenomenological results from experiments were evaluated and it was found that connections that meet the mentioned criterion can initially be considered as gas-tight [21].

In addition to oxidation, **galvanic corrosion** can influence the long-term behavior of current-carrying connections. Galvanic corrosion occurs when two different conductor

Table 5.8 Material combinations at risk of galvanic corrosion [58, 59, 61]

Material pairing metal 1/metal 2	Potential difference in mV
Al/Cu	580
Al/Ag	770
Al/Ni	620
Al/Sn	300
Ag/Sn	470
Ag/Ni	150
Sn/Ni	320
Cu/Sn	260
Cu/Ag	190
Cu/ Ni	40

Gray shaded are material combinations at risk

materials are connected, which are far apart in the electrochemical series. Both metals are electrically conductive with each other via the a-spots. If both conductors are wetted with an electrolyte, galvanic corrosion begins. The less noble metal acts as an anode, is oxidized, and transfers to the electrolyte. The electrons produced in this reaction move via the a-spots to the nobler metal, which acts as a cathode, and enter the electrolyte there. At this point, water and oxygen are reduced and hydroxide is formed. As a result, the less noble metal is continuously consumed and deposited on the surface of the nobler metal as a hydrated oxide. The higher the potential difference, the stronger the galvanic corrosion (Table 5.8). In addition, the surface ratio of both reaction partners is crucial. Galvanic corrosion is intensified at the anode if the area of the anode is smaller than that of the cathode [59, 60]. Galvanic corrosion must be considered in technical applications from a potential difference of >350 mV [61].

Galvanic corrosion can reduce the metallic contact area of a connection, thereby increasing the connection resistance. Since an electrolyte is necessary for the reaction, which covers both reaction partners, the effect of galvanic corrosion is practically excluded in the interior (<50% RH) and at temperatures above 100 °C [61]. Under open-air atmosphere, near the sea or in vehicles on salted, wet roads, galvanic corrosion must be particularly considered, as it can strongly influence the long-term behavior of the connection. By "sealing" the connections with plastic housings, paints or plastic coatings (after joining), galvanic corrosion can be greatly reduced.

5.1.4 Wear

Wear describes the aging of current-carrying connections due to relative movements, which cause changes in the surface topography or the surfaces and contact properties. The following wear mechanisms can act due to these relative movements [63]:

1. Abrasion: Wear of softer surfaces by hard particles or hard surface elevations
2. Adhesion: Formation of adhesive bridges (micro cold welding), which tear open with sufficiently large offset movements in or near the contact and thereby detach cold-hardened particles from the surface
3. Tribochemical reactions: Removal of protective film layers by relative movement. The underlying material then reacts again with components of the atmosphere, e.g., through oxidation. This leads to a continuous material removal.
4. Surface fatigue: Crack formation in the contact surfaces due to temperature change or mechanical stress (embrittlement). The coating can partially break out and oxygen reaches the underlying material.

Wear can generally be distinguished into fretting and sliding wear (Fig. 5.25). The aim is to define application-specific value ranges for the amplitude of the relative movement between the contact partners and the acting contact normal force, in which certain wear mechanisms preferentially occur depending on the contact materials [63].

Through **fretting**, micro-peaks and corrosion layers are removed. Adhesion acts as a wear mechanism. If the offset movements are very small, the adhesive forces lead to a purely elastic deformation of the a-spots. If this is the case, a hysteresis is created in the force-displacement diagram (Fig. 4.16). Depending on the contact force, the hardness of the contact materials, and the static friction between the contact partners, plastic deformation also occurs from a certain offset amplitude (Fig. 5.26).

In plug-in connections with contact lamellas, larger relative movements can also be absorbed by elastic deformation. If the relative movements become too large, the a-spots start to tear and abrasion begins. Due to the small relative movements during fretting

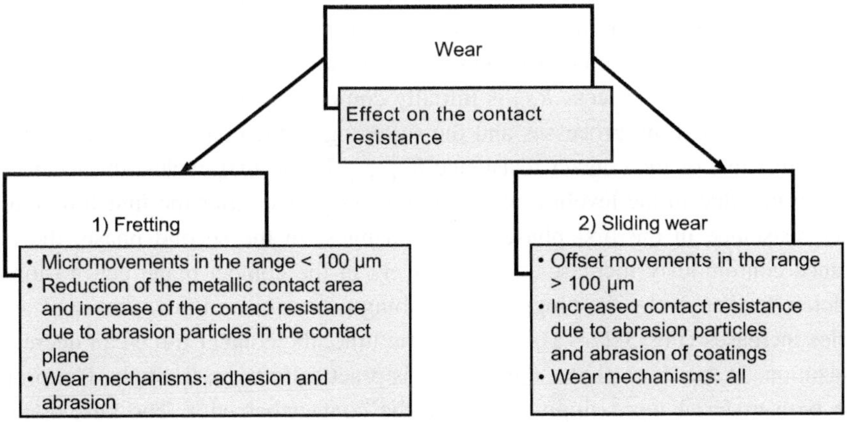

Fig. 5.25 Effect of fretting wear on the long-term behavior of current-carrying connections

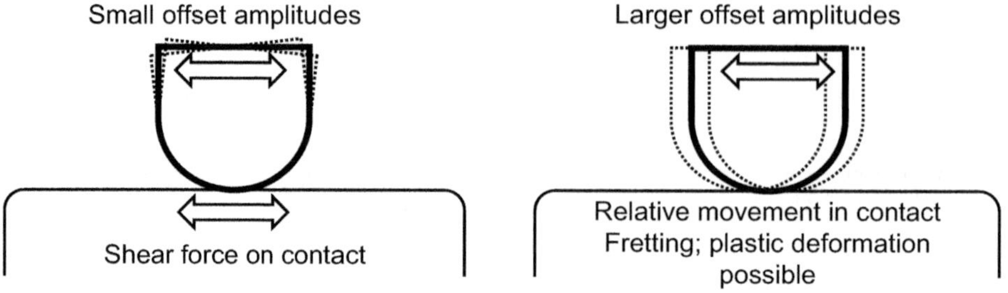

Fig. 5.26 Elastic deformation and breaking of the a-spot at larger amplitudes

(typically < 100 μm), the resulting abraded and subsequently cold-hardened particles remain between the contact partners. This intensifies the abrasion and increases the wear [62, 63]. If there are enough abrasion particles between the contact partners, the wear decreases, as the contact partners each shift against the movable abrasion particles. In the further course, particles are pushed out of the contact area and the mechanical stress distribution becomes uneven. As a result, the contact area wears out, especially in the areas of the highest mechanical stresses [63]. Closed mathematical models for describing the influence of fretting on the course of the contact resistance of connections are not known, so the effect must be determined in experimental investigations [62, 63].

Even larger relative movements lead to sliding of the contacts, and **sliding wear** becomes dominant. In sliding wear, the roughness of the surfaces of the contact partners is leveled and the abrasion particles are pushed out of the contact area. This leads to the removal of the contact material and can be slowed down by suitable lubrication, which also protects the surface from chemical reactions and reduces the coefficient of friction between the contact partners. In sliding wear, all wear mechanisms can occur, but three phases are distinguished. In the first phase, the running-in phase, the undamaged surfaces of the contact partners rub against each other for the first time. The contact resistance after the first plugging process R_{K0} is initially comparatively high (Fig. 5.27—a)). This decreases due to settling processes and the softening of the coating after a few hours to a minimum value of the contact resistance $R_{K0\ min}$. In the first cycles, the wear is comparatively low due to the leveling of the roughness peaks. After the first few hours, the contact resistance in the first phase hardly changes. In the second phase, the contact resistance continuously increases, as film layers in the contact plane can no longer be completely broken up by the plugging or rubbing processes and the number of abrasion particles increases (Fig. 5.27—a)). At the same time, the contact force can decrease due to relaxation. In the third phase, the coating is practically worn through. The properties of the base material increasingly dominate the contact behavior. The base material of plug-in connections, copper, aluminum or their alloys, usually ages faster through chemical reactions than the coating material. The degradation of the a-spots is thus dominated

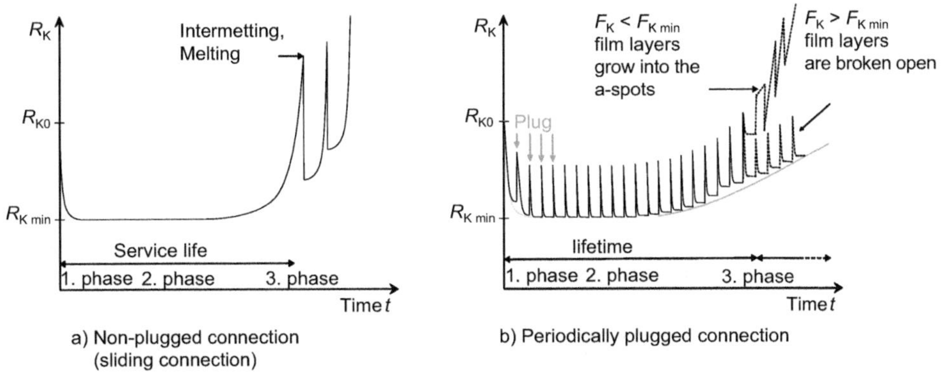

Fig. 5.27 Contact resistance depending on the effect of frictional wear of silver coated sliding and plug-in connections (schematic) **a**) non-plugged connection or sliding connection **b**) periodically plugged connection

in the third phase by the tribochemical wear and the abrasion, until the connection fails (Fig. 5.27a) [18, 64].

When plugging cycles are carried out, the existing contact force is particularly crucial for the contact behavior. If the contact force F_K is greater than the critical minimum contact force $F_{K\,min}$, the existing film layers are broken up during plugging and the contact resistance remains low (Fig. 5.27—b)). The third phase is characterized by the fact that the contact resistance is temporarily reduced by the plugging process, but remains at a very high level. Plugging typically increases the contact resistance to values significantly greater than $R_{K\,min}$ but less than or equal to $R_{K\,0}$. If film layers are present, they are broken up by plugging and rubbing (Fig. 5.27b).

If the contact force F_K decreases below the critical value $F_{K\,min}$, the plugging process still partially breaks up film layers and initially reduces the contact resistance. However, the remaining contact force is not sufficient to prevent the growth of oxide and film layers into the contact point [7, pp. 128, 139].

The mechanisms of frictional wear dependent on the speed of movement. The Stribeck curve (Fig. 5.28) explains this behavior. At the beginning of a movement, static friction must be overcome. In movements in the range of μm, minimal relative movement without breaking up the a-spots is possible due to the elasticity—both of the materials surrounding the contact and within the contact itself. For larger offset movements and slow speeds, boundary lubrication or solid body friction occurs. The coefficient of friction is comparatively high, as metal rubs directly on metal. At higher speeds, metal partly rubs on a solid or liquid lubricant, which is referred to as mixed lubrication. At even higher speeds, there is no longer any metallic contact, which is called hydrodynamic or full-film lubrication[13].

Typically, the contact resistance during movement is higher than at rest, as the a-spots are broken up and, depending on the speed and lubrication, interlock less strongly than

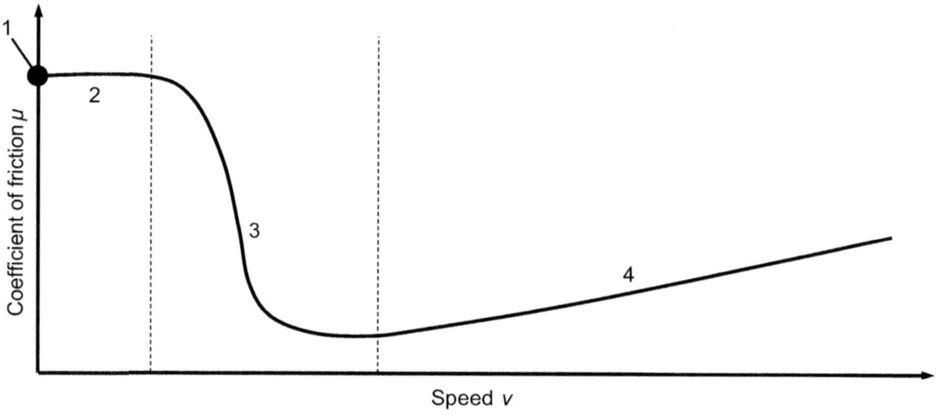

Fig. 5.28 Friction coefficient μ dependent on speed 1) static friction 2) boundary lubrication/ solid body friction 3) mixed lubrication 4) hydrodynamic/full-film lubrication

in the resting state. Sliding wear cannot yet be fully described in models. However, there are empirical-analytical models with which the wear volume in sliding wear can be calculated. This can be used to estimate when a coating is worn through and a significant increase in contact resistance can be expected. The basis for this is the wear formula according to Archard [91]. He calculates the wear volume V_r with the specific wear coefficient k_r determined from experiments, depending on the contact normal force F_K and wear distance s_r as follows:

$$V_r = k_r F_K s_r \tag{5.28}$$

From the wear volume V_r, the removed height of a concact surface due to sliding wear h_w can be calculated if the contact geometry is known. With the simplified assumption of a ball-plate geometry, as often occurs in plug-in connections and switching contacts, this can be done analytically.

The size of the contact area after the first joining of a single-point contact A_w can initially be calculated simply with Eq. (4.21). The associated depth of plastic impression h_{pl}, assuming a hemispherical contact partner, depends on its radius of a sphere r_K and the contact hardness H_K (Table 4.9, Eqs. (5.29 and 5.30)). For line contacts, the same area (Fig. 5.29) is assumed, but with the average contact length l (cylinder-plate) [65]:

Point contact

$$h_{pl} = r_K - \sqrt{{r_K}^2 - \frac{F_K}{H_K \pi}} \tag{5.29}$$

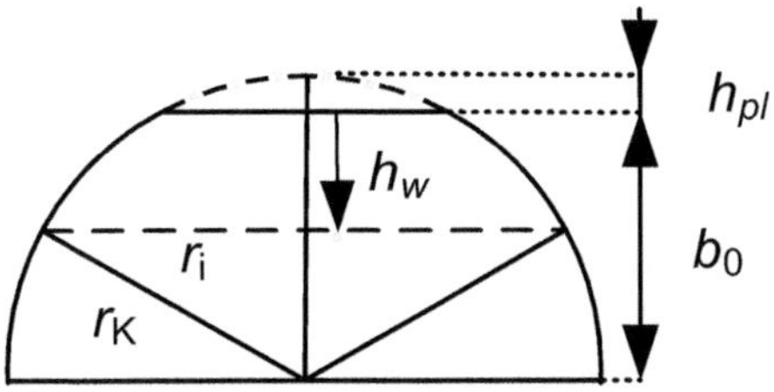

Fig. 5.29 Model concept for frictional wear

Line contact

$$h_{\text{pl}} = r_{\text{K}} - \sqrt{{r_{\text{K}}}^2 - \left(\frac{F_{\text{K}}}{H_{\text{K}}2l}\right)^2} \tag{5.30}$$

Assuming that the wear volume is continuously removed with the specific wear coefficient k_r during the wear process, the wear distance s_r for point or surface contact, depending on the wear depth h_w, results as follows:

Point contact [65]

$$s_{\text{r}} = \frac{\pi}{k_{\text{r}} F_{\text{K}}}\left(-\frac{1}{3}h_w^3 + (r_{\text{K}} - h_{\text{pl}})h_w^2 + \left(2r_{\text{K}}h_{\text{pl}} - h_{\text{pl}}^2\right)h_{\text{w}}\right) \tag{5.31}$$

Line contact

$$s_{\text{r}} = \frac{l}{k_{\text{r}} F_{\text{K}}}\left(\begin{array}{l} r_{\text{K}}^2 \operatorname{ar}\cos\left(1 - \frac{h_{\text{w}}+h_{\text{pl}}}{r_{\text{K}}}\right) - \left(r_{\text{K}} - (h_{\text{w}} + h_{\text{pl}})\right)\sqrt{2r_{\text{K}}(h_{\text{w}} + h_{\text{pl}}) - (h_{\text{w}} + h_{\text{pl}})^2} \\ -\left(r_{\text{K}}^2 \operatorname{ar}\cos\left(1 - \frac{h_{\text{pl}}}{r_{\text{K}}}\right) - (r_{\text{K}} - h_{\text{pl}})\sqrt{2r_{\text{K}}h_{\text{pl}} - h_{\text{pl}}^2}\right) \end{array}\right) \tag{5.32}$$

The specific wear coefficients k_r combines the processes that occur during the first and second phase of wear (Table 5.9) [62, 65, 92, 93].

This is necessary because a more accurate wear model for the first phase can only be determined with a precisely known surface profile [65]. This is very rarely the case in practical applications. Experiments have identified the main influencing factors for sliding wear, in addition to the contact material, as temperature, contact force, and movement speed.

Comparative studies show that sliding wear increases by at least a factor of (2 … 10) at operating temperatures which are higher than room temperature. Thus, the expected lifetime of electrical contacts, in terms of wear, is reduced by a factor of (2 … 17) at typical operating temperatures of (55 … 105) °C. Therefore, the operating temperature must be considered as a significant influencing factor in lifetime studies. A slow movement speed means that the abrasion particles have more time to form oxidation products during a fixed number of cycles, thereby intensifying the effect of wear. Lubrication significantly reduces the coefficient of sliding friction. As a result, suitably lubricated con-

Table 5.9 Specific wear coefficient k_r depending on material, movement speed v and amplitude, valid for amplitudes >100 μm and sphere/flat geometries [62, 65, 92, 93]

Contact Pairing	ϑ_{Test} in °C	F_K in N	Speed v in mm/s	Amplitude in mm	k_r in mm³/(N mm)
Ag-Ag lubricated	25	4	0.15	2.2	$<<1.0 \cdot 10^{-9}$
Ag-Ag lubricated	55	4	0.15	2.2	$1.5 \cdot 10^{-9}$
Ag-Ag lubricated	65	7	15 … 100	0.4 … 0.6	$(1.4 \ldots 1.8) \cdot 10^{-11}$
Ag-Ag lubricated	65	7	1.2	0.3	$8.0 \cdot 10^{-9}$
Ag-Ag lubricated	30	7	20	60	$<<1.0 \cdot 10^{-11}$
Ag-Ag lubricated	100	7	20	60	$6.3 \cdot 10^{-11}$
Ag-Ag bare	25	3	0.4	0.2	$2.0 \cdot 10^{-8}$
Ag-Ag bare	55	4	0.15	2.2	$1.5 \cdot 10^{-8}$
Ag-Ag bare	65	3	49	0.4	$1.43 \cdot 10^{-8}$

tacts can withstand at least ten times more friction cycles compared to unlubricated ones [62, 90, 92, 93].

Solid lubricants also exist for moving contacts (Table 4.8). These can be incorporated directly into a silver matrix in the form of a dispersion layer. Compared to unlubricated contacts, significantly reduced friction coefficients are achieved with both solid lubricants and lubricating greases, thereby reducing wear due to abrasion (Fig. 5.30, Table 4.8). Depending on the type of solid lubricant particle, up to (3 … 18) times more cycles are achieved before wear-through than with unlubricated silver. When using lubricating greases, for example for sliding contacts in gas-insulated switchgear (GIS), even significantly higher cycle numbers of up to 20,000 and more can be achieved.

Transferring the respective experimental results to other geometries is difficult, as the wear coefficient captures the overlapping effects of individual wear mechanisms. Due to the different wear mechanisms that influence each other, there are currently no comprehensive numerical or analytical models. To make statements about friction wear, experimental investigations on the same or similar contact geometries, with the same contact forces, coating structures, used lubricating greases, surrounding atmosphere, and operating temperatures are necessary. For high resistance to a wear-through of the coating, a low contact force, low operating temperatures, and especially a high layer thickness are advantageous. For a fixed number of plug-in cycles, a high movement speed and thus a

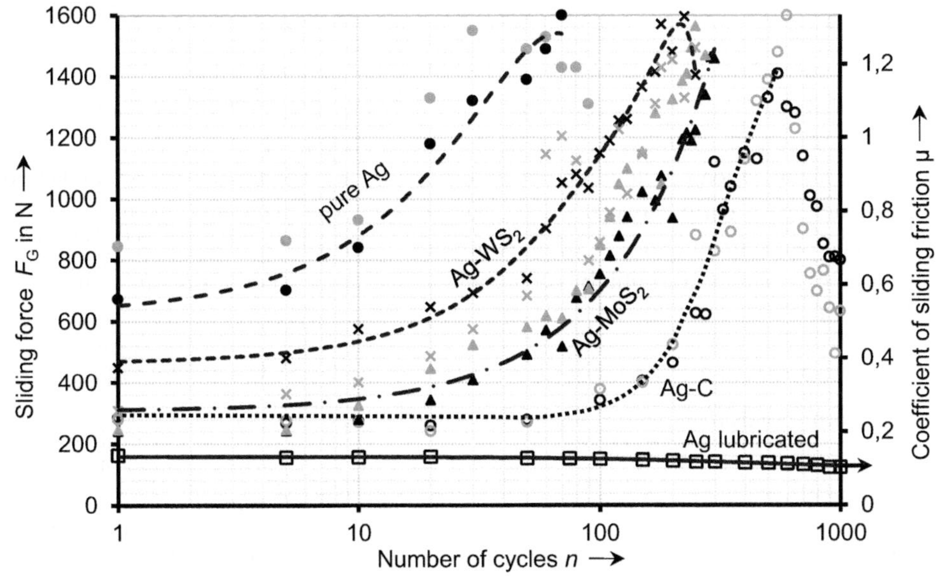

Fig. 5.30 Sliding force and sliding friction coefficient of a GIS tube, which is mounted on two axially loaded coil springs, depending on the coating used on the coil springs—conductor tube and contact carrier are coated with pure silver, sliding path per cycle is 120 mm [90]

short duration in which these cycles are performed is advantageous. However, this can hardly be influenced in practical applications.

In screwed and compression connections with contact partners made of different materials, mechanical shear stresses occur in the contact plane due to different thermal expansion, such as with copper and aluminum. A sufficiently high contact force can prevent relative movement, so that friction wear can only occur at a low contact force. In bimetallic screwed connections with hard conductor materials or in connections with hard coating materials, a displacement movement in the a-spots in the μm range could occur due to different thermal expansion coefficients. If the contact surfaces are hard, such as with nickel, and the interlocking and deformation of the microstructures of both contact partners during joining is low, mechanical shear stresses can lead to relative movements and thus to the breaking of a-spots or the detachment of coatings [66]. Experiments on screwed connections with nickel-phosphorus coated busbars made of aluminum under constant and changing temperature loads support this theory [5]. Due to the numerous influencing parameters of friction wear, the boundary conditions such as the contact force in the initial state, the force reduction in operation, and the effect of formed intermetallic phases must be examined very closely. Further investigations are necessary to better understand the effects and impact on screwed and compression connections.

Example

Connector with silver-plated socket and silver-plated plug lubricated

A connector with an insertion force of 100 N and 20 parallel contacts is used in a switchgear with cylindrical conductors made of aluminium ($\alpha_T = 23.1 \cdot 10^{-6}\,K^{-1}$). Due to the daily load cycle of energy consumption, the temperature of the conductors fluctuates twice daily between −20 °C and 115 °C. The maximum conductor length between two support points is 10 m. A lifetime of 20 years is required. The total friction path is calculated. Initially, each cooling or heating phase results in a change in length of:

$$\Delta L = \alpha\, L_0\, \Delta\vartheta = 23.1 \cdot 10^{-6} \cdot 10\,\mathrm{m} \cdot (115\,°\mathrm{C} - (-20\,°\mathrm{C})) = 31.18\,\mathrm{mm}$$

This corresponds to a total friction path during the lifespan of 20 years of:

$$s_r = 4 \cdot 365 \cdot 20 \cdot 0.01871\,\mathrm{m} = 29200\ \mathrm{cycles} \cdot 0.03118 \frac{\mathrm{m}}{\mathrm{cycle}} = 910.5\,\mathrm{m}$$

The worn layer height is calculated for point contacts according to Eq. (5.31). Contacts for GIS/GIL applications are galvanically silver-coated. The contact force per contact point in the example is 50 N, which is higher than the contact force in the tests in Table 5.9. The expansion speed is 0.0014 mm/s. Therefore, wear coefficients in the range ($1.5 \cdot 10^{-9} \ldots 6.3 \cdot 10^{-11}$) mm³/(N mm) are calculated. Assuming a friction coefficient of $\mu_G = 0.1$. The friction path until wear-through is calculated as:

$$s_r = (128 \ldots 3048)\ \mathrm{m}$$

The calculated friction path until wear-through depends on which comparative values for the friction wear coefficient are known or assumed from the literature. Iteratively, the required layer thickness can be calculated to achieve a real friction path of 910 m. Accordingly, a layer thickness of (10 … 61) µm is necessary. If in doubt, friction wear tests on the real system should reduce the range of the calculation. ◀

5.1.5 Electromigration

Electromigration is a process in which particles J_D migrate. Unlike diffusion due to concentration differences (Sect. 5.1.2), the driving force is created by an existing electric field and the transfer of momentum from charge carriers to atomic cores [73, 74].

$$J_D = -D \cdot grad\, n_a + \frac{D\, n_a}{k\, T} F_{EW} \tag{5.33}$$

$$F = (Z_e + Z_w) \cdot \rho \cdot E = Z^* q \cdot \rho \cdot J \tag{5.34}$$

The resulting diffusion current J_D is, with negligible atomic density concentration gradient grad $n_a = 0$, i.e., a homogeneous structure, dependent on the absolute temperature T, the diffusion coefficient D and the driving force F_{EW} (Eq. (5.33)). The driving force results from the effect of the electric field on the charge of ionized defects Z_e and from the effect of the charges Zw, which are created by the transfer of momentum from electrons to the atomic cores ("electron wind"). In another representation, these components are combined into the effective charge of the atomic cores Z*q, and there is a direct dependence on the current density J and the specific electrical resistance ρ. At very high current densities, forced movements of metal ions occur due to collisions between electrons and the metal lattice, which is also referred to as electron wind [67, 68]. A significant movement of the metal ions occurs from a current density of $(10^3 \ldots 10^4)$ A/mm^2. However, this current density is not reached in conductors of energy technology. Only in areas of constrictions, such as in thin conductor layers or a-spots, which are very well cooled to achieve these current densities, can electromigration occur [69]. Atoms in the metal lattice then oscillate due to the thermal energy around their rest position. The distribution of energy to the individual atoms follows a Boltzmann distribution. This results in a proportion p of atoms that have a higher energy E_A.

$$p = e^{-\frac{E_A}{k \cdot T}} \tag{5.35}$$

If the energy of the atoms is higher than that of the potential well, they can leave it. This process is also referred to as self-diffusion (Sect. 5.1.2). If a force now superimposes on the atoms, which is caused by collisions with electrons moving in the electric field, material can be transported in a directed manner. In this case, direct current leads to a directed material transport, in which atoms are dragged in the direction of the anode at the cathode, creating vacancies [70]. The number of lattice defects in the area of the narrow point increases accordingly. With alternating current, the material transport is alternating and the number of lattice defects increases. This locally increases the electrical resistance in the material and thus the contact resistance [71]. The electromigration of the metal ions is a diffusion-controlled process, which particularly takes place along the current density gradient at grain boundaries or large lattice defects (Fig. 5.31).

Defects arise that can lead to cracks in the area of a-spots. The necessary current density for migration processes in a-spots made of aluminum is in the range between 10^4 A/mm^2 and 10^6 A/mm^2 [68] (Fig. 5.32). For copper, current densities of $2 \cdot 10^4$ A/mm^2 are given at a temperature of 105 °C [72]. With these material-specific current densities S_{EM}, a critical current I_{EM} can be estimated, from which electromigration in the a-spots takes effect, if the true metallic contact area is approximately known.

$$I_{EM} = S_{EM} \cdot A_W \tag{5.36}$$

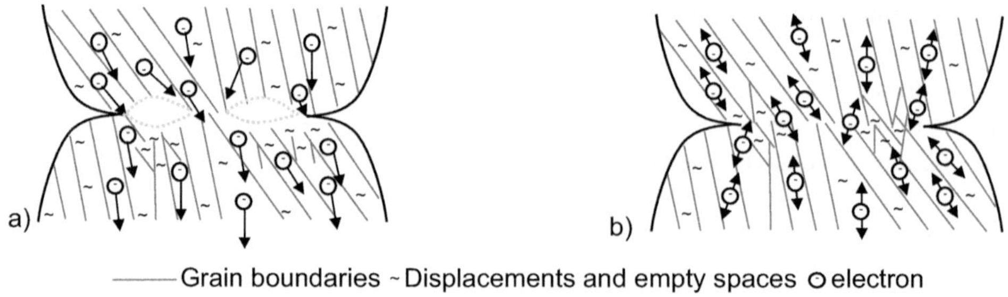

Fig. 5.31 Effect of electromigration in the microcontact with **a**) direct and **b**) alternating current

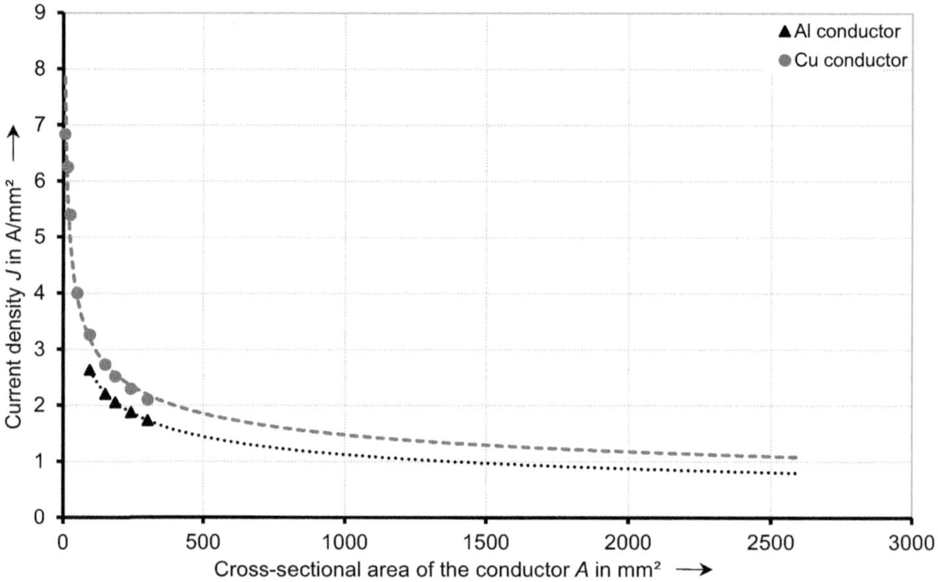

Fig. 5.32 Permissible current density depending on the conductor cross-section using the example of cables laid freely in air with a permissible limit temperature of 90 °C

Example: Critical current I_{EM} of a connection with copper conductors

In a screwed connection with busbars of the geometry (40 × 10) mm^2 made of copper and a connection length of 40 mm, subtracting the borehole results in an apparent contact area of 1477 mm^2. Assuming that 1% of this apparent contact area is true contact area, this results in 14.8 mm^2. This results in a critical current, from which electromigration can become effective, at a temperature of 105 °C of:

$$I_{EM} = 2 \cdot 10^4\,\mathrm{A/mm^2} \cdot 14.8\,\mathrm{mm^2} = 2960\,\mathrm{kA}$$

These currents are not reached in practical operation. ◀

Electromigration is a physical mechanism that is particularly observed in microelectronics and power electronics. In order to estimate the failure of a component, the phenomenological BLACK's equation is used, which describes the mean failure time *MTF* of a component (resistance increase by 5 … 20% of a homogeneous conductor track) depending on the current density J and the absolute temperature T [73, 75].

$$MTF = A^{*} J^{-n} e^{\frac{E_a}{kT}} \tag{5.37}$$

From numerous experiments on homogeneous conductor tracks, the proportionality factor A^* and the activation energy E_a were calculated for various materials (Tables 5.10 and 5.11). The exponent n can be assumed to be $n = 1$ for current density less than 10^4 A/mm^2.

In electrical conductors, the current density decreases with increasing cross-sectional area, as the heat-emitting surface decreases in relation to the cross-sectional area (Fig. 5.32).

The current densities in the conductors are thus much too low for electromigration. In connections, the current is strongly confined in the a-spots, thereby increasing the current

Table 5.10 Selected research results on electromigration behavior

Material	Current density in A/mm^2	Exponent n	Temperature in °C	MTF in h	E_a in eV	Prop.-Factor A^*	Source
Al 99.9	$1 \cdot 10^4$	2	375	0.524	1.03	0.72	[84]
	$0.96 \cdot 10^4$	2	265	50.6	1.03	0.72	
	$2 \cdot 10^4$	2	260	110	1.14	0.72	[80]
AlCu0.5 … 4	$2 \cdot 10^4$	2	260	72	1.12	0.75	
Al(O)	$2 \cdot 10^4$	2	260	75	1.11	0.95	
Cu (pure)	$6 \cdot 10^4$	2	200	26	1.02	1.2 … 1.5	[79]
	$3 \cdot 10^4$	2	300	7	1.1		[81]
	$7 \cdot 10^4$	2	300	3	1.14		
	$8 \cdot 10^4$	2	300	2.5	1.14		
	$1 \cdot 10^5$	2	300	0.55	1.1		
CuAg0.8	$2 \cdot 10^4$	2	300	130	1.2	1.3 … 2.2	[77]
	$3 \cdot 10^4$	2	350	68	1.3		
	$3 \cdot 10^4$	2	300	52	1.2		
	$2 \cdot 10^4$	1	350	26	0.67		
	$2 \cdot 10^4$	1	320	55	0.67		
	$2 \cdot 10^4$	1	290	106	0.67		
Ag (pure)	$2.8 \cdot 10^4$	2	150	0.13	0.82	1.7	[78]

Table 5.11 Activation energy E_a for various mechanisms of electromigration [73, 76, 82, 83]

Diffusion mechanism	Activation energy E_A in eV		
	Aluminium	Copper	Silver
Lattice	1.2 … 1.4	2.1 … 2.3	1.91 … 1.97
Grain boundary	0.4 … 0.5	1.2	0.99
Surface	0.28	0.74 … 1.09	0.51

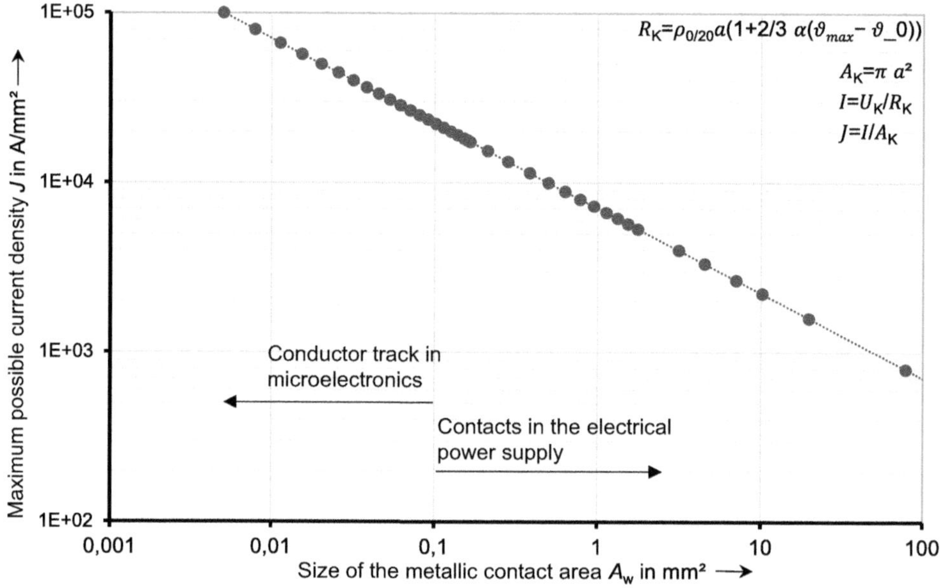

Fig. 5.33 Maximum possible current density in an electrical contact when reaching the melting temperature—connecting conductors at room temperature

density. The internal heating of the contact limits the maximum possible current load and density (Fig. 5.33). The current density achieved in contacts when they reach the melting temperature while the connecting conductors remain at room temperature is calculated based on the simplifications in Sect. 4.4.2. A realistic maximum current density in electrical contacts is in the range of (10^3 … 10^4) A/mm^2.

For the activation energies given in Table 5.10 and temperatures up to 120 °C, mean failure times due to electromigration of several thousand hours to years are calculated using Eq. (5.37) (Table 5.12).

The greatest influence on the speed of the processes during electromigration is the temperature. However, very high temperatures are only reached in the operation of current-carrying connections when the contact resistance is very high and thus the metallic contact area is very small. For connections with conductors made of aluminium and

Table 5.12 Calculated mean time to failure MTF (resistance increase by (5 … 20)%) for assumed current densities in the contact points

Material/ Parameter	Aluminium		Copper		Silver	
Current density J in A/mm^2	$1 \cdot 10^4$ Connector	$1 \cdot 10^3$ Busbar	$1 \cdot 10^4$ Connector	$1 \cdot 10^3$ Busbar	$1 \cdot 10^4$ Connector	$1 \cdot 10^3$ Busbar
Temperature ϑ in °C	120	120	120	120	120	120
MTF in h	117670	$11.8 \cdot 10^6$	$1.93 \cdot 10^6$	$19.4 \cdot 10^6$	562	56208

copper, this is only the case when they are already considered to have failed. For connections with silver-plated contact surfaces, an influence of electromigration on ageing due to operating temperatures in the range (120 … 180) °C is possible. However, in studies on connectors with silver-plated conductors at a temperature of up to 180 °C over 16,000 h, no critical condition was found [88]. This is probably due to the fact that the connections are force-fitted. The material migration due to electromigration in the contact area is compensated by the spring elements and the resulting elasto-plastic deformation, so that the metallic contact area is not significantly reduced. Also, the influence of additional defects forming in the material near the a-spot does not lead to a significantly higher contact resistance. Studies on screwed or compresion connections in electrical energy technology have so far also shown no significant influence of electromigration on long-term behaviour. No significant differences in the course of contact resistance were observed in tests in the heat chamber and with DC current flow [2, 5, 8, 19, 62, 63].

The findings so far, obtained from practically used force-fitted connections in electrical energy technology, show that no failure of a connection could be attributed to the effect of electromigration. In the case of heavily aged connections, it is conceivable that the influence of electromigration can become effective. However, the contact resistances are then already so high that the high thermal stress on the connection when loaded with the rated current also intensifies the other ageing mechanisms. Since the effect of electromigration depends exponentially on the temperature and approximately quadratically on the current density, and mainly occurs at grain boundaries under continuous current load, these influences should be investigated application-specifically [69].

5.2 Interaction of Aging Mechanisms and Resulting Lifetime Characteristics

The considered mechanisms of aging of contacts and connections act simultaneously and overlap. The course of aging and accordingly that of the connection resistance depend on the dominantly acting aging mechanisms. Thus, possible aging mechanisms can already

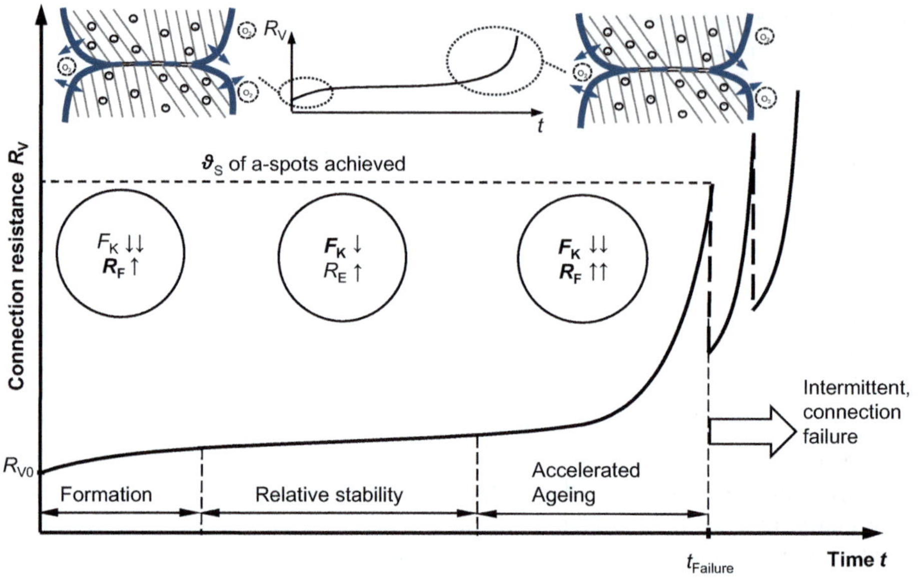

Fig. 5.34 Lifespan characteristic of connections with contact partners made of the same materials [11]

be identified by the course of the connection resistance. Typical lifetime characteristics arise depending on the type of connection, conductor and/or coating materials, and operating conditions. For stationary connections with contact partners made of the same material, aging is dominated by force reduction and chemical reactions (Fig. 5.34).

The typical lifetime characteristic of such connections is characterized in the formation phase by the fact that the mechanical tension in the contact surface is reduced by settling and the primary area of force reduction, and small-scale s-spots break open or are interrupted by growing film layers. This behavior is particularly typical for connections with conductors made of aluminum, where the connection resistance slightly increases in the first operating hours. In the phase of relative rest, the connection resistance changes little, even if the contact force continues to decrease. Only when a minimum force is undercut, e.g., F_K F_{Kmin}≤, further a-spots break open, the gas tightness in the contact surface is reduced, and the chemical reactions (Sect. 5.1.3) are intensified. This reduces the metallic contact area, the temperature increases, and the aging processes are accelerated until the connection fails.

In the case of bimetallic connections, the interdiffusion (Sect. 5.1.2) can be so dominant that the course of the lifetime characteristic significantly differs from that of a connection with the same materials of the contact partners. In the phase of formation and relative rest, the gradient of the connection resistance is higher and can lead to a faster failure of the connection (Fig. 5.35).

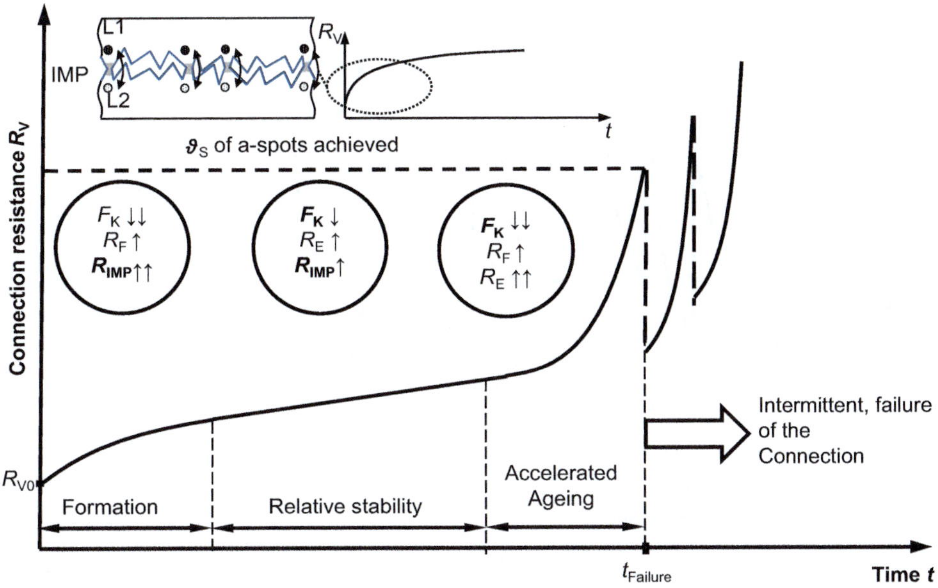

Fig. 5.35 Lifetimen characteristic of bimetallic connections dominated by interdiffusion [11]

In this case, the intermetallic phases grow very quickly at the beginning and can increase the resistance in the a-spots. However, the effect is less than with film layers, as the specific electrical resistance of the considered intermetallic phases is at most about 37 times that of the conductor material (Sect. 5.1.2), while the specific electrical resistance of oxides is about a factor of $10^8 \ldots 10^{11}$ higher than that of the conductor materials. Subsequently, the phase growth slows down as the concentration gradient decreases due to the compensation of concentration differences. This can be compensated to some extent by the increasing temperature in the connection if the current load remains the same. If the contact force is very small, the a-spots break open more easily due to the brittle intermetallic phases (Sect. 5.1.2). Shear forces due to cyclic thermal loading and different thermal expansion coefficients of the contact partners intensify this effect (Sect. 4.2.4).

If chemical reactions, especially oxide layers, dominate the long-term behavior, the gradient of the connection resistance is very steep already in the first operating hours and the connection can fail within a very short time in the range of hours to a few days (Fig. 5.36).

This behavior can be attributed to the fact that even thin oxide layers lead to an additional skin resistance. When a sufficient thickness of the film layer is reached, the significantly higher volume resistance comes into effect. This is several orders of magnitude higher than the material resistance of the contact materials and can be considered electrically insulating. It has a very strong influence on the contact and thus on the connection

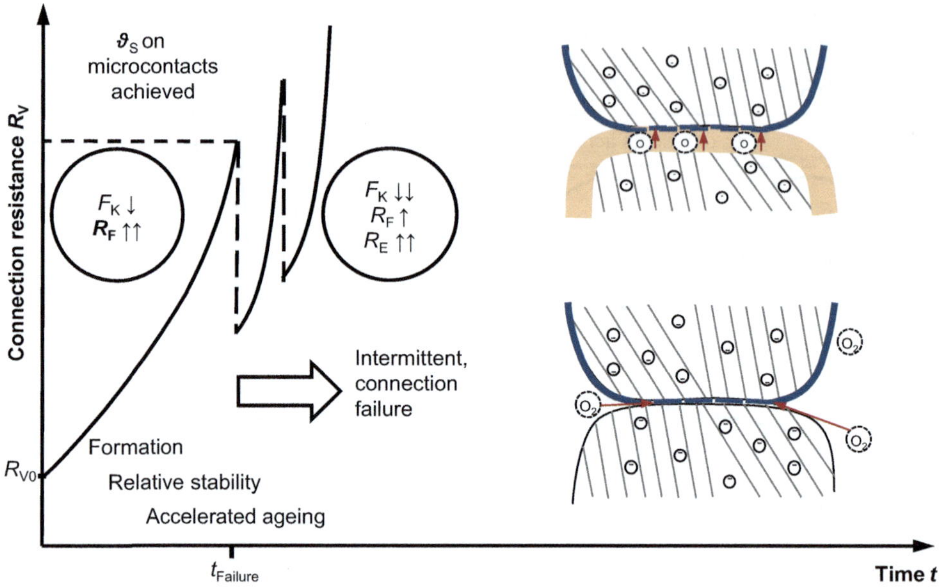

Fig. 5.36 Lifetime characteristic of bimetallic connections dominated by film layers [11]

resistance (Sect. 5.1.3). The growth of film layers into the a-spots will intensify with a low contact force.

From the preceding chapters and the overview of typical lifetime characteristics, it becomes clear that all properties essential for the function of a current-carrying connection are strongly temperature-dependent. As a result, the connections can only be operated safely within a certain temperature range. There is an upper and sometimes also a lower limit temperature. If the upper limit temperature is exceeded, failure can occur, for example due to mechanical failure or excessive thermally induced/accelerated aging of the contacts. The decisive factor is not only the absolute value of the temperature, but also its duration of exposure. If the lower limit temperature is undercut, it can happen, for example, in lubricated plug-in and slide connections that the viscosity of the lubricating grease used decreases and the plug and slide forces become too great for the coupled drive. The lower limit temperatures are often determined by the climate at the location of the device. Limit temperatures are defined in standards for numerous devices and systems and their components (Fig. 5.37).

These limit temperatures take into account a very specific use case and depend heavily on it, as numerous influencing factors affect the lifetime and permanent function of the connection. In Table 5.13, factors influencing the aging mechanisms are described. These determine a limit temperature of the connection, which must not be exceeded in continuous operation. Explanations of the aging mechanisms can be found in Sect. 5.1.

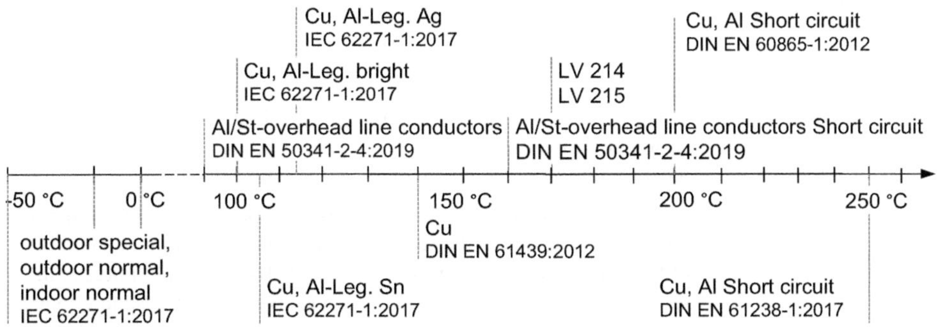

Fig. 5.37 Permanent and short-term permissible limit temperatures of connections and contacts according to standards

5.3 Material Combinations

In numerous experiments as part of research work, the life cycle curve for various material combinations and types of connections was studied depending on the temperature. Tests were stored and evaluated in heating cabinets or with current flowing through them for several years. A simplified overview could be developed from this, which does not claim to be complete and should not serve as a substitute for an individual application-specific qualification of material pairings (Table 5.14). The goal is to make statements for the user about which material pairings are fundamentally more stable or unstable depending on the type of connection, without addressing very specific questions of application and the resulting requirements for, for example, coating systems, material strengths and more. The preceding Sects. 5.1 and 5.2 illustrate the complexity of the topic.

The following universally valid statements can be derived from the overview. In properly dimensioned stationary current-carrying connections such as screwed and compression connections, where the contact materials of the contact partners are the same, long-term stable operation can be assumed if the limit temperatures are adhered to and the contact surfaces are carefully pretreated (Sects. 3.1 and 3.2). The lifetime curve then follows that shown in Fig. 5.34. For moving connections, such as plug-in connections, special boundary conditions apply, as new a-spots are formed with each movement. Furthermore, the contact force is comparatively low, so that stable operation is generally not possible without a coating under normal air atmosphere.

The following notes on stationary current-carrying connections with conductor materials made of aluminum and copper, aluminum and silver, and aluminum and tin should be observed. If coatings are used, the limit temperature of the coating material must also be taken into account (Sect. 3.2).

Table 5.13 Influencing factors that particularly affect the limit temperatures of current-carrying connections

Influencing factor	Explanations	Example
Base material of the contact partners	The mechanical strength decreases at higher temperatures, aging mechanism force reduction	According to [IEC 60439-1], copper conductors are mechanically stable up to 140 °C
Atmosphere	Atmospheres with aggressive ingredients can attack the surface of the contact partners; aging mechanism chemical reactions	Bare copper may be used in oxidizing atmosphere (e.g., air) up to 75 °C for (plug-in) contacts, in non-oxidizing atmosphere (e.g., insulating gas SF6) up to 115 °C
	The relative humidity RH influences chemical reactions and frictional wear	Galvanic corrosion in humid environment and reduction of the friction coefficient in silver layers at RH > 50 %
Surface, or coating material of the contact partners	Aging mechanisms chemical reactions, diffusion	A silver plating protects copper surfaces from oxidation. Silver-plated copper (plug-in) contacts may therefore be operated permanently at 115 °C according to (IEC 62271-1). Bare ones only at 75 °C.
Adjacent components	Conductors are usually used with insulation(s) for which limit temperatures also apply	Class A insulating materials (e.g., epoxy resin insulators in GIS) may be permanently loaded up to 105 °C
	High ambient temperatures in the installation space (ISO 6722-1:2011-10)	Connections near components with high heat dissipation in vehicles (LV 215-1:2013)
Duration of load	Over the lifetime, lower temperatures are usually permissible in normal operation than in short-term loads that can occur in the event of a fault	DIN EN 50341-2-4:2019 Limit temperatures for Al/St conductor cables: 80 °C rated operation, 160 °C short-circuit case
Frictional wear	The hardness of a material is temperature-dependent. Soft materials lead to higher frictional wear	For plug-in connections, the lifetime is lower by a factor of (2 … 10) when frictional wear occurs at operating temperatures of (55 … 105) °C than at room temperature (Sect. 5.1.4)
Lubrication	Lubricating greases frequently used for plug-in and sliding contacts have a lower (viscosity, lubricating ability) and upper limit temperature (chemical stability, dripping)	Lubricating greases can harden and lead to a failure of a connection [88], Sect. 6.3.3) Oils can be displaced from the contact area and the lubricating effect is lost [8]

Table 5.14 Reliability of material pairings depending on the type of connection

Conductor material 1	Conductor material 2	Long-term stability (yes ◆ green, conditional ◆ yellow, no ◆ red, no knowledge ◇): Stationary contacts – Screwed connection	Stationary contacts – Compression connection	Moving contact – Plug-in connection[a]	Remark	Source
					For all types of connections: careful pretreatment of the contact surface and assembly according to instructions	
Aluminium	Aluminium	yes	yes	no	Pay attention to the softening temperature of the aluminium	[5, 12, 85]
Copper	Copper	yes	yes	conditional	Consider the annealing temperature of copper Only pre-treated and possible in non-oxidizing atmospheres or very high contact forces for connectors	[2, 12, 89]
Aluminium	Copper	conditional … no	conditional … no	no	Gas tightness required over the entire lifetime—Oxidation (Sect. 5.1.3)	[8, 19, 21]
Aluminium	Tin	no	yes … conditional	no	Oxide formation from tin oxide to aluminium oxide possible (Sect. 5.1.3)	[11, 21, 35, 36]
Aluminium	Silver	no	conditional … no	no	Oxygen permeability of silver → Oxidation of aluminium (Sect. 5.1.3)	[19, 21, 35, 89]
Aluminum	Nickel	yes	no knowledge	no	Consider abrasive pretreatment of the nickel surface and high hardness difference between Al and Ni	[5, 11]
Copper	Tin	yes … conditional	yes	conditional … no	Conditional solubility due to diffusion-related material closure, reassembly after revision by IMP critical	[2]

(continued)

Table 5.14 (continued)

Conductor material 1	Conductor material 2	Long-term stability yes ◆ conditional ◆ no ◆ no knowledge ◇			Remark	Source
		Stationary contacts		Moving contact	For all types of connections: careful pretreatment of the contact surface and assembly according to instructions	
		Screwed connection	Compression connection	Plug-in connection[a]		
Copper	Silver	◆ (yes)	◆ (yes)	◆ (conditional)	Only pretreated (Cu) and long-term stable in non-oxidizing atmosphere for plug-in connections	[2]
Copper	Nickel	◆ (yes)	◇	◇	Consider pretreatment of Nickel	[11]
Tin	Tin	◆ (yes)	◆ (yes)	◆ (yes)	Contact adhesion (Sect. 5.1.1), formation of whiskers possible; plug-in connection: observe limited number of cycles	[2, 12]
Silver	Silver	◆ (yes)	◇	◆ (yes)	Contact adhesion possible from 100 °C (Sect. 5.1.2)	[2, 12, 63, 89]
Nickel	Nickel	◆ (yes)	◇	◆ (conditional)	High contact hardness → High contact forces for good contact behavior	[11, 12]
Tin	Silver	◆ (yes)	◇	◆ (conditional) … ◆ (no)	Conditional solubility due to diffusion-related material closure, remounting after revision by IMP critical For plug connections continuous resistance increase, thermally unstable after a few 100 h	[32, 34, 86],

(continued)

Table 5.14 (continued)

Conductor material 1	Conductor material 2	Long-term stability yes ◆ (green) conditional ◆ (yellow) no ◆ (red) no knowledge ◇			Remark	Source
		Stationary contacts		Moving contact	For all types of connections: careful pretreatment of the contact surface and assembly according to instructions	
		Screwed connection	Compression connection	Plug-in connection[a]		
Tin	Nickel	◆ (green)	◇	◆ (red)	Long-term tests on screwed connections over 16,000 h at 140 °C Due to the high differences in hardness of both materials and the low contact force, long-term stable operation is not possible with plug connections	Tests unpublished
Silver	Nickel	◆ (green)	◇	◆ (red)	Investigation on screwed connections at 115 °C over 5000 h High initial value scattering and not long-term stable for contact forces up to 10 N in plug connections	[87]

[a]applies to plug connections for transmitting high currents (≥50 A)

Aluminum–Aluminum

Connections with conductors made of aluminum can be used in a long-term stable manner up to the softening temperature of the conductor material (Chap. 3—Table 3.4) with appropriate pretreatment and design. Technically pure aluminum can be used permanently up to a temperature of approx. 90 °C and alloyed aluminum up to 140 °C depending on the type. The main aging mechanisms in aluminum are force degradation and oxidation. For moving contacts, aluminum is not usable without a coating, as rapidly growing oxide layers lead to high contact resistances. In addition, the contact behavior of aluminum-aluminum connections is poor at low contact forces.

Copper–Copper

Connections with conductors made of copper have a higher annealing temperature than aluminum and can therefore be used at permanently higher temperatures. The annealing temperature of technically pure copper, like all pure metals, depends on the degree of deformation. Slightly deformed material states of copper, such as the R250 state, can be used permanently up to a temperature of approx. 160 °C. States like R300 or with even higher strength can only be used permanently at a temperature less than 140 °C, as annealing due to recrystallization occurs above this temperature (Chap. 3—Table 3.4). Even low-alloy copper materials, for example with silver, have higher operating temperatures.

Bimetal

When different materials are contacted with each other, in addition to the aging mechanisms of force reduction, oxidation as part of the chemical reaction, self-diffusion, and frictional wear in plug-in connections, interdiffusion and galvanic corrosion also occur. Whether a long-term stable operation of these connections is possible depends heavily on the properties of the conductor and coating materials. The lifetime curve can follow a different course than for connections with the same materials of the contact partners, as shown in Figs. 5.35 and 5.36. The current state of knowledge for non-long-term stable material pairings will be explained in more detail below.

Connections with conductors made of aluminum and copper often show very poor long-term behavior. This has been repeatedly attributed to interdiffusion and thus growing intermetallic phases (IMP) in the past. In recently published works, it was shown that the influence of the IMP at operating temperatures below 200 °C is low and the measured resistance change cannot be explained with it [8]. Based on these findings, it was shown that oxidation in particular has a decisive influence. If a gas-tight contact level is reached over the lifetime, long-term stable operation is conceivable [21]. This is shown by tests on screwed connections with conductors made of aluminum and copper stored at 140 °C in an inert atmosphere (Fig. 5.38).

How gas tightness can be ensured in screwed connections (necessary surface condition, minimum contact force, optimal distribution of mechanical stress) has not yet been conclusively clarified. Different approaches show inconsistent results [21].

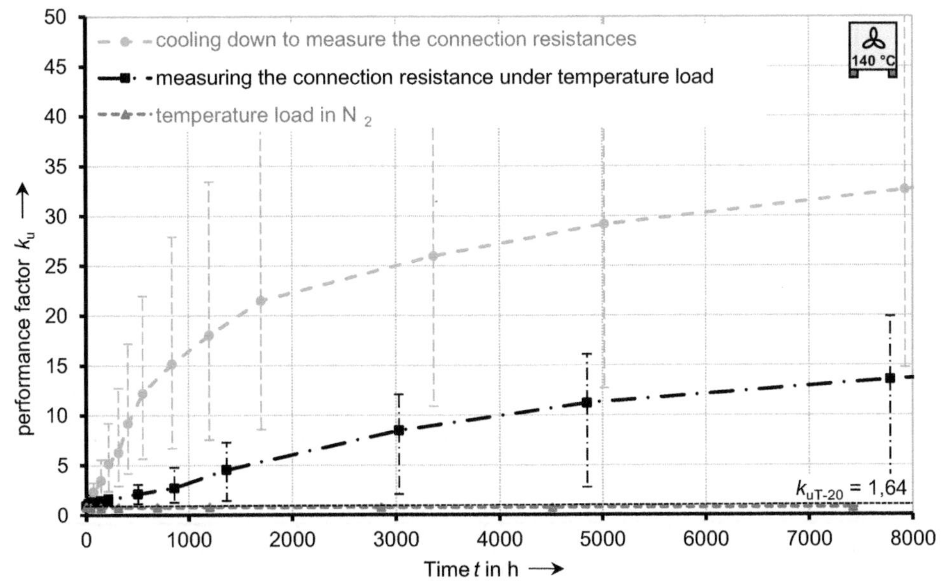

Fig. 5.38 Performance factor k_u dependent on time t of screwed connections with busbars made of aluminum and copper as well as washers in air and inert atmosphere. Connection resistances measured at a temperature of 140 °C or after cooling at room temperature [21]

In compression connections, the access to atmospheric oxygen can be greatly reduced by the press grease, the high deformation of the contact partners, and the correspondingly large-scale a-spots. The influence of oxidation on the long-term behavior of aluminum-copper compression connections is therefore significantly less than with classic screwed connections.

Aluminum–Tin

The material combination of aluminum and tin is not long-term stable in screwed connections. Tin oxide adhering to the tin surface transforms into aluminum oxide in the joined state, so that the contact resistances increase in a short time, sometimes already at room temperature in a few hours. This has been shown in numerous experimental investigations (Fig. 5.39). The effect was also demonstrated in an inert atmosphere, so that the influence of the oxygen from the environment of the connection could be excluded [21, 36]

During the joining process , a large plastic deformation of the joining partners is usually achieved by compression connections. Tin oxide layers and, to some extent, the tin coating are displaced or tear open. If the oxide layers tear open, the chemical reaction between the tin oxide and the aluminum surface is prevented. If the coating tears open,

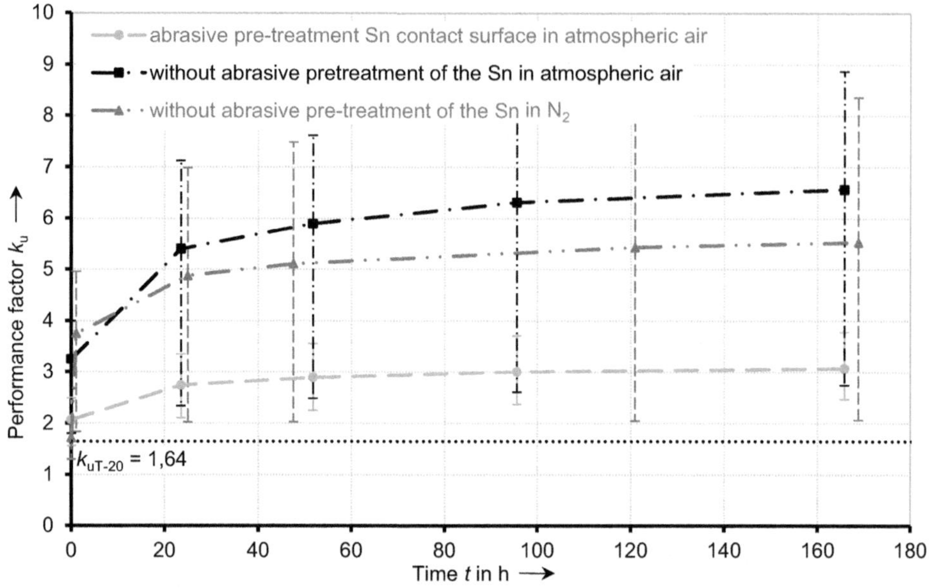

Fig. 5.39 Performance factor dependent on time in air and inert atmosphere of screwed connections with busbars made of aluminum and tinned copper stored at room temperature [21]

contact with the base material is established. In both cases, stable a-spots can be formed, which enable long-term stable operation. In addition, the press or contact grease used provides a sealing function against the atmosphere. This material combination cannot be used for plug-in connections, as the low contact forces and the solid and insulating oxide layers on the aluminum prevent reliable operation.

Aluminum–Silver

Connections with conductors made of aluminum and silver-plated conductors are not long-term stable in atmospheric air. Oxygen diffusing along the grain boundaries of the silver leads to the oxidation of the contact surface of the contact partner made of aluminum. This significantly increases the contact resistance in a short time. The speed of the process strongly depends on the length of the diffusion path. Since this effect only occurs in atmospheric air and could not be observed in an inert atmosphere, it has been shown that the oxygen penetrates into the connection from the outside (Fig. 5.40) [21].

As with compression connections with contact partners made of aluminum and copper or tin, the access to atmospheric oxygen can be greatly reduced by the press grease and the high deformation of the contact partners and the correspondingly large a-spots. A tearing open of the silver coating due to the high deformation already during the joining of longitudinal compression connections is also suspected as a cause for a partially long-term

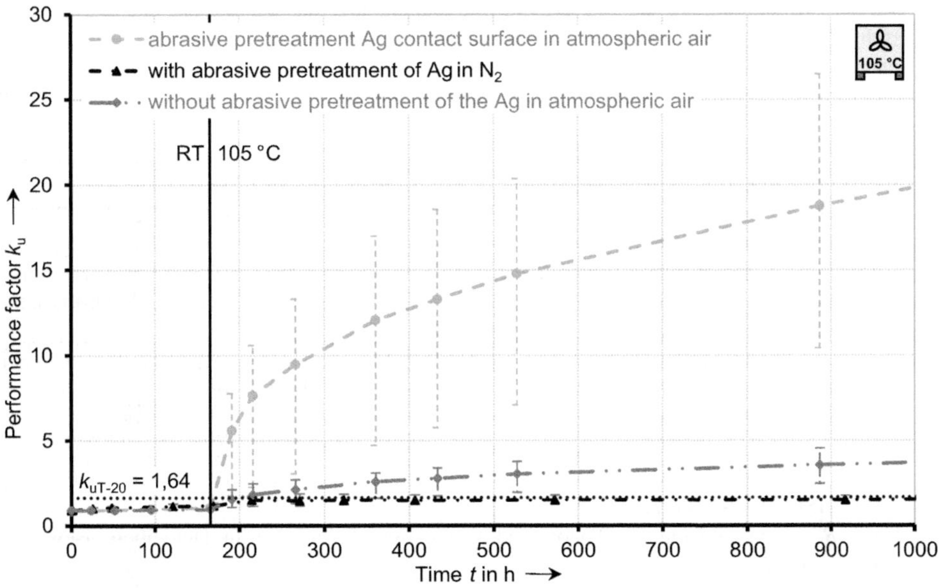

Fig. 5.40 Performance factor dependent on time in air and inert atmosphere of screwed connections with busbars made of aluminum and silver-plated copper stored at room temperature (RT) and subsequently in a heating cabinet at 105 °C [21]

stable behavior [94]. For plug-in connections, this material combination cannot be used for the same reasons as for the combination of aluminum and tin.

References

1. *Bürgel, R.*: Handbuch Hochtemperatur-Werkstofftechnik. Vieweg & Sohn Verlag 2006.—ISBN: 978-3-528-23107-1.
2. *Schlegel, S.*: Langzeitverhalten von Schraubenverbindungen mit Stromschienen aus Reinkupfer in der Elektroenergietechnik unter besonderer Berücksichtigung der Temperatur. Dissertation, TUD 2011.
3. *Bergmann, R.*: Zum Langzeitverhalten des Widerstands elektrischer Stromschienen-verbindungen. Dissertation, TU Dresden, Fortschritt-Berichte VDI Nr. 195. VDI-Verlag 1996.—ISBN 3-18-319521-6.
4. *Großmann, S.; Löbl, H.; Böhme, H.*: Contact lifetime of Connections in electrical power systems. 16th International Conference on Electrical Contacts, Loughborough (UK), September 7–11, 1992, pp. 129–134.
5. *Fuhrmann, T.*: Kontakt- und Langzeitverhalten von stromführenden Schraubenverbindungen mit vernickelten und versilberten Leitern aus Aluminiumwerkstoffen. Dissertation, TU Dresden, 2019.—ISBN 978-3-86780-636-7.
6. *Larson, F.; Miller, J.*: A Time-Temperature Relationship between for Rupture and Creep Stresses. Trans. ASME, pp. 765–775, 1952.

7. *Stennett, N. A.; Campell, D. S.*: Normal force reduction: A variable activation energy process. 16th ICEC, Loughborough (UK), pp. 191–197, 1992.
8. *Pfeifer, S.*: Einfluss intermetallischer Phasen der Systeme Al-Cu und Al-Ag auf den Widerstand stromtragender Verbindungen im Temperaturbereich von 90 °C bis 200 °C. Dissertation, TU Dresden, 2015.
9. *Revie, W. R.; Uhlig H. H.*: Corrosion and Corrosion Control: An Introduction to Corrosion Science and Engineering, 4. Aufl. Hoboken, NJ: Wiley, 2008.—ISBN 978-0-471-73279-2.
10. *Heumann, T.*: Diffusion von Metallen – Grundlagen, Theorie, Vorgänge in Reinmetallen und Legierungen. Springer-Verlag 1992—ISBN 978-3-642-86413-1.
11. *Schlegel, S.*: Stromführende Verbindungen und Leiterwerkstoffe der Elektroenergietechnik – Theorie zum Kontakt- und Langzeitverhalten von Schraubenverbindungen mit Flächenkontakten. Habilitation, TU Dresden, 2019.—ISBN 978-3-86780-614-5.
12. *Dreier, S.*: The Impact of Films on the Long-Term Behavior of Stationary Electrical Connections and Contacts in Electric Power Systems. Dissertation, TU Dresden, 2015.
13. *Vinaricky, E.*: Elektrische Kontakte, Werkstoffe und Anwendungen – Grundlagen, Technologien, Prüfverfahren. 3. Auflage, Springer Verlag, 2016.—ISBN 978-3-642-45427-1.
14. *Rieder, W.*: Elektrische Kontakte: Eine Einführung in ihre Physik und Technik. VDE Verlag; 2000.—ISBN 978-3-80072-542-7.
15. *Holm, R.*: Electrical Contacts: Theory and Application, 4. Aufl. Berlin, Heidelberg, New York: Springer, 1967.
16. *Simmons J. G.*: Generalized Formula for the Electric Tunnel Effect between Similar Electrodes Separated by a Thin Insulating Film. Journal of Applied Physics 34 (1963), pp. 1793–1803.
17. *Slade, P. G.*: Electrical Contacts: Principles and Applications, Second Edition. NewYork, Basel: Crc Press Inc, 2013.—ISBN 978-1-4398-8130-9.
18. *Braunovic, M.*: Electrical Contacts-Fundamentals, Applications and Technology. Taylor & Francis Group 2006.—ISBN 1-57444-727-0.
19. *Schneider, R.*: Langzeitverhalten geschraubter Stromschienenverbindungen in der Elektroenergietechnik. Dissertation, TU Dresden, 2012.
20. Bericht zum IGF-Vorhaben (AIF 17860BG): Auswirkung von Oxiden innerhalb galvanischer Beschichtungen auf die Zuverlässigkeit und das Langzeitverhalten elektrischer Verbindungen der Energietechnik. 2017.
21. *Oberst, M.*: Einfluss von Sauerstoff auf die Alterung stromführender Bimetall-Verbindungen. Dissertation, TU Dresden, 2021.—ISBN 978-3-86780-655-8.
22. *Funamizu, Y.; Watanabe, K.*: Interdiffusion in the Al-Cu system, Trans. Japan Institute of Metals 12 (1971), pp. 147–152.
23. *Wulff, F. W.; Breach, C.D.; Stephan, D.; Dittmer, K.J.*: Characterisation of intermetallic growth in copper and gold ball bonds on aluminum metallisation, Electronics Packaging Technology Conference, 2004, pp. 348–353.
24. *Krautz, H.*: Untersuchungen zur Diffusion und Phasenbildung im Dünnschichtsystem Kupfer-Aluminium. Dissertation, TU Dresden, 1989.
25. *Pierre Villars (Chief Editor), PAULING FILE in*: Inorganic Solid Phases, Springer Materials (online database), Springer, Heidelberg (ed.) Springer Materials.
26. *Dongmin, S.; Bin, W.; Roderick, M.; Shan, Y.; Tingju L.*: First-principles studies of Al-Ni intermetallic compounds. Journal of Solid State Chemistry 182, 2009, pp. 2664–2669.
27. *Moris, K.*: Review: reaction synthesis processing of Ni—Al intermetallic materials. Materials Science and Engineering A299, 2001, pp. 1–15.
28. *Wang, Y.; Liu, Z.-K.; Chen, L.-Q.*: Thermodynamic properties of Al, Ni, NiAl, and Ni3Al from first-principles calculations. Acta Materialia 52, 2004, pp. 2665–2671.

29. *Mitra, R.*: Intermetallic Matrix Composites: Properties and Applications. 1st Edition, Woodhead Publishing, 2017—ISBN 978-0-85709-346-2.
30. *Creydt, M.*: Diffusion in galvanisch aufgebrachten Schichten und Weichloten bei Temperaturen zwischen 23 °C und 212 °C. EMPA-Publikation 19/1971.
31. *Blair, H. D.; Pan, T-Y., Nicholson, J. M.*: Intermetallic Compound Growth on Ni, Au/Ni and Pd/Ni Substrates with Sn/Pb, Sn/Ag and Sn Solders. Electronic Components and Technology Conference, Seattle (USA), 25–28.05.1998.
32. Schlussbericht (IGF-Vorhaben 17EWBG): Zuverlässigkeit und Sicherheit von Betriebsmitteln in Niederspannungsverteilstationen für die Anwendung im Bereich der Windkraft- und Photovoltaikanlagen (Sicherheit von WK- und PV-Anlagen). 2019.
33. *Yao, Y.; Fry, J.; Keer, L.M.*: The Wiedemann-Franz-Lorenz relation for lead-free solder and intermetallic materials. Acta Materialia 61, 2013, pp. 1525–1536.
34. *Oberst, M.; Schlegel, S.; Großmann, S.; Willing, H.; Freudenberger, R.*: Einfluss von intermetallischen Phasen der Systeme Ag-Sn und Ni-Sn auf den elektrischen Widerstand. 25. Albert Keil Kontaktseminar, Karlsruhe, 9.–11.11.2019.—ISBN 978-3-8007-5083-2.
35. *Oberst, M.; Schlegel, S.; Großmann, S.*: Influence of Oxygen on the Aging of Electrical Joints with One Bare and One Coated Aluminum Contact Member. IEEE Holm Conference on Electrical Contacts, Albuquerque, NM, 2018, pp. 262–269.
36. *Oberst, M.; Hildmann, c.; Schlegel, S.*: Deterioration and Breakdown Mechanisms in force-fitted current-carrying Connections between Aluminum and Tin. 66th IEEE Holm Conference on Electrical Contacts, San Antonio, 2021, pp. 1–7.—ISBN 978-1-6654-2107-2.
37. *N. N. Dzektser und V. V. Izmailov*: On the Life-Time of Stationary Contacts. 16th International Conference on Electrical Contacts, Loughborough, UK, 1992, pp. 175–180.
38. *Meyerhofer, D.; Ochs, S. A.*: Current Flow in Very Thin Films of Al_2O_3 and BeO. Journal of Applied Physics 34 (1963), pp. 2535–2543.
39. *Pfeifer, S.; Oberst, M.; Willing, H.*: Auswirkung von Oxiden innerhalb galvanischer Beschichtungen auf die Zuverlässigkeit und das Langzeitverhalten elektrischer Verbindungen in der Energietechnik (ODiEn). Abschlussbericht AiF-Projekt 17860 BG, 2017.
40. Der Silberspiegel: The Ellingham diagram is giving the free energy of formation of metal oxides and the corresponding oxygen partial pressure at equilibrium. Labels in English (30.03.2016), https://commons.wikimedia.org/wiki/File:Ellingham_Richardson-diagram_english.svg (Stand: 02.02.2018).
41. *Buresch, I.*: Oberflächenbeschichtungen für Steckverbinderkontakte bei höheren Einsatztemperaturen. 24. Albert Keil Kontaktseminar, Karlsruhe, 2017.—ISBN 978-3-8007-4449-7.
42. *J. H. Tripp; S. M. Garte*: The Gas-Tightness of Separable Base Metal Electric Contacts. IEEE Transactions on Components, Hybrids, and Manufacuturing Technology, Bd. CHMT-4, Nr. 1, pp. 85.92, 1981.
43. *Bergwerff, J.*: The Interaction of O2 with Silver—A survey on the molecular processes in the passage of O2 through a silver membrane. Diplomarbeit, Universität Utrecht [Online].14.07.2016.
44. *Scher, H.; Zellen, R.*: Critical density in percolation processes. The Journal of chemical Physics, vol. 53, pp. 3759–3761, Juni 1970.
45. *N Ravi Chandra Raju, et al.*: Physical properties of silver oxide thin films by pulsed laser deposition: effect of oxygen pressure during growth. Journal of Physics D: Applied Physics 42, 2009.
46. *Zähr, J.*: Einfluss des Oberflächenzustandes auf das thermische Fügen von Aluminium. Dissertation, TU Dresden, 2010.
47. *Fischmeister, H.; Rönnquist, A.*: The Oxidation of Copper: A Review of Published Data. Journal of the Institute of Metals 89, 1960–61.

48. *Campbell, W. E.; Thomas, U. B.*: The Oxidation of Metals. Transactions of the Electrochemical Society 91, 1947, pp. 623–640.
49. Gmelin-Handbuch der anorganischen Chemie, Bd. 60: Kupfer, Teil D: Elektrische Eigenschaften der Kupferoxide, 8. Aufl. Weinheim: Verl. Chemie, 1963.
50. Gmelin-Handbuch der anorganischen Chemie, Bd. 46: Zinn, Teil C 1: Verbindungen mit Wasserstoff, Sauerstoff, Stickstoff und Halogenen, 8. Aufl. Weinheim: Verl. Chemie, 1972.
51. *Samsonov, G. V. (Hrsg.)*: The Oxide Handbook, 2. Aufl. New York: IFI/Plenum, 1982.—ISBN 0-306-65177-7.
52. Gmelin-Handbuch der anorganischen Chemie, Bd. 35: Aluminium, Teil B: Die Verbindungen des Aluminiums, 8. Aufl. Berlin: Verl. Chemie, 1934.
53. Landolt-Börnstein: Zahlenwerte und Funktionen aus der Physik, Chemie, Astronomie, Geophysik und Technik, 6. Auflage, Band II-Teil 5 (Transportphänomene, Kinetik, Homogene Gasgleichgewichte), Springer Verlag, Berlin-Göttingen-Heidelberg, 1968.
54. Landolt-Börnstein—Group III Condensed Matter, Volume 41C, Non-Tetrahedrally Bonded Elements and Binary Compounds I, Silver sulfide (Ag2S) crystal structure. Berlin, Heidelberg: Springer, 1998.
55. *Boggs, W. E.*: The Oxidation of Tin: III. The Mechanisms of Oxidation of Pure Tin and Their Dependence on Time and Oxygen Pressure. Journal of the Electrochemical Society 108, 1961, pp. 124–129.
56. *Leja, E.; Korecki, J.; Krop, K.; Toll, K.*: Phase Composition of SnOx Thin Films Obtained by Reactive D.C. Sputtering. Thin Solid Films 59, 1979, pp. 147–155.
57. Informationszentrum Technische Keramik. Brevier Technische Keramik. Lauf: Fahner Verlag, 3. überarb. Auflage 1999.
58. *Roos, E; Maile, K.*: Werkstoffkunde für Ingenieure: Grundlagen, Anwendung, Prüfung, 3. Aufl. Berlin, Heidelberg: Springer, 2008.—ISBN 978-3-540-68403-9.
59. DIN EN 3197 (Ausg. 2011-03): Luft- und Raumfahrt – Konstruktion und Installation elektrischer und optischer Verkabelung in Luftfahrzeugen.
60. DIN EN 50343 (Ausg. 2014-09): Bahnanwendungen – Fahrzeuge – Regeln für die Installation von elektrischen Leitungen.
61. IEC/TR 60943 (Ausg. 2009-03): Guidance concerning the permissible temperature rise for parts of electrical equipment, in particular for terminals.
62. *Blumenroth, F.*: Zum Langzeitverhalten von Steckverbindungen mit Schraubenfedern in Anlagen der Elektroenergietechnik. Dissertation, TU Dresden, Shaker Verlag, 2010.—ISBN 978-3-8322-9506-6.
63. *Lücke, N.*: Zum Langzeitverhalten elektrischer Steckverbindungen mit Kupfer-Beryllium-Kontaktlamellen. Dissertation, TU Dresden, TUDpress 2004.—ISBN 978-3-9443-3175-1.
64. *Czichos, H.; Habig, K.-H.*: Tribologie-Handbuch. 3. Aufl.: Vieweg+Teubner Verlag 2010.—ISBN 978-3-8348-0017-6.
65. *Song, J.; Yuan, H.; Schinow, V.*: Berechnungsmodelle für die Vorhersage von Verschleiß und Durchrieb der Oberflächenschutzschicht von elektrischen Kontakten mit einer sphärischen/ebenen Reibpaarung. 24. Albert-Keil-Kontaktseminar, Karlsruhe, 2017.
66. *Bond, N. T.; McGeary, F. L.*: Nickel Plating for Improved Electrical Contact to Aluminum. IEEE TRANSACTIONS ON INDUSTRY APPLICATIONS, Vol. IA-9, No. 3, 1973.
67. *Aronstein, J.*: Electromigration Failure of Aluminum Contact Junctions. 41st IEEE Holm Conference on Electrical Contacts. Montreal, 1995, pp. 10–16.
68. *Aronstein J.*: AC and DC Electromigration in Aluminum Contact Junctions. 18th International Conference on Electrical Contacts, Chicago, 1996, pp. 311–320.

69. *Roth, W.; Eden, K.*: Elektromigration – Ein Schadensmechanismus in mikroelektronischen Schaltungen. Werkstoffe und Korrosion 44, VCH Verlagsgesellschaft mbH, 1993, pp. 130–136.
70. *Arzt, E.; Kraft, O.; Spolenak, R.; Joo, Y.-C.*: Physical Metallurgy of Electromigration: Failure Mechanisms in Miniaturized Conductor Lines. Zeitschrift für Metallkunde. Carl Hauser Verlag, München, 1996, pp. 934–942.
71. *Runde, M.; Hodne, E.; Totdal, B.*: Current-induced Aging of Contact Spots. 35th IEEE Holm Conference on Electrical Contacts, Chicago, 1989, pp. 213–220.
72. *Thiele, M.; Lienig, J.*: Elektromigration Der Feind auf dem Chip. Electroniknet.de, 24.01.2012—http://www.elektroniknet.de/elektronik/halbleiter/der-feind-auf-dem-chip-85242.html.
73. *Mildenberger, O.*: Einführung in die Mikrosystemtechnik: Grundlagen und Praxisbeispiele. Vieweg-Verlag, 2000.—ISBN 3-528-03891-8.
74. *Ho, P. S.; Kwok, T.*: Electromigration in metals. Reports on Progress in Physics. Volume 52, 1989, pp. 301–348.
75. *Thiele, M.; Lienig, J.*: Fundamentals of Electromigration-Aware Integrated Circuit Design. Springer-Verlag, 2018.—ISBN 978-3-1833-8121-0.
76. *Höft, H.*: Informationselektronik: Elektrische Kontakte. Werkstoffe, Einsatzbedingungen, Zuverlässigkeit. 1. Auflage., VEB-Verlag Technik, 1980.
77. *Strehle, S.*: Cu(Ag)-Legierungsschichten als Werkstoff für Leiterbahnen höchstintegrierter Schaltkreise – Herstellung, Gefüge, thermomechanische Eigenschaften, Elektromigrationsresistenz. Dissertation, 2006.
78. *Wirtz, C.*: Elektromigrationsverhalten von nanostrukturierten Silber-Leiterbahnen. Diplomarbeit, Universität Duisburg Essen, 2008.
79. *Meyer, W.-K.*: Electromigration of Damascene copper for IC interconnect. Oregon Health & Science University, 2004.
80. *Böhm, J.*: Legierungseinflüsse auf die Elektromigration in Submikrometer Aluminium-Leiterbahnen. Diplomarbeit, Universität Stuttgart, 2001.
81. *Walther, T.*: Elektromigrationsuntersuchungen an der Grenzfläche zwischen Kupferleiterbahn und Kupferdiffusionsbarriere. Studienarbeit, TU Dresden, 2012.
82. *Schultz, l., Freudenberger, J.*: Physikalische Werkstoffeigenschaften. Script. Leibniz Institute for Solid State and Materials Research Dresden, Germany, 2019.
83. *Ming Tan, C., Roy, A.*: Investigation of the effect of temperature and stress gradients on accelerated EM test for Cu narrow interconnects. Thin Solid Films 504, pp. 288–293.
84. *Aronstein, J; Hare, T.K.*: AC and DC electromigration failure of aluminum contact junctions. IEEE Transactions on Components and Packaging Technologies 28, 2005, pp. 701–709.
85. *Hildmann, C.*: Zum elektrischen Kontakt- und Langzeitverhalten von Pressverbindungen mit konventionellen und Hochtemperatur-Leiterseilen mit geringem Durchhang. Dissertation, TU Dresden, 2016.—ISBN 978-3-00-056211-2.
86. *Oberst, M.; Schlegel, S.; Großmann, S.; Willing, H.; Freudenberger, R.*: Impact of the Formation of Intermetallic Compounds in Current-Carrying Connections. IEEE Transactions on Device and Materials Reliability, vol. 20, no. 1, 2020, pp. 157–166.
87. *Oberst, M.; Hildmann, C.; Schlegel, S.*: Einfluss von Grenzflächenreaktionen auf das Kontakt- und Langzeitverhalten von kraftschlüssigen Verbindungen mit unterschiedlichen Werkstoffkombinationen. 26. Albert-Keil-Kontaktseminar, Karlsruhe, 2022.
88. *Gatzsche, M.*: Elektrisch-thermisches Betriebs- und Langzeitverhalten hochstromtragfähiger Kontaktelemente. Dissertation, TU Dresden, Shaker Verlag, 2016.—ISBN 978-3-84404-894-0.
89. *Großmann, S.*: Zur Gestaltung und zum Betriebsverhalten von Steckverbindern der Elektroenergietechnik. Dissertation, TU Dresden, 1988.

90. *Israel, T.; Oberst, M.; Hildmann, C.; Schlegel, S.*: Selbstschmierende Beschichtungen für Hochstrom-Steckverbindungen. 26. Albert-Keil-Kontaktseminar, Karlsruhe, 2022.
91. *Archard, J. F.*: Contact and rubbing of flat surfaces. Journal of Applied Physics, vol. 24, no. 8, 1953, pp. 981–988.
92. *Hagen, B.; Kufner, T.; Freudiger, G.*: Grundlegende Untersuchungen an tribologischen Systemen zur optimierten Auslegung von Steckverbindern in der Elektroenergietechnik., 24. Albert-Keil-Kontaktseminar, Karlsruhe, 2017.
93. *Yuan, H.; Song, J.*: Fretting Corrosion Behavior of Various Contact Surfaces under Diverse Operat-ing Conditions. IEEE Holm Conference on Electrical Contacts, 2019. pp. 19–26.
94. *Ramonat, A.*: Untersuchungen zum elektrischen Kontakt- und Langzeitverhalten von Fügetechnologien mit zylindrischen Leitern aus Aluminium. Dissertation, TU Dresden, 2018.

Designing Current-Carrying Connections 6

The procedure described in this chapter aims to translate the requirements for designing current-carrying connections for operation in electrical power systems into a constructive design and assembly instructions in six clearly defined steps (Fig. 6.1).

Based on the requirements, a significant aspect of the constructive design of the connection is determined in each of the successive steps. In principle, later steps may require an adjustment in a previous step. This is primarily the case in the design, continuous operation, and short-term operation steps. For example, a high short-circuit current may require additional contact points or increased contact force. To illustrate the application of the guideline, three example connections are designed step by step in Sect. 6.7.

For each step of the design, it is crucial that the requirements for the connection are clearly defined. In practice, many core features are determined by the device containing the connection and especially the conductors to be connected. The connection is a "means to an end", but has a significant influence on the reliability of the overall system. An explicit specification for the connection therefore often does not exist and many requirements, such as the continuous current carrying capacity, are dictated by the device.

6.1 Type of Connection

Before the actual design, the type of connection must be selected. Which type of connection can be used depends primarily on the requirements for assembly and operation. The following two steps need to be clarified:

S. Schlegel et al., *Contact and Long-Term Behavior of Current-Carrying Connections in Electrical Power Engineering*, https://doi.org/10.1007/978-3-662-69644-6_6

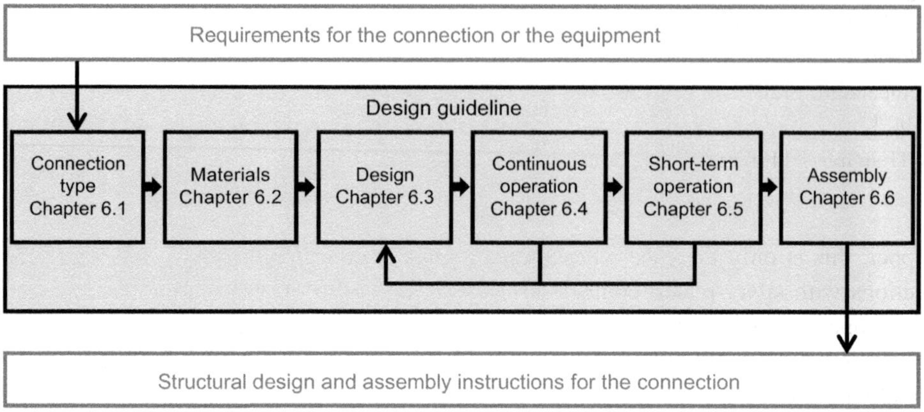

Fig. 6.1 Guideline for designing current-carrying connections

- Definition of the **main connection** based on the requirements for operation and assembly: selection from the types of connections plug-in connection, sliding connection, compression connection, screwed connection or alternative joining methods (Fig. 6.2).
- Definition of the **auxiliary connection**: Depending on the conductors to be connected, it may be necessary to add further connections in order to technically realize the main connection (Fig. 6.3).

The main connection must fulfill all required functions of the connection, such as fixed, movable or separable. Auxiliary connections are always necessary when two conductors cannot be directly connected to each other through the main connection (Table 6.1). If two cable ends are to be connected separably via a plug-in connection, auxiliary connections, for example as a compression connection between plug and cable 1 and between socket and cable 2, are required.

When selecting the type of connection, in addition to the requirements for function and assembly, the necessary space requirement, the costs of the chosen type of connection taking into account the quantity and the available suppliers, as well as the costs for the required tools and their maintenance must be considered.

6.2 Materials

6.2.1 Conductor and Coating Material

For both contact partners, the conductor and optionally the coating material must be selected. The conductor materials of the contact partners are usually predetermined by

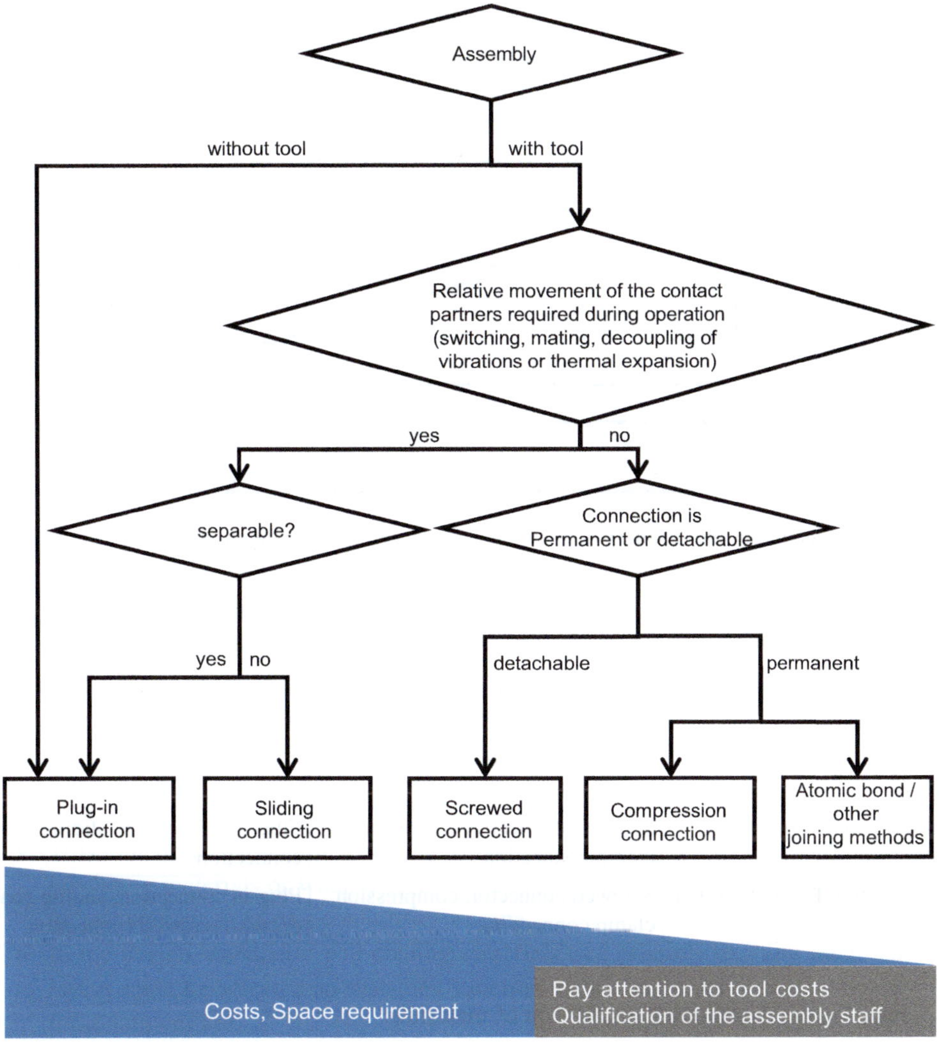

Fig. 6.2 Selection of the main connection

the electrical conductors to be connected or the device to be connected. A special feature are connectors, where a different material can be chosen for the connector body itself than for the conductors to be connected (e.g. plug and socket on a cable, press fitting for connecting two overhead line conductors).

6.2.2 Special Features in Corrosive Environments

If connections are used in corrosive environments such as direct rainfall or salt fog, galvanic corrosion can occur (Sect. 5.1.3). This must be taken into account when two metals with a large potential difference in the electrochemical series (guideline value >350 mV [61]) are conductively wetted at the transition point by an electrolyte such as ionized water. The less noble metal with the lower standard potential is continuously dissolved. In current-carrying connections, two areas are mainly affected (Fig. 6.4-a):

- The contact level in a bimetallic connection such as nickel-aluminum connections
- In the case of coated contact partners, the transition from the uncoated, less noble conductor material to an area coated with precious metal

Ideally, areas affected by galvanic corrosion should be avoided by using a suitable material combination, e.g., by connections with the same conductor materials or the same coating materials on the contact partners. If this is not practically possible, the areas should be sealed against direct conductive wetting (Fig. 6.4-b)), e.g., with a grease coating (short term) or a paint coating (long term).

The ratio of the surfaces of the less noble to that of the noble contact partner influences the corrosion resistance. The current density of the ion current from the less noble to the noble contact partner is smaller, the larger the surface of the less noble wetted with electrolyte is in relation to that of the noble contact partner. A low current density means that the material removal is much slower.

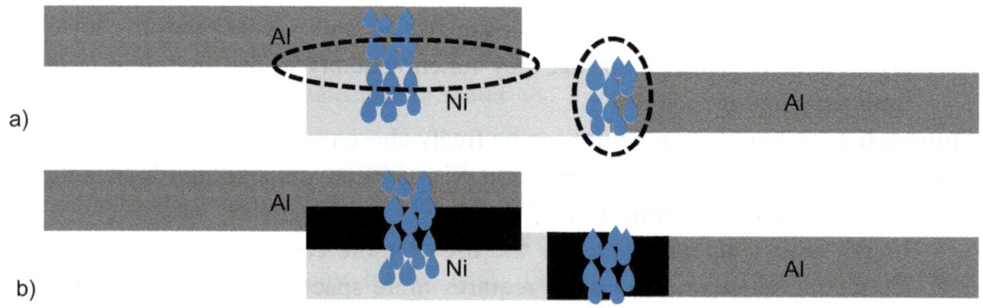

Fig. 6.4 **a**) Areas of a connection affected by galvanic corrosion when wetted with an electrolyte, each with a bare and a nickel-plated contact partner made of aluminum **b**) Reducing galvanic corrosion through a paint coating

Table 6.2 Permissible and non-permissible combinations of materials for the corrosion resistance of grounding systems, surface ratio large to small ≥100:1, selection from [18, 19]

Material with large surface → Material with small surface ↓	Steel	Galvanized steel	Stainless steel	Copper	Tinned copper
Steel	ok	ok	not ok	not ok	not ok
Galvanized steel	ok (Zinc removal)	ok	not ok	not ok	not ok
Stainless steel	ok	ok	ok	ok	ok
Copper	ok	ok	ok	ok	ok
Tinned copper	ok	ok	ok	ok	ok

From this, two extreme cases can be derived:

- Less noble contact partner with a very small surface → rapid corrosion, critical (e.g., unalloyed steel screw in a screwed connection with copper busbars)
- Less noble contact partner with a very large surface → slow corrosion, acceptable for many practical applications (e.g., copper busbars on steel frame)

The case of different material pairings with different sized surfaces is particularly relevant for grounding systems. There, many material pairings are permissible due to the large surface ratio of at least 100:1 (surface of the less noble metal to that of the noble metal) in terms of corrosion resistance (Table 6.2).

6.3 Design

The current-carrying connection must be designed in such a way that a sufficiently high and long-term stable contact force is generated, so that the contact resistance is small enough to meet the electrical and mechanical requirements for the continuous (Sect. 6.4) and the short-term load (Sect. 6.5) over the lifetime of the connection. Each type of connection has specifics that are discussed in the subchapters. The design recommendations in this chapter can be understood as a "basic version" that covers most use cases. In Sect. 6.4 and 6.5, it is then checked whether the requirements for continuous operation and short-term operation are met. In certain cases, a design iteration may be necessary for special requirements (Fig. 6.1).

6.3.1 Screwed Connection

Screwed connections are one of the most common types of connections in the electrical power supply network and have inherent advantages for the electrically conductive connection of conductors such as:

- Detachability and reassembly
- Joining different uncoated and coated conductor materials
- No heat input during the joining process
- Easy assembly
- Cost-effective

When mechanically designing an electrically conductive screwed connection, the scheme presented by *Kloos* and *Thomala* or the VDI 2230 can generally be followed [20, 21]. In ten calculation steps, the necessary preload force is systematically determined, taking into account an arbitrarily directed operating force (Fig. 6.5). The following influences are considered:

- Loss of preload due to settling processes, even after thermal expansion or relaxation processes at higher temperatures (non-ferrous metals such as Cu and Al not considered)
- Loss of clamping force due to operational forces
- Required clamping force to prevent the separation gap from opening, the automatic loosening, and to achieve gas tightness in the separation gap
- Variation of the assembly preload force

The electrical properties of the screwed connection have not been considered in this scheme so far. They must be considered separately. For this purpose, an additional step is added, which must be taken into account when determining the required minimum clamping force (Fig. 6.5—R3) and includes the following (Fig. 6.6):

- Determining the overlap length of the conductors
- Defining the hole pattern
- Designing and pretreating the contact surface of the conductors
- Selection of suitable joining elements
- Determining the necessary minimum clamping/contact force or average mechanical stress in the contact plane between the conductors

With the steps for mechanical (Fig. 6.5) and electrical design (Fig. 6.6) of screwed connections, the essential practical use cases can be covered. This is explained in Sect. 6.7.1 using an example.

6.3.2 Compression Connections

When designing compression connections, the system, consisting of, for example, press sleeve and tool, must be selected for the conductor to be connected. Initially, a technology and a process can be selected based on the conductor(s) to be connected according

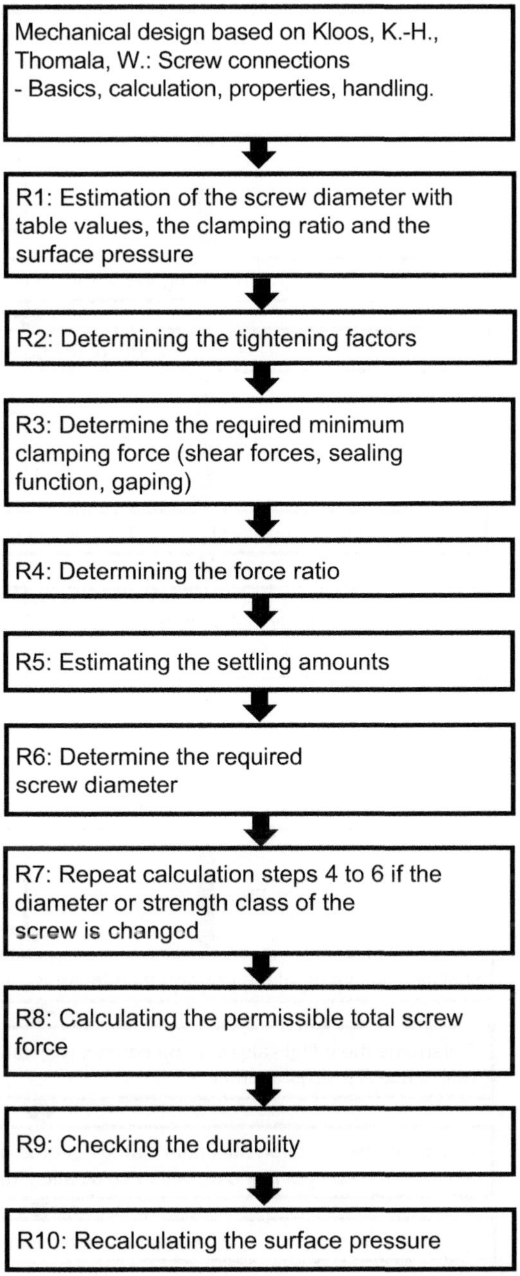

Fig. 6.5 Calculation steps for the mechanical design of a screw connection according to [20]

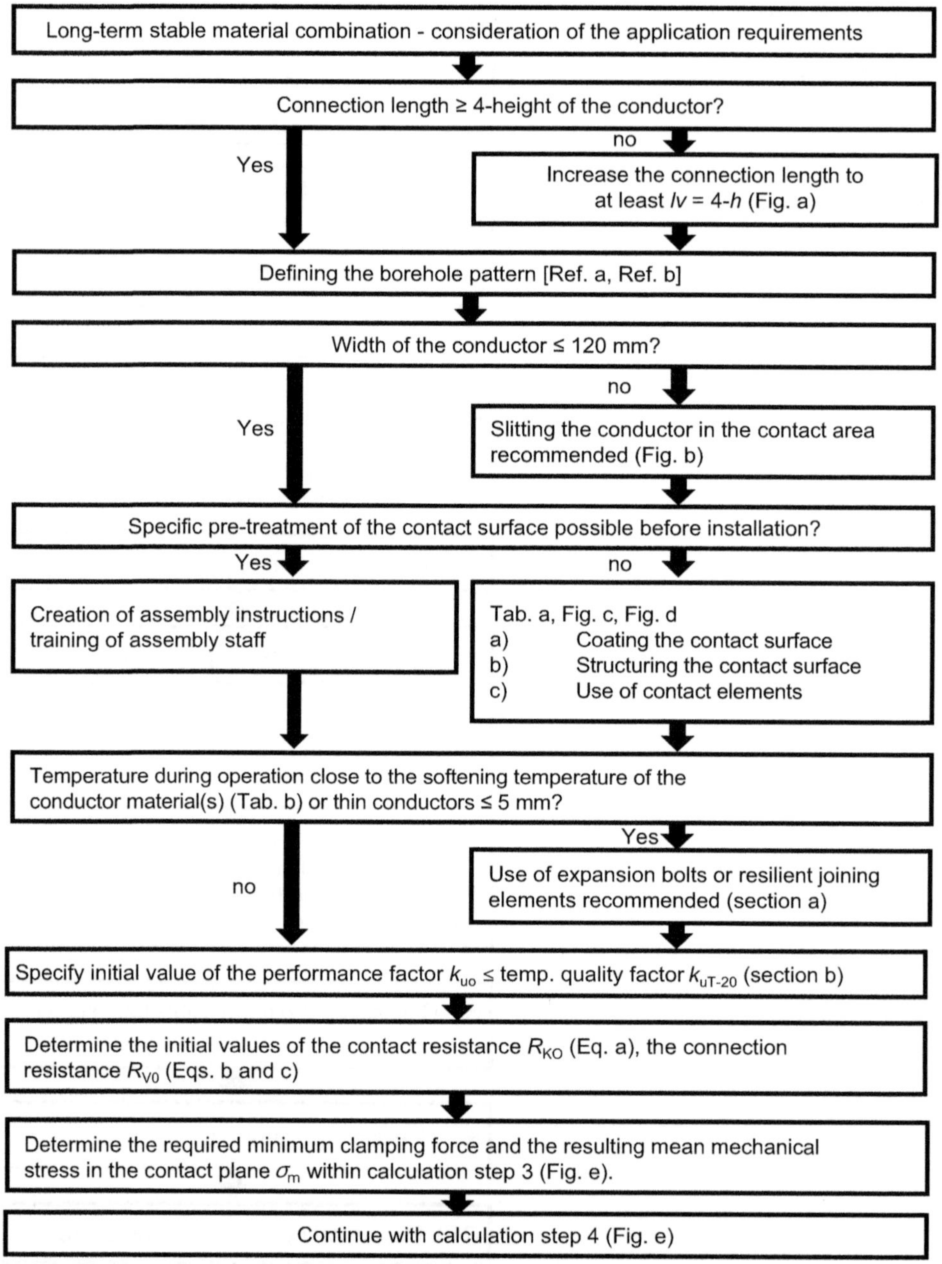

Fig. 6.6 Procedure for the electrical design of a screwed connection(detailed listing of step R3 in Fig. 6.5) Ref. a, Ref b = [20, 21], Fig. a = Fig. 4.58, Fig. b = Fig. 4.63, Fig. c = Fig. 4.61, Fig. d = 4.62, Fig. e = Fig. 6.5, Table a = Table 4.20, Table b = Table 3.4, Eq. a = Eq. 4.30, Eq. b = Eq. 4.3, Eq. c = Eq. 4.66

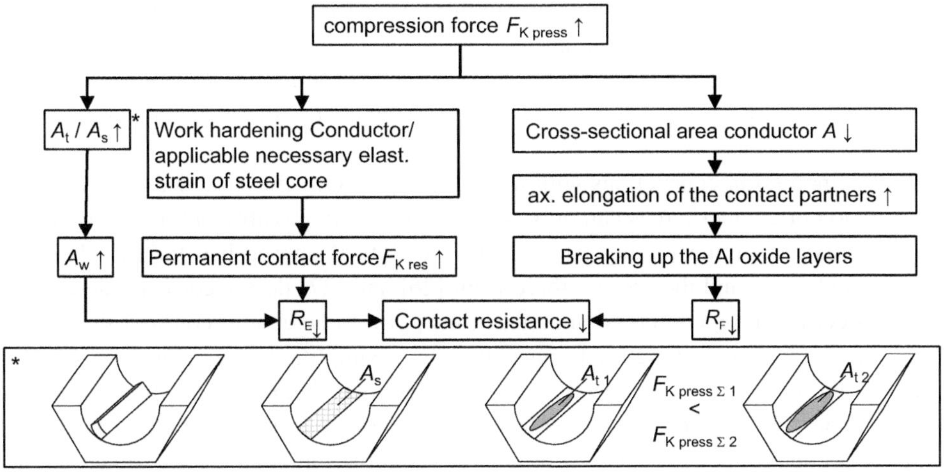

Fig. 6.7 Influence of the force effect in press connections on the electrical contact behavior

to Table 4.6. Furthermore, the press sleeve and the tool can be matched to each other using simple geometric parameters, e.g., based on the guideline values in Table 4.7. In principle, when designing compression connections, the dependence of the contact resistance as a function of the contact force should be taken into account in the same way as for screwed connections. The relationships between mechanical and electrical contact behavior are specific for compression connections, especially due to the deformation of the contact partners compared to other types of connections (Fig. 6.7).

The electrical contact behavior of the entire connection is also dependent on the distribution of the force on the individual contact surfaces and the resistance-force characteristic specific to respective point, line, or surface contacts (see Sect. 4.3.1). With a higher contact force during assembly, the load bearing contact area increases and usually the remaining contact force. Furthermore, any existing oxide layers on the contact partners can break up during joining due to the shear stresses that occur [24]. In compression connections, a constructively determined deformation is always achieved. The force effect in the connection is thus a dependent variable. The resulting contact forces are determined by the design and the material selection of the connection as well as by the conductors to be connected and the chosen press process (Sect. 4.2).

Accordingly, the structure and the material(s) of the conductor are the essential input variables for the dimensioning of a compression connection. For each compression connection to be designed, the optimization of the contact force in the connection and thus the electrical contact behavior is in the foreground, which can be achieved with numerous “adjusting screws”. These are primarily the material and the chosen compression procedure .

A simplified, mechanical interpretation of compression connections with two concentric conductors is analytically possible using the method according to DIN 7190-1:2017

[25]. The result of the calculations is a uniformly distributed average mechanical stress in the contact surface, which is always somewhat lower than that from FEM calculations. In the contact surface, the mechanical stress distribution is inhomogeneous, and in particular, stress peaks in the edge areas are not depicted with the analytical calculation [11].

For compression connections with multi-, fine-, or ultra-fine stranded conductors, the distribution of contact forces can only be determined and possibly optimized by numerical calculations of the joining process. The analytical approaches for calculating the mechanical stress and the contact force from [26] and [27] do not correctly describe the force effect for these compression connections [24]. To be able to consider the effect of the contact force in compression connections in a simplified way, geometric substitute sizes of the compression connections have established themselves in the design (Sect. 4.2.2, Table 4.7). Crimp connections can be designed in a simplified way based on the degree of compression V. For circumferential compression connections, which are used in particular for connecting multi-stranded conductors, the compression and fit factor are used for the design (Table 4.7). The force input into the press sleeve and from the press sleeve into the conductor is comparatively uniform. Accordingly, high fit factors can be set in circumferential compression connections. Furthermore, compression factors of at least 1.1 should be aimed for. The forming degrees in circumferential compression connections are approximately in the range of 0.2 … 0.3. Thus, comparatively hard materials can also be used for the press sleeve in circumferential press connections if the required forming degree can be achieved with these. Higher forces can be achieved in the connection during joining with harder materials of the press sleeve. An evaluation of the compression and fit factors, which result for compression connections with compression cable lugs according to DIN 46235:1983 for copper conductors of class 2 and with the compression dies according to DIN 48083:1985, shows that the mentioned criteria are well met (Fig. 6.8). Especially when conductors with large cross-sections need to be connected, only lower compression factors can be achieved due to design constraints.

For impression compression connections, which have primarily established themselves for connecting fine and ultra-fine stranded conductors, lower fit factors are chosen (Fig. 6.8). This is partly due to the lower fill factors of the conductors. The compression sleeves accordingly have larger outer diameters than those of circumferential compression connections. The wall thickness is about the same compared to the press sleeves for circumferential compression connections, so that sufficient material is provided for the forming process. Locally much higher forming degrees are produced, so that in particular the materials of the press sleeves must have a very high ductility. The compression factor of impression compression connections can no longer be determined analytically.

For the correct choice of the connection length of compression connections, the basic relationships according to Fig. 4.35 (p. 73) must be considered. Furthermore, attention must be paid to long-term stable material combinations (Table 5.14). For the adapted design of compression connections, the necessary partial steps are described in Fig. 6.9.

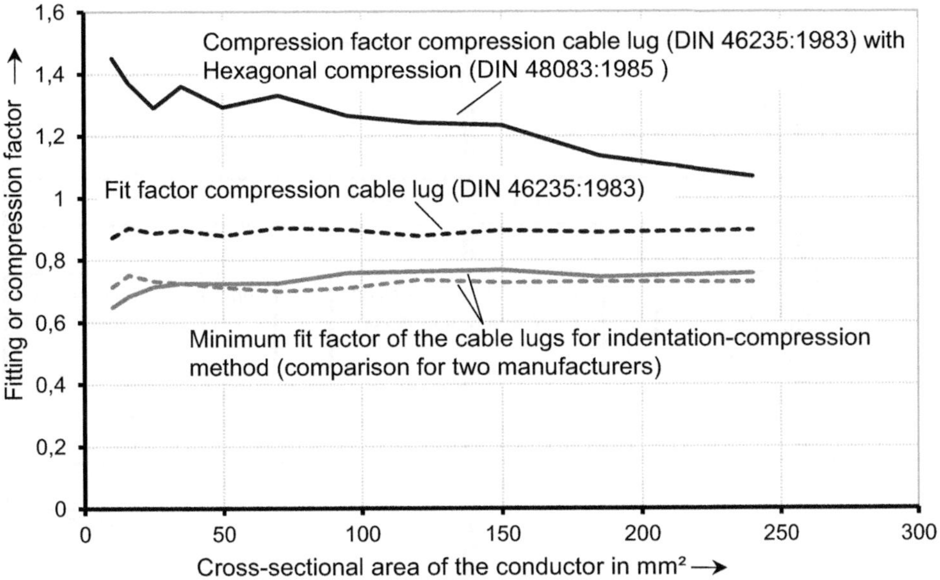

Fig. 6.8 Compression and fit factors of compression connections for Cu conductors of class 2 and class 5

6.3.3 Plug-in and Sliding Connections

Only in plug-in and sliding connections is a relative movement of the contact partners permissible during operation. Other advantages typically include low insertion forces and tool-free assembly. Therefore, they are used in almost all areas of electrical power transmission, such as plug in and sliding connections in switches, switchgear, and gas-insulated lines, as well as in connections that often have to be plugged in by hand, in household and industrial sockets, charging ports for electrically driven vehicles and aircraft as well as ships. The design of plug-in and sliding connections should follow the scheme shown in Fig. 6.11.

In principle, all the technologies listed in Fig. 6.10 are suitable for transmitting high currents in plug-in connections. The technologies mainly differ in the number of contact points and their contact force. For example, reducing the number of contact points while increasing their contact force reduces contact resistance but increases frictional wear. Conversely, a high number of contact points with lower contact force increases the contact resistance of individual contact points and reduces the current carrying capacity per contact point, but also reduces frictional wear. Depending on the application and requirements, the respective advantages and disadvantages must be weighed up when selecting the technology [7] (Table 6.3).

Connectors with contact elements (a) to (c) have the advantage that the plug and the socket are technologically simple to manufacture. The necessary spring deflection for

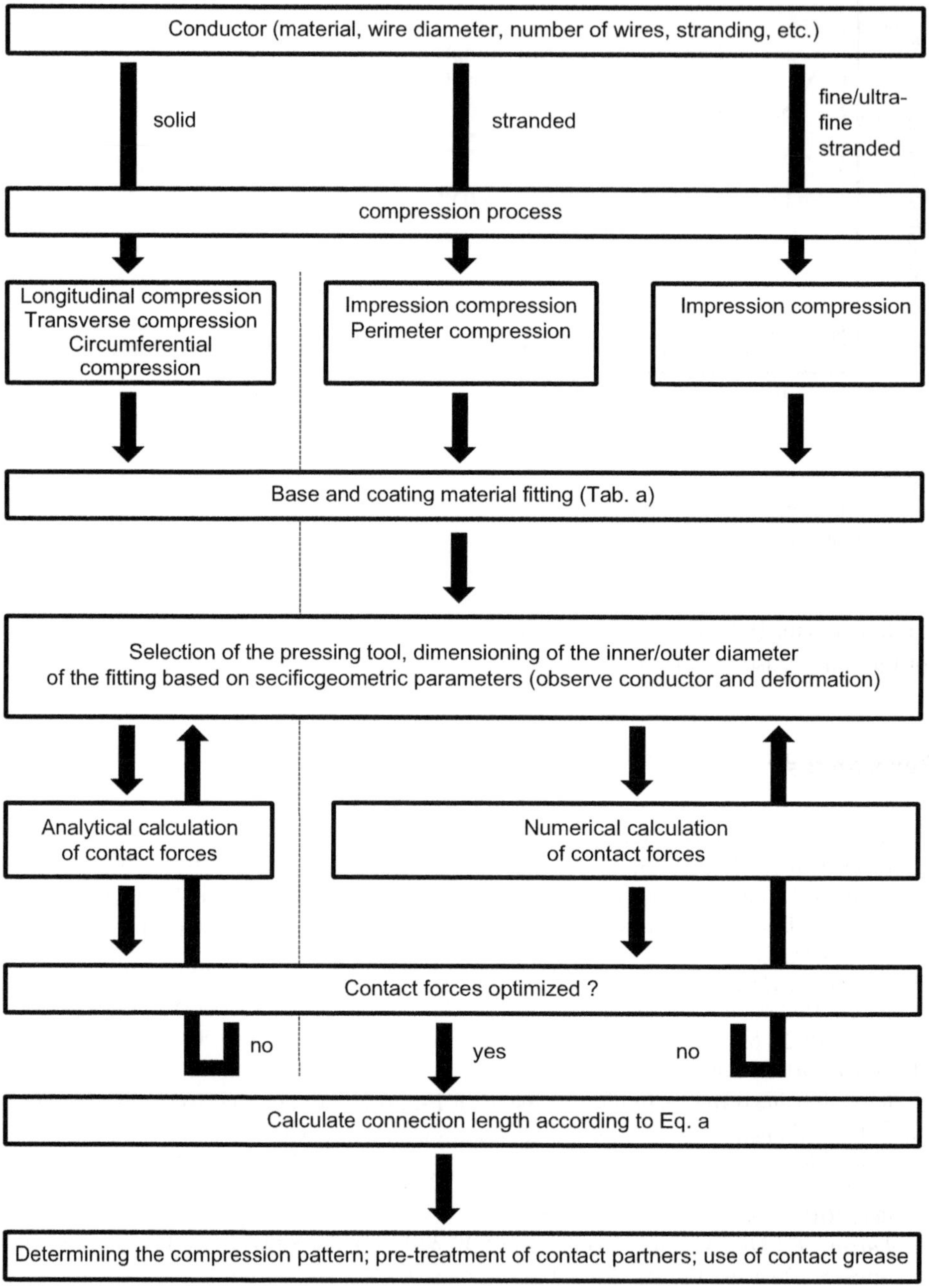

Fig. 6.9 Procedure for the adapted design of press connections. Table a = Table 5.14, Eq. a = Eq. 4.49

Fig. 6.10 Options for applying contact force in connectors: **a**) radially stressed helical spring, **b**) slotted socket with finger contact, **c**) contact element according to the leaf spring principle (elliptical or parabolic spring), **d**) contact element according to the torsion spring principle with and without carrier band, **e**) finger contacts with additional compression spring for applying a contact force

good contact properties and the geometry for accommodating the contact elements is usually specified by the manufacturer of the contact elements. Due to the variable length of most contact elements, these can be used in round and flat installation for various connector diameters and thus current ranges. Small quantities can also be produced economically. A disadvantage of contact elements is that there are at least two series-connected transition points between the contact element and the plug as well as the contact element and the socket. Contact elements in the form of lamella contacts (b) and (c) are available in versions with functional separation. In this case, the current mainly flows over louvers with high electrical conductivity and the contact force is generated by a high-strength carrier band. This functional separation has advantages in relaxation behavior, especially at high temperatures, and at the same time reduces the material resistance of the lamella contacts.

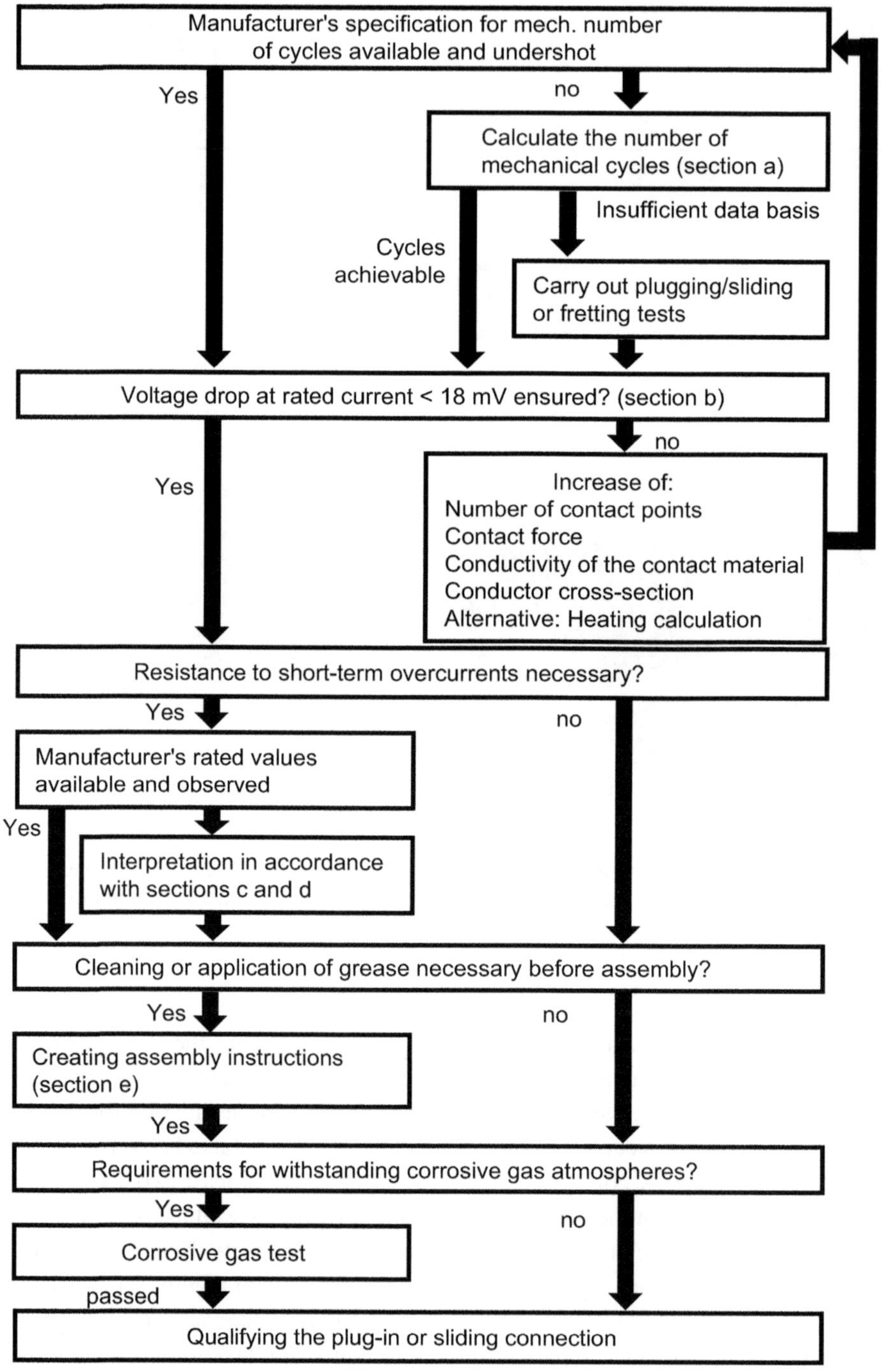

Fig. 6.11 Procedure for the electrical design of a plug-in and sliding connection. Sect. a = Sect. 5.1.4, b = 4.1.3, c = 4.4.2, d = 4.5.3, e = 6.6

Table 6.3 Selection of the type of connector for various applications

Application	Electric power supply, industrial plants		Electrical transportion		Special cases
	High and medium voltage	Low voltage	Charging connector	On-board network	e.g. particle accelerators [12], test facilities
Types of connectors	a) Coil spring b),c) Lamella contact e) Finger contact with pressure spring	a) Coil spring b),c) Lamella contact e) Finger contacts	a)-c) Contact elements d) Slotted socket contact	b) Contact elements; made of conductor material; shaped spring contacts similar to e)	a) Coil spring b),c) Lamella contact
Installation location	Power, grounding, and disconnect switches, expansion compensation, assembly connector		Plug and socket at charging stations and in vehicle/aircraft, ship	Connection of individual components of the wiring harness	Expansion compensation, Assembly connector
Explanation	a)-c) small quantities, e) very large quantities (low voltage); particularly high contact force necessary for circuit breakers and grounding switches		d), e) high quantities, easy to manufacture a)-c) high number of parallel contacts, therefore less frictional wear		a)-c), due to small quantities, particularly high short-term currents

Slotted sockets with finger contacts (b) are inexpensive and can be produced in large quantities. However, they are only designed for one diameter and are therefore not scalable. Compared to contact elements, the number of parallel contact points is usually lower. They are used in applications such as charging connectors and industrial sockets.

With finger contacts with additional springs (e), the contact force can be easily adjusted by the spring constant of the spring, and very high contact forces of several 100 N are also possible. A disadvantage is that the is design tailored for an specific application and the comparatively large installation space, especially with high contact forces. Compared to lamella contacts and coil springs, the connections usually have a lower number of parallel contact points. This type of connector is used, for example, for earthing switches in generator circuit breakers.

The rated values of continuous and short-circuit current carrying capacity of contact elements given by the manufacturers usually refer to the use of galvanically silver-plated plugs and sockets made of pure copper. If galvanically silver-coated aluminum is used instead, the current carrying capacity is reduced to approximately (75 … 80)% compared to technically pure copper. This is also referred to as the derating factor. When using galvanically silver-coated brass, the derating factor is approximately 60% based on technically pure copper. [50]

6.3.3.1 Geometry

In plug-in connections, a spring deflection s, which results from the height of the gap between the plug and socket h and the type of force application, must be defined (Fig. 6.12). Based on the spring characteristic curve (Fig. 4.16, p. 46), a sufficiently high spring force is generated. When using contact elements, the manufacturer specifies the recommended contact force and the required spring deflection. For special solutions, for example with slotted socket contacts, this relationship must be calculated or measured in advance.

The geometry of the contact element's groove in the plug or socket, the insertion, as well as the inlet geometry on the opposite side of the contact element are often also

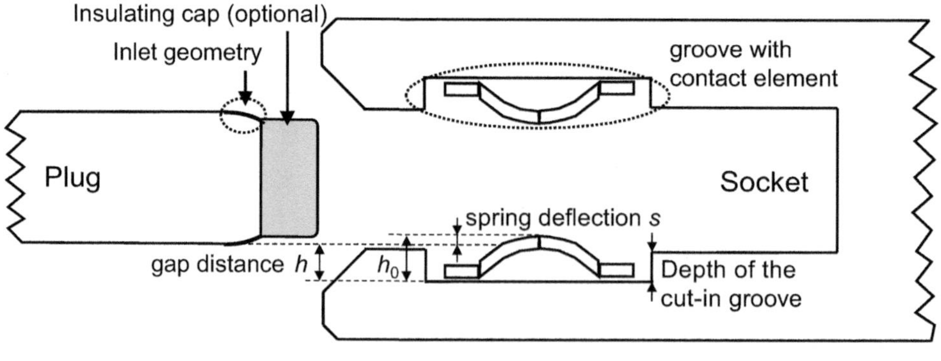

Fig. 6.12 Longitudinal section through a round plug-in connection with contact element (schematic)

specified by the manufacturer of the contact element. The groove holds the contact element mechanically firm and the inlet geometry ensures a steady spring-in when plugging in, as hard edges would cause a high insertion force and possibly damage the coating. The required plug-in and pull-out forces can be estimated based on the considerations given in Sect. 4.2.3 and with the corresponding sliding friction coefficients (Table 4.8) taking into account the inlet geometry of the plug and socket. Insulating caps can be used for touch protection and to avoid direct contacting between the plug and the socket, away from the intended contacts (or contact elements) between the plug and socket. For other types of applying contact force, the considerations for defining dimensions are qualitatively similar.

The contact force must allow a sufficiently low contact resistance (Sect. 4.3.1). At the same time, a large contact force increases frictional wear. In Sect. 5.1.4, simple estimations for this are described. In manufacturing, particular attention must be paid to the tolerances that influence the gap dimension, as even a small change in the gap dimension of (50 … 100) μm can significantly influence the contact force depending on the type of force application.

6.3.3.2 Lubrication

Lubrication is necessary when the required number of plug-in or friction cycles is high, as then the coating of unlubricated contacts can be worn through and the electrical contact properties can deteriorate significantly (see Sect. 6.6.2). For unlubricated silver coatings, typically a maximum of a few 100 plug-in cycles are achieved, with lubricated silver coatings more than 10,000 plug-in cycles are possible. Tin coatings are mainly used unlubricated. About 20 plug-in cycles are achieved until wear-through. With mixed material pairings, usually only slightly more plug-in cycles than with the tribologically poorer of the two materials are achieved. Another reason for lubrication can be that the plug-in and pull-out forces are too high for the application and need to be reduced.

Currently, physically applied silver layers [7] and dispersion layers based on silver with incorporated solid lubricant particles [5, 6] are being developed, which have a lower frictional wear than electroplated pure silver coatings. These coatings can have advantageous properties for certain applications compared to classically unlubricated or lubricated galvanic silver coatings, but due to the development status, a qualification test (Chap. 7) on the final product is necessary.

6.3.4 Further Joining Methods

The following presents the advantages and disadvantages of alternative types of connections and joining methods that have already been examined for their use as current-carrying connections (Table 6.4). The most important electrical properties are summarized in brief form and corresponding literature sources are given, which contain detailed information.

Table 6.4 Examples of alternative types of connections/joining methods

Type of Closure	Type of Connection	Principle	Advantages/Disadvantages	Literature Sources
Force-fit/ Form-fit	Clinching	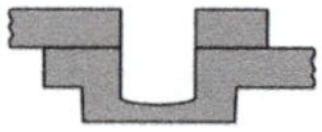	*Advantages:* • No heat input during joining • Direct joining of Al and Cu possible • No additional joining elements • $k_u \leq k_{uT\text{-}20}$ possible *Disadvantages:* • High deformation of the contact partners • Assembly not possible on the construction site	[9, 10]
	Semi-hollow punch riveting (optimized)		*Advantages:* • No heat input during joining • No abrasive pretreatment of the contact surface before joining • $k_{u0} = 1.6 \ldots 2.2$ possible (Al-Al) *Disadvantages:* • High deformation of the contact partners (softening possible under temperature load—strong force reduction) • Assembly not possible on the construction site	[9, 10]
	Electromagnetic Forming (EMU)	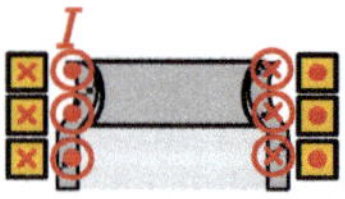	*Advantages:* • No pretreatment of the contact surfaces before joining • Cold welding of the contact partners possible during joining • No heat input during joining *Disadvantages:* • Adapted component geometry in the joining zone necessary • Expensive plant technology • Assembly not possible on the construction site	[11] see also Sect. 6.3.2

(continued)

Table 6.4 (continued)

Type of Closure	Type of Connection	Principle	Advantages/Disadvantages	Literature Sources
	Longitudinal Pressing (see Sect. 4.2.2)	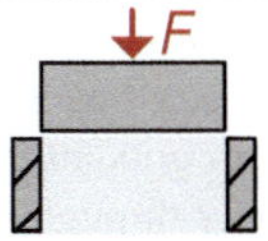	*Advantages:* • Cold welding of the contact partners possible during joining (Al-Al) • No pretreatment of the contact partners necessary due to frictional relative movement *Disadvantages:* • Possibly greasing of the contact surfaces to reduce the pressing forces • Assembly not possible on the construction site	[11] see also Sect. 6.3.2
	Press fitting (Sect. 4.2.2)	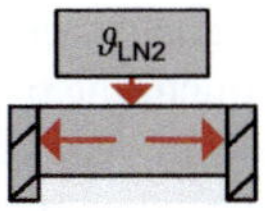	*Advantages:* • No pretreatment of the contact partners required • No heat input during shrinkage by cooling *Disadvantages:* • Excess ~ contact force limited by thermal expansion coefficients of the contact partners • Occupational safety when handling LN_2	[11] see also Sect. 6.3.2
Material joint	Soldering	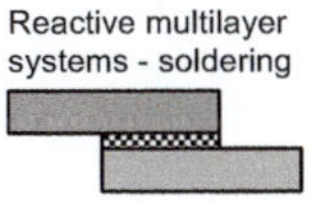	*Advantages:* • Low heat-affected zone during joining • Large true contact area *Disadvantages:* • Intermetallic phase in the contact zone • Force during the joining process between the contact partners necessary • Pretreatment of the contact surfaces required	[12, 13]

(continued)

Table 6.4 (continued)

Type of Closure	Type of Connection	Principle	Advantages/Disadvantages	Literature Sources
	Welding	Friction stir welding, Resistance welding, Ultrasonic welding, Laser welding, Cold gas welding	*Advantages:* • Large contact areas • Very good electrical conductivity through material joint *Disadvantages:* • Formation of intermetallic phases during joining in bimetallic connections • Partly high heat input during joining	[14, 15]
	Gluing		*Advantages:* No heat input during joining *Disadvantages:* • Special process necessary • Structuring of the contact surface required • Special pretreatment of the contact surface of particular importance	[16, 17]

The presented types of connections and joining methods can be used as main or auxiliary connections depending on the application. The overview does not claim to be complete and refers to variants that have already been considered in individual studies or extensive works. These are pointed out in the literature references.

6.4 Continuous Operation

In general, an electrically conductive connection must be stable in the long term. In continuous operation or with regularly occurring maximum load, the permissible limit temperature (Sect. 5.2) must not be exceeded so that no accelerated aging occurs. The basis for this is that the type of connection used (Sect. 6.1) and the selected materials (Sect. 6.2) are adapted to the permissible limit temperature (Chap. 3 and Sect. 5.3) and the connection is designed accordingly (Sect. 6.3). Current-carrying connections are usually spatially compact and thermally closely coupled with the connecting conductors. Therefore, the thermal dimensioning of the connecting conductors is of central importance. The connection itself also has a significant influence on the temperature, as any additional power loss that occurs in the connection is conducted into the contact partners and usually also into the connecting conductors. For this reason, the connecting conductors are usually thermally "influenced" over a length of several 10 cm by the connection (see concept of the thermal spatial constant [30]). This chapter describes the thermal dimensioning of the connection for continuous operation depending on the type of connection. In general, two basic cases can be distinguished regarding the limit temperature of the connection:

- **"same temperature":** Connection and connecting conductors have the same limit temperature
- **"cooler connection":** The limit temperature of the connection is lower than that of the connecting conductors. This is the case when the type of connection would age faster with the temperature of the connecting conductors during operation. For example, a lower limit temperature (75 °C) is specified for copper contacts in low and high voltage switchgear than for the copper conductor itself (140 °C) (see IEC 62271-1:2017 [41] and IEC 61349-1:2020 [42]). In practice, this means that additional measures must be taken to cool the connection (Sect. 6.4.1, 6.4.2 and 6.4.3). Alternatively, it is always possible to cool the entire system by using larger conductor cross-sections and not fully utilizing the higher possible limit temperature of the conductors.

With a temperature rise test or a thermal calculation, it can be ensured that the limit temperatures are adhered to. Depending on the application, this is prescribed in test standards (Sect. 7.2). If this is not required and the connection is designed according to the diagrams shown, the heating test can possibly be skipped and instead the long-term behavior can be tested (Fig. 6.13).

6.4.1 Screwed Connection

Case "same temperature"

In this case, the requirements for continuous operation are already met if the connection is designed according to Sect. 6.1, 6.2 and 6.3. and the connecting conductors comply with the limit temperatures. The decisive factor here is the temperature equivalent performance factor $k_{uT\text{-}20}$ (Table 4.21). If the connecting conductors are "too hot", their temperature must be reduced by larger conductor cross-sections, higher electrical conductivity ("copper instead of aluminum") or better cooling (e.g., paint or active cooling). The thermal dimensioning of the connecting conductors is not discussed in detail here. The method according to *Böhme* [30] or the use of tabulated values for typical conductor dimensions, for example in [18], is recommended.

Case "colder connection"

To cool the connections to a lower temperature than that of the connected conductors, additional heat must be dissipated from the connection. Ideally, the connection resistance can be reduced to decrease the resulting power loss. The following measures can be applied.

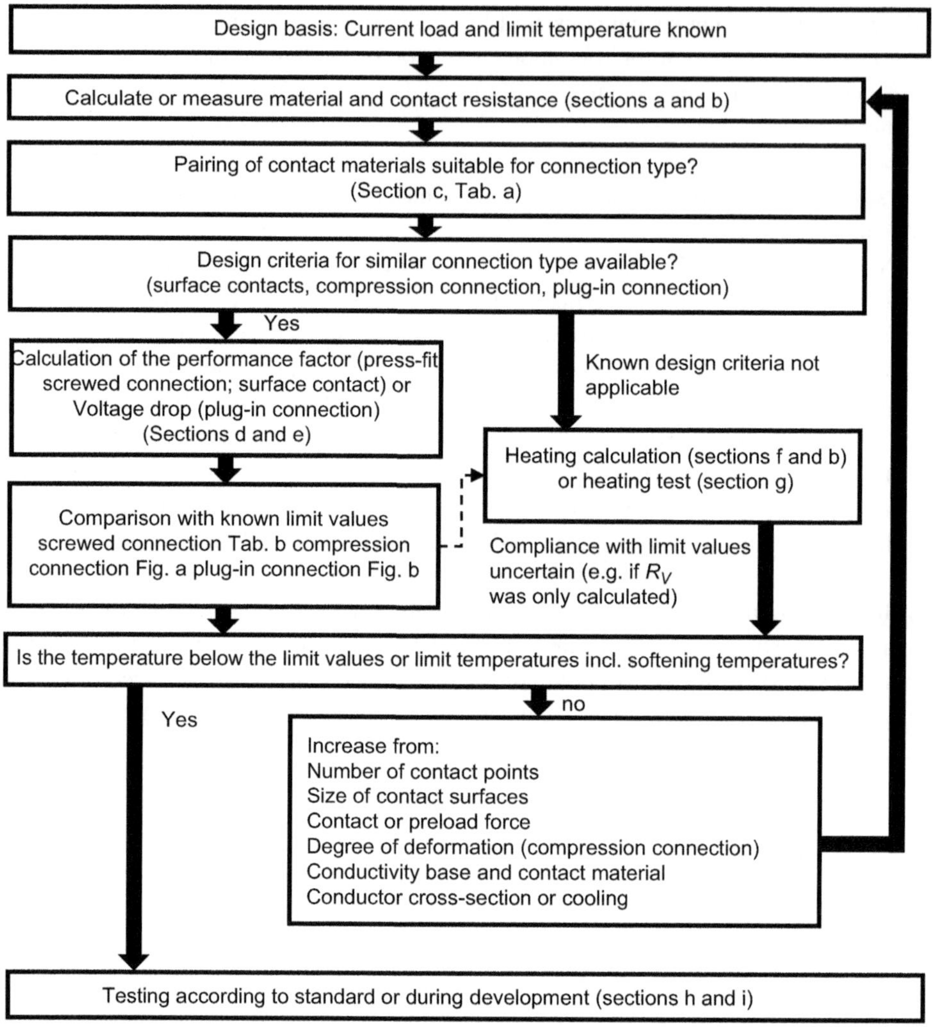

Fig. 6.13 Procedure for designing a connection for continuous operation. Sect. a = Sect. 4.3, b = 4.5, c = 5.3, d = 4.1.2, e = 4.1.3, f = 4.4, g = 7, h = 7.2, i = 7.3; Table a = Table 5.14, Table b = Table 4.21, Fig. a = Fig. 4.75, Fig. b = Fig. 4.7

- Increase the overlap length of the contact partners
- Use of a larger conductor cross-section
- Increase the surface area of the connection by using joining elements or attaching a heat sink
- Increase the emissivity of the surface by, for example, painting

The efficiency of the measures should be checked by thermal dimensioning using calculation methods (e.g., according to *Böhme* [30].) or a heating test.

6.4.2 Compression Connections

Press Compression connections designed according to Sect. 6.1, 6.2 and 6.3 typically have performance factors significantly less than one. Typical guideline values are:

- Complete fitting for overhead line conductors $k_u < 0.75$ (DIN EN 61284:1997, limit value after aging test)
- Cable fitting $k_u \approx 0.7 \ldots 0.8$

Due to the massive connector bodies, the temperature at the connection is usually lower or, approximately the same as at the connected conductor (see concept of thermal location constant, Sect. 5.4.3.2 in [5]).Nevertheless, the following two cases can also be distinguished here.

Case "same temperature"
Compression connections designed according to Sect. 6.1, 6.2 and 6.3 meet this requirement if their performance factor is not greater than the temperature equivalent performance factor $k_{uT\text{-}20}$ for e.g., rotationally symmetric conductors (Fig. 4.75).

Case "colder connection"
This case occurs especially with compression connections for overhead line conductor . The compression connections themselves must not reach the operating temperatures of the overhead line conductors in order to limit the force reduction. In this case, the connector must be chosen to be sufficiently long and with an appropriate diameter to achieve a lower power loss through the larger conductor cross-section and to increase the heat-dissipating surface in the area of the connector, thus reducing the temperature.

6.4.3 Plug-in and Sliding Connections

Generally, for plug-in connections, the type of contact element and the number of parallel contacts should be chosen so that the contact resistance is small enough that the voltage drop across the plug-in connection at rated current is at most (14 … 18) mV (category "typical design plug-in connections" in Table 4.3, Fig. 4.7). This corresponds to an overtemperature of around (3 … 5) K relative to the solid contact partners plug and socket. Due to the small cross-section of the contact elements, the power loss density in the area of these and the contacts is generally increased. The additional power loss is conducted into the plug and socket as well as into the conductors connected to them (Fig. 4.76 and 4.81). This leads, in most practical applications, to the entire plug-in connection being slightly warmer than the unaffected connected conductors.

6.4.3.1 Case "same temperature"

First, the number of contacts or contact elements should be chosen so that the voltage drop U_V across the connection at rated current is a maximum of 18 mV (see Sect. 4.1.3). Many manufacturers provide the connection resistance of contact elements in the data sheets. This value can be used for a first estimate. The design for $U_V \leq 18$ mV often still results in a slightly higher temperature at the plug-in connection compared to the unaffected connecting conductors. It depends on the application whether this is acceptable. If this is not the case, the overtemperature can be reduced by the following measures:

- Increase the number of parallel contacts or contact elements
- Use contact elements of a different design with lower resistance
- Modifications to the plug and socket to generate less power loss in the connection area overall and to better dissipate the power loss of the contact elements
  - Increase the conductor cross-sections
  - Use conductor material with higher electrical and thermal conductivity ("copper instead of aluminum")

Ideally, the effectiveness of these measures is checked with a thermal calculation or a simple temperature rise test to find the optimum balance between effort and benefit.

6.4.3.2 Case "cooler Connection"

This case is only possible for high-current applications with additional measures. The measures described in the case "same temperature" should be checked first. If these are not sufficient, active cooling, for example with a liquid cooling medium as close as possible to the plug-in connection, can be used.

6.5 Short-term Load

Many electrical devices must be able to carry significantly increased currents for a short time, e.g. in the event of a short circuit, which causes a strong thermal and mechanical load, especially in the connections with often reduced cross-section compared to the main current path. A short-term current ten times larger than normal operation causes a hundred times the power loss density, as well as a hundred times the current forces. For the reliability of the overall system, it is crucial in this case that all devices can carry the occurring short-term current without damage until the fault is switched off. Devices for the electrical energy supply system therefore have an important requirement, a rated short-time current I_k, which is typically 10 to 20 times larger than the rated operating current and must be safely carried for short-circuit durations t_k of up to a few seconds [41, 42]. Generally, the following characteristics of the short-time current are decisive for the highest stress on the connection:

- Thermal stress of the conductor material in the connection: RMS value and duration, for example $I_k$and t_k in switchgear [41]
- Thermal stress of the contacts, mechanical stress: highest instantaneous value, for example $i_p$in switchgear [41]

The aim of the design for short-term current load is to design the connection in such a way that it is not significantly damaged itself and also does not damage adjacent components. A design according to the Sect. 6.1, 6.2 and 6.4 designed connection already has a basic short-term current carrying capacity. For increased requirements, this can essentially be increased by the following measures:

Reducing the maximum temperature by a smaller contact resistance

- Increasing the contact force
- Creating additional parallel contact points
- Use of a contact material with higher electrical conductivity or better contact behavior

Reducing the maximum temperature by a smaller material resistance and a larger heat capacity

- Using a connection conductor with a larger cross-section
- Use of a conductor material with better electrical and thermal properties
- Use of contact elements with a larger cross-section

Increased resistance to current forces:

- Increasing the contact force
- More stable attachment of the connection conductors, e.g. by shorter support distances
- Using conductors with increased bending stiffness and/or strength
  - Cross-sections with more area moment of inertia
  - Materials with higher strength: e.g. copper alloys instead of pure copper, copper instead of aluminum

To determine whether a particular connection has sufficient short-term current carrying capacity, either

- the mechanical and thermal stress can be calculated and compared with the strength values of the materials (Sect. 6.5.1 and 6.5.2) or

- Results of tests can be evaluated (Sect. 6.5.3). Under certain conditions, it is possible to transfer results from similarly designed connections or with other current and time characteristics

Generally, when designing for short-term current load, the entire connection with connected conductors, possibly existing contact elements and other adjacent current-carrying connections must be considered, as these are thermally and mechanically coupled. In addition, the influence of other conductors in the vicinity must be investigated, as electromagnetic forces can act between them (Sect. 4.2.4).

- Electromagnetic forces on connected conductors have a reaction on the connection
- In force-fit connections, the effective contact force and thus the contact resistance can dynamically change due to these forces
- In multipole and multiphase systems, an inhomogeneous current distribution can occur due to the proximity effect on the circumference of the connection

As a result,different stresses on the connections in the individual phases of a multi-conductor system are possible. Furthermore, the stress on the individual connections on the circumference may be inhomogeneous, i.e. there are areas that are significantly higher electrically and/or mechanically loaded than others [43].

6.5.1 Step A: Determining the mechanical load

According to Sect. 4.2.4.1 and 4.2.4.2, the contact force resulting for the connection can be calculated. First, the electromagnetic forces through the conductors and, in the case of very highly loaded point contacts, additionally the lift-off forces are to be determined. The peak value of the current i_p should be noted. If the contact force acting in the static case is significantly (more than 10%) reduced by the acting currents, the dynamically changing contact force and its feedback on the contact resistance should be considered in step B (determining the thermal stress). Under no circumstances should the resulting contact force go against or below zero, then a lifting of the contact partners can occur and an arc can be created.

6.5.2 Step B: Determining the thermal load

Calculation of Temperature Distribution
For point contacts, simplified analytical considerations for calculating the welding limit current and thus the permissible load current for a given limit temperature for load durations up to approximately 0.1 s are available (Sect. 4.4.2). For longer load durations,

the inhomogeneous, transient heating of the connecting conductors must be taken into account, for example by numerical calculations (Sect. 4.5.3).

Strengths and Limits

For mechanically loaded conductors, as well as mechanically supporting parts of the contact elements, the maximum calculated temperature should be compared with the material-dependent short-term limit temperatures. Guidance is provided by [38] and Chap. 3:

- Copper conductor (solid or stranded): 200 °C
- Steel (solid or stranded): 300 °C

However, it is also possible to use deviating short-term limit temperatures determined from empirical values, for example determined by the adhesion of the coating or adjacent insulating materials [1]. In certain systems with less mechanical stress, it may also be permissible to allow limit temperatures that are close to the melting temperature [40].

For maximum temperatures in the contacts calculated with the voltage-temperature relationship, it is permissible in most cases for short-term loads to reach or exceed the softening temperature (softening voltage) (Sect. 4.5.3.2). However, it should remain well below the melting temperatures (melting voltage), as otherwise welding of the contact points occurs and a transition to the arc can occur even at slightly higher currents.

Benefit of Manufacturer Specifications

Especially with contact elements with manufacturer-specific design, there are often design values that are specified by the manufacturer based on tests or calculations. These specifications can also be used for the design, as long as comparable conditions prevail in the target application, especially for the materials of the connecting conductors (typically copper) and their coating (typically silver). It should be noted that the design values $I_{\mathrm{k\,per}}$ are usually related to a period unit (for example a louver or a turn). They must be converted to the number n of actually existing period units. In this case, a correction factor k_{per} may need to be taken into account:

$$I_{\mathrm{K}} = k_{\mathrm{per}} \cdot n \cdot I_{\mathrm{K\,per}} \tag{6.1}$$

It should be taken into account that the number n of period units is one of the most effective parameters for the design of plug-in and sliding connections. Thus, by adding another contact element, the short-term current carrying capacity can be significantly increased. As a first approximation, $k_{\mathrm{per}} = 1$ can be calculated. However, in plug-in connections with inhomogeneous current distribution over the circumference or contact elements distributed over the length, a reduction may also occur [39].

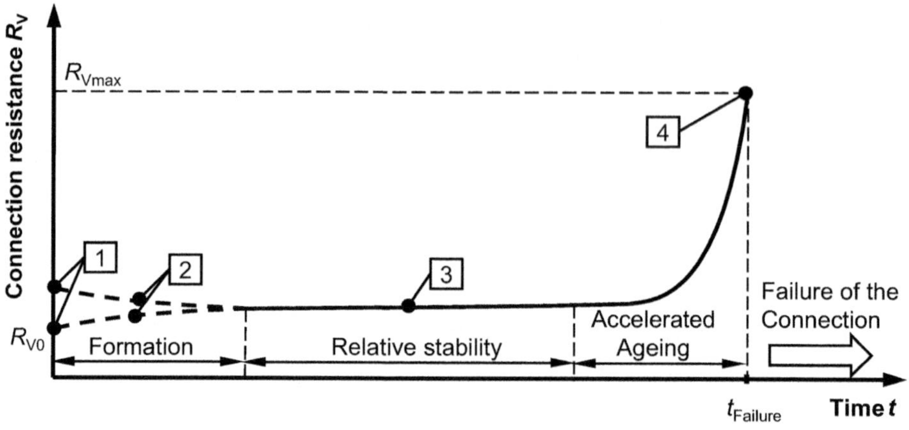

Fig. 6.14 Possible times of occurrence of a short circuit and the associated contact resistance of a silver-coated plug-in connection and temperature of the connection ϑ_V at the time of the short circuit

- Case 1: Short circuit immediately after assembly of the plug-in connection: $R_V \approx R_{V0}$; $\vartheta_V = \vartheta_0$ (corresponds to type testing)
- Case 2: Short circuit during the first operating hours: $R_{V\ min} < R_V < R_{V0}$; $\vartheta_0 < \vartheta_V < \vartheta_{stationary}$. The higher ϑ_V, the lower R_V
- Case 3: System in stationary state: $R_V \approx R_{V\ min}$; $\vartheta_V = \vartheta_{stationary}$ (e.g. 105 °C)
- Case 4: System at the end of its life: $R_K = R_{KA} = (1.2 \ldots 2)\ R_{K0}$; $\vartheta_K \geq \vartheta_{stationary}$

Lifetime Considerations

An electrical short circuit can occur at any time during the lifetime of a connection. Typically, the contact resistance should not be significantly higher during the lifetime than after assembly. Therefore, when designing for short-term current load, it should be decided for which state of the contact this is done. In standardized short-circuit tests of connections with contact elements, only the new condition is tested. In compression and screwed connections for stranded conductors, short-circuit tests take place after or during a duration test (Sect. 7.2) (Fig. 6.14).

6.5.3 Alternative: Design Based on Short-Term Current Tests

A direct test of the connection to be designed with the desired load (current, time) provides the highest security. Due to the great effort for such tests, it may be sensible to transfer the results of already existing tests to other conditions.

Transfer to comparable contact systems using the example of a plug-in connection

For plug-in connections, tests with the same contact element and the same or stronger spring deflection, as well as connection conductors with the same conductor and coating material, are generally transferable. If in the reference test with the short-term current $I_{k\,ref}$ a different number n_{ref} of period units (or webs, turns) was used, the current can be scaled accordingly in the first approximation:

$$I_K = \frac{n}{n_{ref}} \cdot I_{k\,ref} \qquad (6.2)$$

Transfer to other short-term currents and other short-term current durations

In the case of short-term current load, the energy that can be transferred via a contact system depends on the heat flows. For connectors with contact elements and point contacts, the permissible currents under short-term load deviate from those determined with the I^2t criterion (Fig. 6.15 and 6.16) [1, 28].

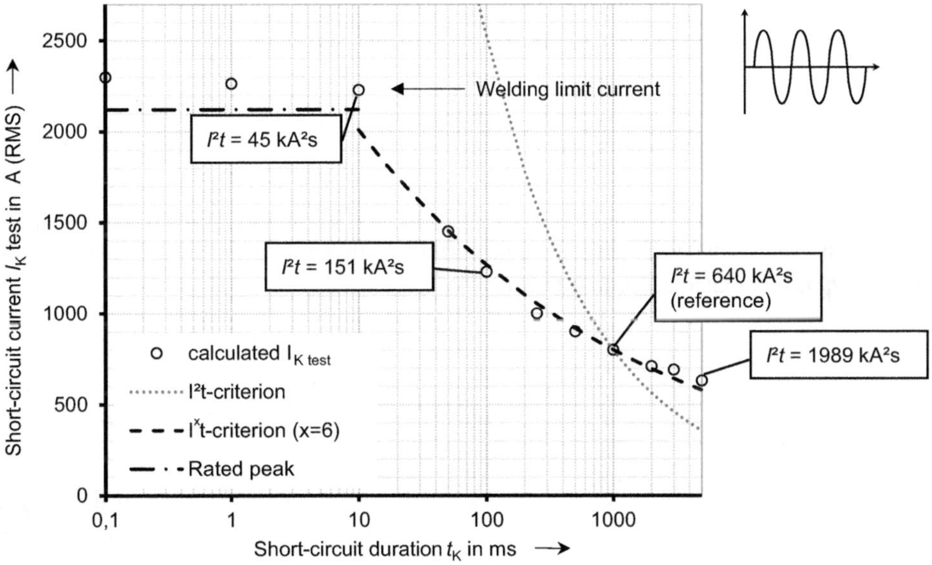

Fig. 6.15 Short-circuit current calculated with the I^2t criterion (starting from the 1-s value) and short-circuit current of an exemplary connector calculated with the I^xt criterion. Values for one period unit ($n = 1$). The actually valid short-circuit currents determined with a calculation model and the welding limit current of the contacts are also plotted

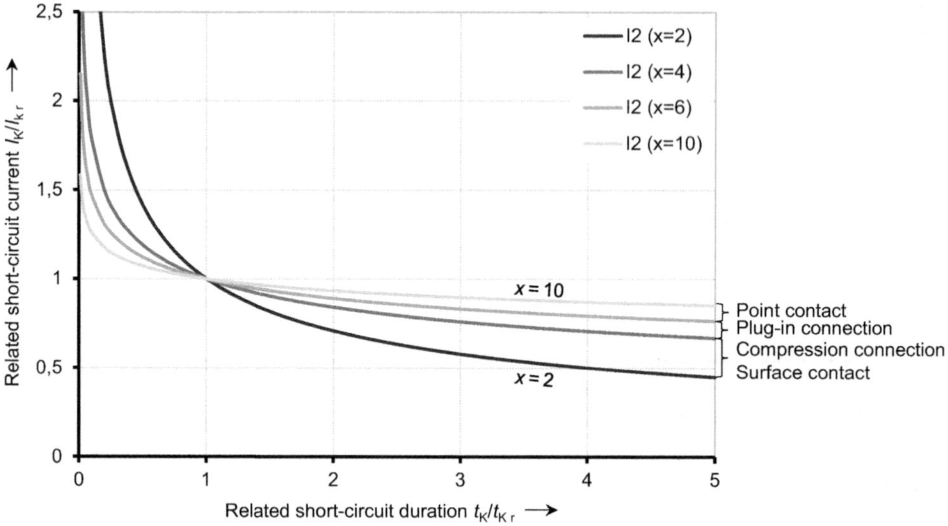

Fig. 6.16 Permissible short-circuit current depending on the load duration

This is because the adiabatic heating of the contacts and connections assumed for the I^2t criterion is generally not fulfilled, as heat is not only stored during the short circuit, but also conducted from the contacts into the connecting conductors. By adjusting the exponents to the I^xt criterion, a much better agreement with the calculated, permissible short-circuit currents is achieved (Fig. 6.16). The exponent x is dependent on geometry and material. It can be calculated directly with Eq. (6.3) if at least two current-time pairs are known. The transferable short-circuit current I_K depending on the real short-circuit duration t_K is calculated according to Eq.(6.4).

$$x = \frac{\ln {}^{t_{K2}}\!/_{t_{K1}}}{\ln \left({}^{I_{K1}}\!/_{I_{K2}}\right)} \tag{6.3}$$

$$I_K = I_{K1} \sqrt[x]{\frac{t_{K1}}{t_K}} \tag{6.4}$$

The I^2t criterion is also formally not valid for screwed and compression connections, as there are a-spots/constrictions in the connection and the current-carrying cross-section changes along the conductor.

The magnitude of the error when applying the I^2t criterion has not yet been determined for compression and screwed connections, but is likely to be smaller due to the smaller differences in cross-section along the current path compared to plug-in connections and point contacts. Maximum tolerable surge currents also exist for compression and screwed connections, which result, for example, from the heating of the a-spots

Table 6.5 Rules for the short-circuit-proof design of high-current connections for other short-circuit durations than the rated short-circuit duration, if Eq. 6.4 cannot be applied, or x is unknown.

	1st condition	2nd condition
Design **Connection can** **I_K carry for t_K if:**	$I_K \leq I_{K\,r}\sqrt{\frac{t_{K\,r}}{t_K}}$ für $t_K \geq t_{K\,r}$ or $I_K \leq I_{K\,r}$ für $t_K \leq t_{K\,r}$	$i_p \leq i_{p\,r}$

(melting) and the effect of electromagnetic forces. If the exponent x for a connection or the time-dependent rated short-circuit current is not known, a short-circuit-proof design can still be carried out for short-circuit durations that deviate from the rated values. If the short-circuit duration is shorter than the rated value, the effective value of the short-circuit current must not exceed the rated short-circuit current. If the test duration is extended, the effective value can be adjusted using the I^2t criterion, as this does not result in increased thermal stress in the contact (Table 6.5—1st condition). Regardless of the actual short-circuit duration, the rated surge current must not be exceeded at any time during the short circuit (Table 6.5—2nd condition). [2–4]

Example

Application of the I^xt criterion to a sliding connection for high-voltage switchgear

The manufacturer of contact elements for sliding connections in a switchgear specifies a rated short-time current of 55 kA for a rated short-circuit duration of 1 s and 42 kA for a short-circuit duration of 3 s. The rated peak current is 140 kA. Short-circuit currents that can be safely carried for a load duration of 0.3 s; 0.5 s and 5 s are sought. Since two current-time pairs are known, the I^xt criterion can be used. The exponent x is calculated according to equation (6.3):

$$x = \frac{\ln \frac{t_2}{t_1}}{\ln \frac{I_1}{I_2}} = \frac{\ln \frac{3\,\mathrm{s}}{1\,\mathrm{s}}}{\ln \frac{55\mathrm{kA}}{42\mathrm{kA}}} = 4.1$$

The controllable short-circuit current for the specified short-circuit durations is calculated according to equation (6.4). The table also lists the short-circuit currents calculated using "I^2t = constant" and according to the design with an unknown parameter x (Table 6.5).

$$I_K = I_{K1} \cdot \sqrt[x]{\frac{t_{K1}}{t_K}} = 55\,\mathrm{kA} \cdot \sqrt[4.1]{\frac{1\,\mathrm{s}}{t_K}}$$

Short-circuit duration t_K in s	Permissible short-circuit current according to $I^x t$ in kA (x = 4.1)	Permissible short-circuit current according to $I^2 t$ in kA	Short-circuit current according to Table 6.5
0.3	73.9	100.4	55
0.5	65.2	77.8	55
1	55	55	55
3	42	42	42
5	37.1	32.5	32.5

The currents calculated for 0.3 s and 0.5 s short-circuit duration using the I^2t criterion are significantly higher than those calculated using the I^xt criterion. They would thermally load the connection more heavily (see Fig. 6.15). The design with an unknown exponent x according to Table 6.5 is safe, as the connection is not thermally stressed more than by the rated current. However, the performance of the connection for short-circuit durations <1 s and >3 s is significantly underestimated compared to the I^xt criterion.

It should be noted that for all short-circuit times, the maximum instantaneous value of the current must not exceed 140 kA, as the contacts can then melt and weld. When using the I^2t criterion, this would be the case even without a pronounced surge current ($\kappa = 1$) at a short-circuit duration of 0.3 s ($\hat{I} = 142$ kA). With the I^xt criterion, it must be checked for a short-circuit duration >3 s whether the connecting conductors do not heat up excessively. ◀

6.6 Assembly and Quality of Contact Surfaces

6.6.1 Pretreatment of the Contact Surface

The quality and thus the quality of a current-carrying connection can often not be checked after assembly, as this is done by personnel who are not trained in this respect or because the appropriate measuring technology for measuring the connection resistance is not available. A robust design and an easily understandable instruction for the assembly personnel is often the only way to influence the quality. In series production, additional measures can be taken to check the quality after assembly (Table 6.6).

The quality of an electrically conductive connection is crucially dependent on reliable and secure assembly with the correct joining parameters. To this end, the manufacturer should create as simple and clear assembly instructions as possible, which include the following points:

- Determining the aids for pre-treating the contact surfaces
- Defining the wear limit of the aids for pre-treatment (replacement intervals)

Table 6.6 Observed state from practice—Ensuring the quality of the assembly of current-carrying connections

Measure	Assembly at the place of use	Assembly in series production
Measure resistance or voltage drop	no/conditional	yes/conditional
Assembly instructions	yes	yes
Training of assembly personnel	no/conditional	yes

- Describing the pre-treatment steps as clearly as possible with pictures and bullet points
- Specifying the time between pre-treatment and assembly
- Necessary tools for assembly
- Assembly parameters (e.g., torque)
- Sequence during assembly (e.g., tightening sequence for multi-screwed connections)
- Special notes due to specific features of the connection type "Test procedure after assembly, e.g., measuring the connection resistance, with indication of the potential measurement points and comparison with limit values."

If possible, the assembly instructions should be presented on a DIN A4 page. The assembly of electrically conductive connections should ideally only be carried out by trained personnel who are informed about the main sources of error and risks in training courses. Failure or a reduced lifetime of connections is in most cases due to errors during assembly. This particularly affects connections with uncoated aluminum conductors (Fig. 6.17). However,pre-treatment is also necessary for copper contact surfaces or for certain coating materials and must be carried out in a material-specific manner (Table 6.7). [33–35]

6.6.2 Lubrication of the the Contact Surface

To reduce friction between the contact partners or to protect the contact surface from atmospheric influences, contact greases or lubricants are sometimes applied to the surface of the contact partners before joining. The composition of these is strongly dependent on the requirements during the joining of the connection and the conditions in operation. Therefore, the contact grease or lubricant must be qualified for the application in each case. Depending on the type of connection, an effect is intended and thus requirements are placed on the contact grease or lubricant, which must be taken into account when used (Table 6.8).

For screwed connections with base contact materials, contact grease can be used to protect the contact surface from oxidation after pretreatment. After joining, the grease is displaced into the cavities between the a-spots and is supposed to shield these from atmospheric influences such as moisture or oxygen, which can lead to oxidation or

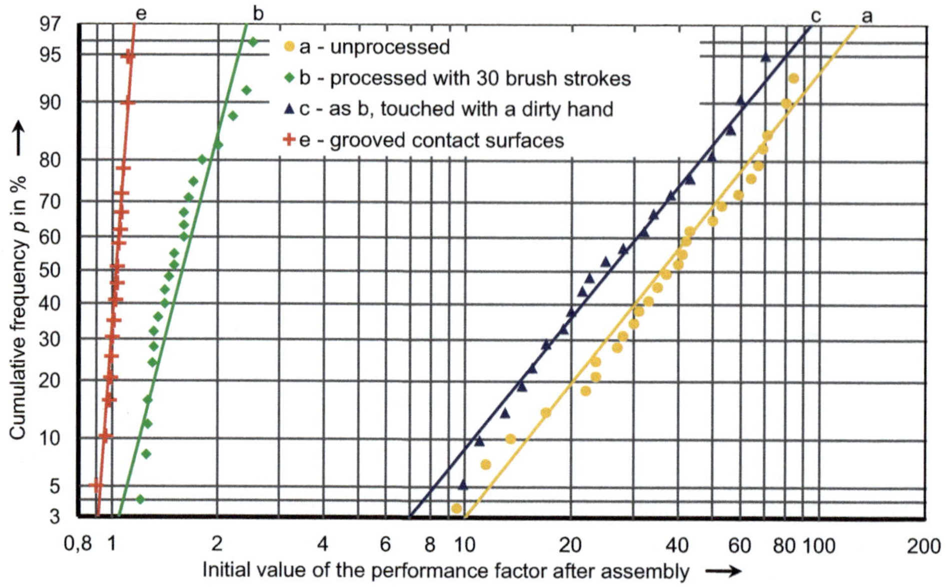

Fig. 6.17 Initial value of the performance factor after assembly of screwed connections with busbars made of Al99.5 with a contact force of $F_K = 18$ kN depending on the condition of the contact surface before assembly [33], data from [49]

Table 6.7 Pre-treatment of contact surfaces/storage times between pre-treatment and assembly

Contact material	Pre-treatment	Storage time after pre-treatment
Copper	• Degreasing the contact surfaces • Breaking up oxide layers and micro-structuring the contact surface with a abrasive fleece or a wire brush	Hours to a few days (depending on the environmental conditions)
Aluminium	• Degreasing the contact surfaces • Breaking up oxide layers and micro-structuring the contact surface with a sharp-edged wire brush	Immediately to minutes
Silver	• None • Possibly greasing in the case of plug-in connections	No specifications, in SO_2-poor atmosphere
Tin	None	No specifications
Nickel	• Degreasing • Breaking up oxide layers and micro-structuring the contact surface with a grinding fleece	Hours to a few days (depending on the environmental conditions)

corrosion, especially if no gas tightness is achieved during joining (Fig. 6.18; see also Sect. 5.1.3). [32, 37]

It should be noted that the contact grease should be applied thinly so that no grease flows or drips out after joining and at operating temperature of the screwed connection. This is particularly important for metal-filled greases to prevent pollution flashover on insulating paths. The contact grease must be qualified for the application and the material combination. Results of investigations have shown that contact greases can not only improve but also deteriorate the long-term behavior of the connection depending on the composition [31]. Also, the adhesion and embedding of dirt particles in the grease before the connection is joined can have a negative impact on the contact and long-term behavior. It is therefore critical to check whether a contact grease needs to be used or not, as connections with the same materials of the contact partners are long-term stable even without contact grease. For screwed connections with different materials of the contact partners such as copper and aluminum, it has been shown that a more stable long-term behavior is possible through the use of a contact grease [32].

In compression connections, contact greases are used that prevent the conductor from being damaged, for example by seizing or cold welding, already during joining. Furthermore, the required press force should be reduced. The addition of conductive or non-conductive metallic particles in the contact grease can increase the abrasion of the contact surfaces during pressing and improve the contact behavior in some applications. The particles are intended to break up oxide layers on the contact surface of the sleeve and the conductor. Contact grease is used as standard for compression connections in outdoor applications to prevent the ingress of moisture or oxygen during operation and thus ensure the corrosion protection of the contact area. In indoor applications of compression connections, it is conceivable that the contact grease can also reduce the ingress of oxygen during operation and achieve a certain protective effect of the contact area against oxidation. Also, by applying a suitable grease immediately after the abrasive cleaning of the contact surface, a certain protection of the pretreated contact surfaces against further oxidation can be achieved.

For plug-in connections, lubricants are often used to improve the tribological behavior. They act directly in the contact points and reduce the insertion and sliding forces (Sect. 4.2.3). Since less abrasion occurs, the frictional wear is reduced (Sect. 5.1.4). There are different types of lubricants. Solid lubricants, which can be incorporated directly into the coating, (Table 3.9, 4.8 and Fig. 5.30) and lubricating greases and oils, which must be applied separately before use. However, not just any lubricants can be used for electrical contacts. The use of oils should be checked especially when heated, as the oil remains in the contact [48]. For solid lubricants, the lubricating effect and the influence on the electrical contact behavior must be checked for the chosen material combination, as only a few types of particles sufficiently fulfill both requirements [45, 46].

Table 6.8 Effects and challenges when using a contact grease or lubricant for current-carrying connections depending on the type of connection

Type of connection	Screw connection	Press connection	Plug connection
Effect	Protection against the penetration of moisture and oxygen, reduced aging due to chemical reactions (corrosion, oxidation) possible when using a grease suitable for the material pairing [31, 32]	• Breaking up of oxide layers during joining by solid particles in the grease • Reducing the necessary press force • Protection against moisture and oxygen in operation • Preventing cold welding during joining	• Reducing the coefficient of friction during plugging and sliding cycles • Increasing the service life by reducing friction wear in operation due to plugging and pulling or sliding cycles • Reducing the tendency to cold weld
Notes for application and qualification	• The amount of grease during application must be defined → Risk of leakage at elevated temperature in operation (possible metal particles in the grease can lead to dielectric problems) • Limiting the operating temperature to the maximum permissible limit temperature of the grease (risk of resinification) • With connections of the same conductor materials and correct pretreatment of the contact surface as well as correct choice of joining parameters (tightness), hardly any effect • Adhesion of particles in the grease before joining due to storage/environment	• The amount of grease during application must be defined • Limiting the operating temperature to the maximum permissible limit temperature of the grease (risk of resinification)	• Increasing the contact resistance depending on the amount of grease • Determining the amount of grease by assembly instructions or factory greasing • Limiting the operating temperature to the maximum permissible limit temperature of the lubricating grease (risk of resinification) [1, 36] • Remaining in contact during the service life must be ensured • Binding of foreign particles especially when plugging over or by particles from the environment

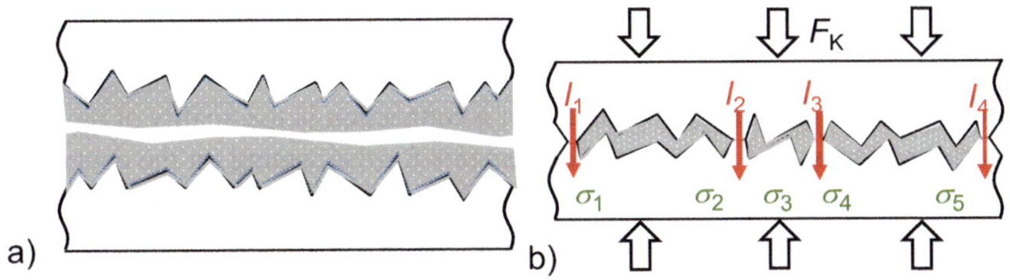

Fig. 6.18 **a**) greased contact surface before joining, **b**) closed connection with contact grease in the cavities and metallic contacts (schematic)

The best tribological behavior of plug-in connections is currently achieved with lubricating greases. In addition to the positive influences of lubricating greases on the operation of plug-in connections, there are numerous challenges that limit their use to certain applications, such as in gas-insulated switchgear. It should be noted that the grease film is applied only thinly to the contact surface in order not to increase the contact resistance too much. Depending on the application, the correct thickness may be so small that no closed lubricating film is visible to the naked eye. For this, an appropriate instruction should be created both when applying the grease in the factory and at the user's site. Furthermore, a lubricating grease must be qualified for the application. In particular, the thermal aging must be considered taking into account the required lifetime of the plug connection, as components of lubricating greases can harden. This has already been proven as a cause of failure [1]. If plug-in connections are operated and plugged in outdoors, a direct effect of rain and moisture as well as dust and dirt on the contact is possible. It is then conceivable that these influencing factors affect the mechanical and electrical properties of the lubricating grease. If not always the same plug and the same socket are used and these are additionally from different manufacturers, the use of lubricating grease is difficult. Film layers can then be transferred and lubricating greases mixed. The resulting effect on the resistance behavior has not yet been sufficiently investigated [47].

Lubricating greases influence the contact resistance depending on the speed of movement. At low speeds of movement, a metallic contact remains and the influence on the contact resistance is also low. However, the contact resistance changes during a movement cycle, as new metallic contacts are constantly being formed and separated [44]. At high speeds and high viscosity of the lubricant, hydrodynamic lubrication occurs without metallic contact. The frictional wear is then very low but the contact resistance is comparatively high (Fig. 5.27, p. 176). This condition should be avoided when a high current load is acting simultaneously with the movement (example: sliding contacts in a circuit breaker, especially under short-circuit load).

6.7 Examples

Three different connections are designed according to the steps 6.1 to 6.6.

6.7.1 Connection 1: Busbar to cable in a low-voltage cable distribution cabinet (LVCDC)

A cable introduced into an LVCDC with a conductor made of Al99.5 of class 2 and a busbar made of Cu-ETP are to be connected. The busbar has the dimensions (50 x 10) mm^2, the cable has a cross-section of 240 mm^2. The permissible limit temperature of the cable is 90 °C due to the used XLPE insulation. The cable outlet has a maximum continuous current of 500 A. In the event of a short circuit, a short-time current of 12.5 kA for 3 s with a peak short-circuit current of 25 kA can occur. In addition, a maximum operating force of $F_{\text{B max}} = 2$ kN can act on the cable.

6.7.1.1 Type of Connection

The main type of connection is selected according to Fig. 6.2:

- Assembly **with tool**s on site is acceptable
- **No relative movement** of the contact partners during operation: cable and busbar are static
- The connection should be **detachable** for maintenance and revision

Therefore, a **screwed connection** is necessary as the main connection. According to Fig. 6.3, it is now checked whether auxiliary connections are necessary. According to Table 6.1, they are not necessarily required for the connection of solid conductors (busbar) and stranded conductors (cable). In step b), it is checked whether the ends of the conductors can be adapted so that the screwed connection between the cable and busbar can be used directly. For this, a clamped connection would have to be made directly in the busbar (see Table 2.1). Since the busbar is an extruded rectangular profile, it is complex to implement the clamped connection directly and an auxiliary connection is needed. For this,the procedure according to Fig. 6.2 is applied again:

- Assembly **with tool** on site is acceptable
- **No relative movement** of the contact partners during operation: cable and busbar are static
- The auxiliary connection should be **permanent**, it is not necessary to open it non-destructively in the event of maintenance or revision.
- The assembly should take place **on site**

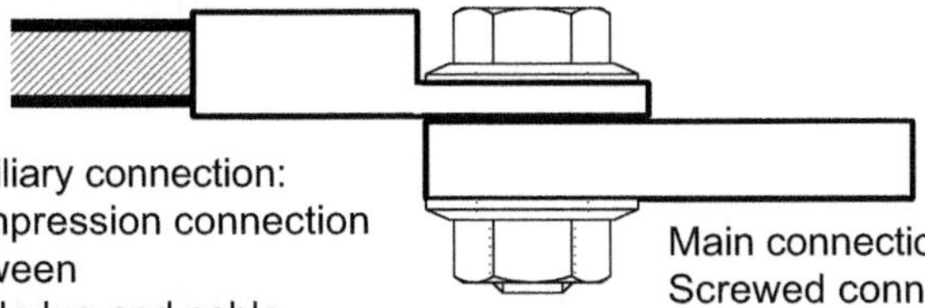

Fig. 6.20 Example connection 1 with the chosen types of connections

In this case, screwed and compression connections are permissible. In the example, we decide for a more cost-effective compression connection in the form of a cable lug, which is joined to the class 2 aluminum conductor via a hexagonal compression connection. The previously selected screwed connection as the main connection is used to connect the cable lug to the busbar (Fig. 6.20).

6.7.1.2 Materials

The conductor material of the busbar is Cu-ETP and that of the conductor in the cable is Al99.5 (stranded, class 2). The material of the cable lug must be chosen within the scope of the design. A cable lug made of Al99.5 is chosen, which is joined to the class 2 aluminium conductor via a compression connection. According to Fig. 6.2, a material pairing of copper and aluminium results between the cable lug and the busbar, which is not long-term stable. Therefore, the cable lug is tinned with an intermediate layer of nickel so that in the screwed connection tin and copper and in the compression connection aluminium and tin/aluminium contact. Both combinations are stable in their respective type of connection (Table 5.14). The flat contact surfaces of copper and aluminium must be pretreated before joining (Sect. 6.6.1). The XLPE insulation of the cable has a maximum permissible operating temperature of 90 °C. This is less than or equal to the permissible limit temperature of the conductor and contact materials used (Table 3.4). Thus, all material-technical prerequisites are fulfilled.

6.7.1.3 Design

The design of the screwed connection (main connection) and the compression connection (auxiliary connection) is described below.

1. Auxiliary connection: compression cable lug (Al99.5 with tinned surface)
   - For 240-mm^2 Al-conductor, a cable lug with a grease-filled sleeve, a connection tab width of 45 mm and a height of 10 mm is selected
   - Hexagonal form inserts with the code number 32 are used as crimping tool (Table 4.6, DIN 48083:1985)
2. Main connection: Screwed connection to the busbar (50 x 10) mm^2

- Connection length: $l_V/h_L \geq 4 \rightarrow l_V \geq 4 \cdot 10$ mm ≥ 40 mm (Fig. 4.58) $\rightarrow$ The length of the connection tab of the cable lug of 45 mm is chosen as the connection length l_V.
- Creation of a procedure for pretreating the contact surfaces of the contact partners (Sect. 6.6.1)
- Operating temperature $\leq$90 °C due to the permissible maximum temperature of the cable insulation
- Total thickness of the contact partners 10 mm + 10 mm = 20 mm > 5 mm $\rightarrow$ no spring loaded joining elements necessary; use of two washers of the normal series (Sect. 4.2.1)
- Determination of the initial value of the contact and connection resistance R_{K0}, R_{V0} and the replacement cross-sectional area A_{Ers} based on the reference conductor (Sect. 4.5.1) with the aim of generating equal power loss in the screw connection and the connected conductor $\rightarrow k_{u0} = 1$
  The resistance of the reference conductor R_{BL} is estimated from the average resistance of the connection tab of the aluminium cable lug and the copper busbar.

$$R_{BL} = \frac{1}{2} \cdot \left(\frac{1}{k_{CU}} \cdot \frac{l_V}{A_{Schiene}} + \frac{1}{k_{Al}} \cdot \frac{l_V}{A_{Lasche\ KB}} \right)$$

$$R_{BL} = \frac{1}{2} \cdot \left(\frac{1}{58\frac{MS}{m}} \cdot \frac{0.045m}{500mm^2} + \frac{1}{36\frac{MS}{m}} \cdot \frac{0.045m}{450mm^2} \right) = 2.16\,\mu\Omega$$

$$k_{U0} = \frac{R_{V0}}{R_{BL}} = 1 \rightarrow R_{V0} = 1 \cdot 2.16\,\mu\Omega = 2.16\,\mu\Omega \qquad \text{(Eq. 6.5)}$$

$$R_M = k_{uI}(A_B) \cdot R_{BL} = 0.73 \cdot 2.16\,\mu\Omega = 1.58\,\mu\Omega \qquad \text{(Eq. 6.6)}$$

$$R_{K0} = R_{V0} - R_M = 2.16 - 1.58\,\mu\Omega = 0.58\,\mu\Omega \qquad \text{(Eq. 6.7)}$$

$$A_{Ers} = \frac{\pi}{4}\left(20.6mm^2 - 13mm^2\right) + \frac{\pi}{8} 20.6mm \cdot 25mm\left[\left(\sqrt[3]{\frac{25mm \cdot 20.6mm}{(25mm + 20.6mm)^2}} + 1\right)^2 - 1\right]$$

$$A_{Ers} = 536\,mm^2 \qquad \text{(Eq. 6.8)}$$

Parameters for Sn-Cu: $k_V = 0.4 \cdot 10^{-5}$, $m_V = 0.59$ (Table 4.11)

$$F_{K0}(N) = \left(\frac{k_V}{R_{K0}(\Omega)}\right)^{1/m_V} \cdot A_{ers}\left(mm^2\right) = \left(\frac{0.4 \cdot 10^{-5}}{0.58 \cdot 10^{-6}\Omega}\right)^{1/0.59} \cdot 536\,mm^2 = 14144\,N$$

(see Eq. 6.9)

- Required contact/clamping force $F_{K0} \approx 14.1$ kN (Electrical design according to Fig. 6.6)
- Pretensioning force $F_V = 27.8$ kN, tightening torque $M_a = 55$ Nm according to VDI 2230 for a static operating force of $F_{B\,max} = 2$ kN and required clamping force of $F_K = 14.1$ kN (Type of load static, assembly with torque wrench, M12 hexagon socket screw of strength class 8.8, total friction value $\mu = 0.12$, mechanical design according to Fig. 6.5)

Resulting borehole size A_B from the calculation for the mechanical design: 13 mm (fine tolerance)

6.7.1.4 Continuous Operation

The required continuous current is 500 A. The XLPE insulated aluminum cable itself, with a conductor cross-section of 240 mm^2 when laid as a single conductor in air according to [18] (in [18] Table 13-45), is suitable for currents up to 600 A. Therefore, the required load current is uncritical.

The limit temperature of the cable is restricted to 90 °C. The limit temperature of the cable lug and the Cu-ETP busbar are higher. Therefore, for the design of continuous operation, the case "same temperature" applies (Sect. 6.4.1 and 6.4.2). This case is already fulfilled by the design according to $k_{u0} = 1$.

6.7.1.5 Short-term load

Main Connection: Busbar-Cable Lug

In the case of short-term current load, the voltage drop across the contact at the maximum amplitude in the first period and at the end of the short-term current is considered to estimate the maximum temperature rise in the contact according to Eq. (4.55). The aim is to consider the most critical case. The voltage drop across the contact between the cable lug and busbar at the moment of short-term current entry for the peak value of the current of $i_p = 25$ kA, under the simplified assumption that the initial temperature of the conductors is 20 °C, is:

$$u_K = R_{K0} \cdot i_p = 0.58\,\mu\Omega \cdot 25\,\text{kA} = 14.5\,\text{mV}$$

This voltage drop leads to an overtemperature in the contact of about 4 K (Eq. 4.55) compared to the connected conductors. Due to the small difference in temperature between the contact and the connected conductors, an exact calculation of the overtemperature according to Eq. (4.57) is omitted.

The thermally effective short-time current of $I_k = 12.5$ kA for $t_k = 3$ s leads to a heating of about 13 K in the unaffected Cu-ETP busbar and a heating of about 144 K in the unaffected Al99.5 cable (Eq. 7.2) with a simplistically assumed initial temperature of both conductors of 90 °C (maximum permissible continuous temperature of the cable).

The temperature at the connection is assumed to be the average temperature of the busbar and the cable. This results in a voltage drop across the contact at the end of the short-time current load, taking into account the temperature coefficient according to (Eq. 4.56) of:

$$u_K = R_{K0} \cdot (\vartheta) \cdot \sqrt{2} \cdot I_K$$
$$u_K = 0.58\,\mu\Omega \cdot \left(1 + 0.0039 \cdot \left(\left(\frac{234^\circ C + 103^\circ C}{2}\right) - 20^\circ C\right)\right) \cdot \sqrt{2} \cdot 12.5\text{kA} = 16.2\text{mV}$$

It becomes clear that the maximum voltage drop of 16.2 mV across the contact is reached at the end of the short-time current. At this voltage drop, the maximum overtemperature in the contact is about 3 K (Eq. 4.55). For both cases, it can be assumed that the maximum temperature in the contact approximately corresponds to the average conductor temperature. Since the cable does not reach 90 °C at a maximum operating current of 500 A and the short-time current annealing temperature for the conductor materials copper and aluminum is not exceeded (Table 3.4), this load is considered uncritical for screwed connections.

Auxiliary Connection: Cable Lug-Cable

The connection is designed for cables made of aluminum or aluminum alloy with a cross-section of 240 mm^2 and is short-circuit proof (Class A). Further data are not known from the test according to standard. The short-circuit current to be safely carried depends on the short-circuit duration and is to be determined. The crimp connection for conductors of low or medium voltage cables is considered short-circuit proof if it has passed the short-circuit test. According to the DIN EN IEC 61238-1-1:2020 standard series, the connection is to be tested with short-circuit currents that lead to a temperature of (250 … 270) °C on the reference conductor. The maximum permissible test duration is 1.5 s. Since, according to the explanations in Sect. 6.5.3, a long test duration corresponds to a "soft" test, this value is assumed. According to Eq. (7.2), a current of at least $I_{K\,r} =$ 23 kA with a short-circuit duration of $t_{K\,r} = 1.5$ s is necessary to heat the conductor from 25 °C to 250 °C. The shock factor κ during the test is not known. At a minimum, $\kappa = 1$ must be assumed. The tested peak current is therefore at least:

$$i_{p\,r} = \sqrt{2} \cdot \kappa \cdot I_{K\,r} = \sqrt{2} \cdot \kappa \cdot 23\,\text{kA} = 32.5\,\text{kA}$$

The compression connection can safely carry a short-circuit current of $I_{K\,r} = 23$ kA with a short-circuit duration of $t_{K\,r} = 1.5$ s. The peak current may be a maximum of 32.5 kA. Thus, the requirement of $i_p = 25$ kA in the application is met.

Since the exponent x of the $I^x t$ criterion is unknown, the same effective and peak current value applies for short current flow durations (Table 6.5). For longer current flow durations, x = 2 can be assumed. This reduces I_K accordingly for a short circuit duration of $t_K = 3$ s to

$$I_{\mathrm{K\,r}}(t_{\mathrm{K}} = 3\,\mathrm{s}) = 23\,\mathrm{kA}\sqrt{\frac{1.5\,\mathrm{s}}{3\,\mathrm{s}}} = 16.3\,\mathrm{kA}.$$

This current is higher than the short-circuit current of $I_k = 12.5$ kA required for a duration of $t_k = 3$ s in the application. Thus, all requirements in practical application are met by the test according to the standard.

6.7.1.6 Pretreatment and Assembly

Create a procedure for the pretreatment of the contact surfaces made of aluminum and copper on the contact partners and the contact element (Sect. 6.6.1). The following points should be included

- Pretreatment:
  - Copper busbar: Degrease with ethanol, microstructure the contact surface with a abrasive fleece or a wire brush
  - Tinned aluminum cable lug: Degrease with ethanol
- Tightening torque: 70 Nm with calibrated torque wrench

6.7.2 Connection 2: Two identical overhead line conductors

A high-temperature overhead line conductor ACSS/TW-265/43 with a diameter of 20.9 mm consists of annealed, trapezoidal aluminum wires with a total cross-sectional area of A = 265 mm^2 and a high-strength steel core with a cross-section of $A_K = 43$ mm^2. Since the overhead line conductor cannot be manufactured in unlimited length, it must be connected in the tension field with an overhead line conductor of the same type, both tensile and non-tensile in current loops. The conductor has a rated current of 1140 A at a permissible rated temperature of 200 °C and a length-related resistance of 103.5 μΩ/m. The conductor is to be used in the 420-kV high-voltage network as a triple bundle, with the maximum continuous current of a circuit being 3150 A. With evenly distributed operating current in the triple bundle, the thermal dimensioning has resulted in a highest operating temperature of 167 °C. In the event of a short circuit, a short-time current of 50 kA can occur for 1 s with a surge short-circuit current of 125 kA.

Fig. 6.21 Example Connection 2: Current-carrying compression connections realized with a compression connector with two identical overhead line conductors

6.7.2.1 Type of Connection

The procedure according to Fig. 6.2 is applied:

- Assembly **with tools** on site is acceptable
- **No relative movement** of the two conductors in the connection during operation
- The connection should be **permanent**, i.e., it is not necessary to open it non-destructively in the event of maintenance or revision.
- The assembly should take place **on site**

In this case, screwed and compression connections are permissible for the current-carrying connection. Since a large number of such connections are to be installed, the tools are established and additional mechanical requirements exist, we opt for the more cost-effective **compression connection**. According to Fig. 6.3, no further auxiliary connection is necessary for the current-carrying connection of two stranded conductors with the compression connection (Fig. 6.21).

6.7.2.2 Materials

Pure aluminium or aluminium alloy materials are suitable for current-carrying hexagonal compression connections with high-temperature overhead line conductors (Table 5.14). When selecting the material for the fitting of the current-carrying connection with a high-temperature overhead line conductors, which is to be joined by hexagonal compression, the aluminium material of the conductor must first be taken into account. In the chosen example, pure, soft aluminium is used for the wires of the outer layers. Due to the high temperature resistance of the steel material for the core, operating temperatures up to 200 °C are permitted for the ACSS conductor. For conventional overhead line conductors with wires made of technically pure hard-drawn aluminium (A199.5), an aluminium alloy with medium mechanical strength can be used for the fitting, for example. To connect HTLS conductors with wires made of soft aluminium, aluminium materials of the fitting with higher mechanical strength are required in order to apply high forces to the conductor already during joining and thus achieve good contact behaviour. The different mechanical strengths of conductor and fitting allow a high press force to be specifically introduced into the conductor during joining, and the remaining contact force, which is established by the elastic spring-back of the conductor, can be optimised. For this reason, the two aluminium materials EN AW-6061 and EN AW-6082 are considered for the compression sleeve in the example considered here with overhead line conductors made of annealed aluminium wires. When dimensioning the fittings, the limit temperature of the materials of up to 150 °C for compression connections must be taken into account [24]. For the material of the fitting, it must be ensured that the required degree of deformation $\varphi_u \approx 0.2 \ldots 0.3$ can be achieved without the material failing locally (e.g. cracking). The use of the fittings under open-air conditions also requires that the material of the fitting has a high corrosion resistance. This is the case with both aluminium materials. For the purpose of corrosion protection, a suitable temperature-resistant contact grease should be used in the compression connection.

6.7.2.3 Design

An optimized current-carrying hexagonal compression connection with the compression dies of size number 34 according to DIN 48083-4:1985 is to be designed. The two aluminum materials EN AW-6061 and EN AW-6082 are considered for the crimp sleeve. The degrees of freedom of the design are the inner and outer diameter of the crimp sleeves. The compression factor K_F and the fit factor p are considered as auxiliary variables.

The contact force during the joining of the connections and the remaining contact force can only be determined with FEM calculations. Depending on the dimensions of the crimp sleeves, the optimal contact behavior was determined experimentally (Fig. 6.22) [24]. It turns out that the optimized crimp connection should have a compression factor greater than 1.1 and a fit factor greater than 0.9. Hexagonal compresssion connections with the material EN AW-6061 have slightly better contact behavior than those with crimp sleeves made of EN AW-6082 with the same design. With the same design of the connection, the force input into the conductor when pressing connections with crimp sleeves made of EN AW-6061 is higher [24]. This material is therefore to be preferred. First, the inverse exponential constant of the electrical current distribution α_{el} is calculated with the specific material resistance of the crimp sleeve of 66.3 $\mu\Omega$/m and a highest specific transverse resistance of 0.2 $\mu\Omega$m (results of the preliminary tests on contact behavior (Fig. 6.22)) with Eq. (4.33):

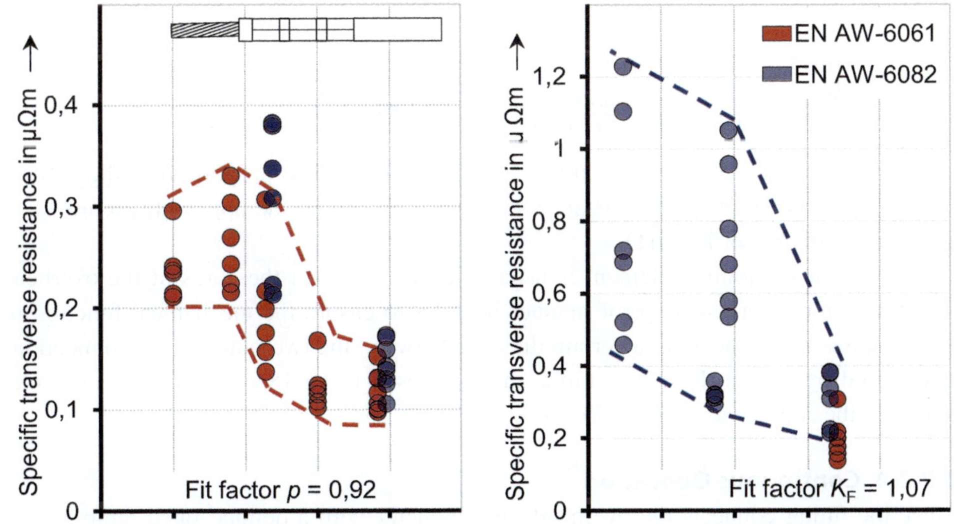

Fig. 6.22 Specific transverse resistance depending on compression and fit factor of hexagonal compression connections with the ACSS/TW-265/43 overhead line conductor

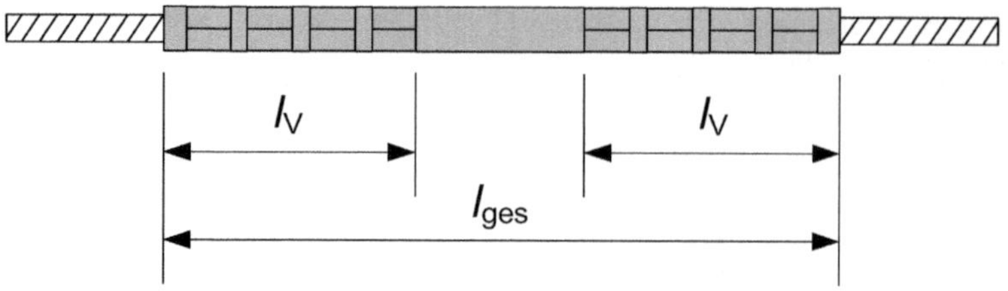

Fig. 6.23 Example Connection 2: Dimensions of the compression conection

$$\alpha_{el} = \sqrt{\frac{103.5\,\mu\Omega/\mathrm{m} + 66.3\,\mu\Omega/\mathrm{m}}{0.2\,\mu\Omega m}} = 29.14\frac{1}{\mathrm{m}}$$

The length of the compression connection can then be determined with Eq. (4.49):

$$l_V > 4/(29,14/\mathrm{m})$$
$$l_V > 13.7\,\mathrm{cm}$$

If a typically 4 cm wide press insert of the characteristic number 34 is used, at least four hexagonal compressions are required. The connector to be designed in the example consists of two compression connections to the two overhead lines conductor with a connection length of $l_V = 16$ cm each (Fig. 6.23).

With Eq. (4.32), a connection resistance R_{V0} of

$$R_{V0} = \frac{103.5\ \mu\Omega/\mathrm{m}}{103.5\ \mu\Omega/\mathrm{m}+66.3\,\mu\Omega/\mathrm{m}} \left[66.3\,\mu\Omega/\mathrm{m}\ 0.16\ \mathrm{m} + \frac{(103.5\ \mu\Omega/\mathrm{m}-66.3\ \mu\Omega/\mathrm{m})}{29.14\,/\,\mathrm{m}} \tanh\left(\frac{29.14/\mathrm{m}\,0.16\ \mathrm{m}}{2}\right)\right] + \frac{103.5\,\mu\Omega/\mathrm{m}+66.3\,\mu\Omega/\mathrm{m}\ \cosh(29.14/\mathrm{m}\,0.16\,\mathrm{m})}{29.14\,/\,\mathrm{m}\ \sinh(29.14/\mathrm{m}\,0.16\,\mathrm{m})} = 9.57\,\mu\Omega$$

is obtained for each of the two compression connections. The proportion of the contact resistance to the connection resistance was calculated using the approach according to Eq. (4.37) with $R_{K0} = 1.68\ \mu\Omega$.

Since a further joining element is usually used to connect the cores of the overhead line conductors in a tensile-prrof manner in order to ensure the mechanical function of the compression connection, a certain distance between the two outer crimp connections is required by design. In the example, this is chosen to be 18 cm, resulting in a total length of the connector of $l_{ges} = 50$ cm.

6.7.2.4 Continuous Operation

Since the fitting connects the overhead lineconductor with a defined rated temperature, the connection temperature is a result of the temperature of the unaffected conductor and the thermal design of the fitting. The limit temperature for the aluminium materials EN AW-6061 and EN AW-6082 for compression connections with high-temperature conductor cables was determined in [24] to be 150 °C. This must be taken into account when thermally dimensioning the fitting. In the example, the highest operating temperature of the unaffected conductor cable is 167 °C. For the 50 cm long connector, a calculation with a thermal network resulted in a highest operating temperature of the connection of 115 °C at maximum temperature of the overhead line conductor. This ensures that the permissible limit temperature of the material for the sleeve is safely maintained.

If the limit temperature is exceeded, the heat-emitting surface must be increased, if necessary, to maintain the limit temperature. An impact on the previous considerations for design must be taken into account. If the limit temperature cannot be maintained, a different material with a higher permissible limit temperature should be chosen.

6.7.2.5 Short-term Load

The load capacity of the fittings with a short-term current is to be tested according to IEC 61284:1997 only for non-tensile fittings of conventional overhead line conductor. So far, this standard does not define any test requirements for the current-carrying fittings of HTLS overhead line conductor. However, application-related test criteria can be determined by using the highest calculated temperature of the HTLS overhead line conductor in the event of a short circuit as a basis. The following estimate is intended to show that it is necessary to carry out a test with a short-term current.

The continuous short-circuit current of the considered overhead line conductor is 50 kA and the maximum transient short-circuit current is 125 kA. Initially, the continuous short-circuit current is considered. With evenly distributed current in the three-bundle, a load of 16.67 kA per overhead line conductor results. The temperature of the overhead line conductor after one second of short-term current load at a starting temperature of 25 °C is according to Eq. (7.2):

$$\vartheta_{\mathrm{KL}} = \frac{1 + 0.004/\mathrm{K}(25^{\circ}\mathrm{C} - 20^{\circ}\mathrm{C})}{0.004/\mathrm{K}} \left(e^{\frac{1.01\,(16667\,\mathrm{A})^2\,0.004/\mathrm{K}\,1\,s}{36\,\mathrm{MS/m}\,910\,\mathrm{J/(kg\,K)}\,2700\,\mathrm{kg/m^3}\,\left(0.000265\mathrm{m}^2\right)^2}} - 1 \right) + 25^{\circ}\mathrm{C} = 75.3^{\circ}\mathrm{C}$$

The temperature of the compression connection is simplistically assumed to be the same as that of the overhead line conductors after the short-term current load. The contact voltage at this load per conductor is then:

$$u_{\mathrm{K}} = R_{\mathrm{K0}}(\vartheta) \cdot \sqrt{2} \cdot I_{\mathrm{K}}$$
$$u_{\mathrm{K}} = 1.68\ \mu\Omega \cdot \left[1 + 0.004 \cdot \left(75.3^{\circ}\mathrm{C} - 20^{\circ}\mathrm{C}\right)\right] \cdot \sqrt{2} \cdot 16.67\mathrm{kA} = 48.4\mathrm{mV}$$

so that with simplistically assumed symmetric heating of the contact partners, a temperature about 36 K higher in the contact is to be expected (Eq. 4.55):

$$\vartheta_{\mathrm{K}} = \sqrt{\frac{(0.0484\mathrm{V})^2}{4 \cdot 2.2 \cdot 10^{-8}\mathrm{V}^2/\mathrm{K}^2} + (75.3^{\circ}\mathrm{C} + 273\,\mathrm{K})^2} - 273\,\mathrm{K} = 111.6^{\circ}\mathrm{C}$$

The resulting contact temperature is uncritical, as it is lower than the temperature in continuous operation.

Assuming the highest operating temperature of the overhead line conductors of 167 °C before the short circuit, a maximum temperature on the overhead line conductors after the short circuit results in:

$$\vartheta_K = \frac{1{+}0.004/\text{K}\left(167^\circ\text{C}{-}20^\circ\text{C}\right)}{0.004/\text{K}}\left(e^{\frac{1.01\,(16667\,\text{A})^2\,0.004/\text{K}\,1\,s}{36\,\text{MS/m}\,910\,\text{J/(kg K)}2700\,\text{kg/m}^3\,\left(0.000265\text{m}^2\right)^2}}-1\right)+167^\circ\text{C}$$

$$= 245.6^\circ\text{C}$$

At the time of the short circuit, the contact voltage due to the effect of the peak current per conductor is

$$u_K = R_{K0}(\vartheta)\cdot i_p$$

$$u_K = 1.68\ \mu\Omega\cdot\left[1+0.004\cdot\left(167^\circ\text{C}-20^\circ\text{C}\right)\right]\cdot 41.67\text{kA} = 111,2\text{mV}$$

Thus, simplifying and neglecting the resistance increase due to internal heating in the contact (Eq. 4.54), a contact temperature about 138 K higher results in

$$\vartheta_K = \sqrt{\frac{(0.1112\text{V})^2}{4\cdot 2,2\cdot 10^{-8}\text{V}^2/\text{K}^2}+\left(167^\circ\text{C}+273\,\text{K}\right)^2}-273\,\text{K} = 305^\circ\text{C}.$$

Just before the end of the steady-state short-circuit current, a contact voltage of

$$u_K = R_{K0}(\vartheta)\cdot\sqrt{2}\cdot I_K$$

$$u_K = 1.68\ \mu\Omega\cdot\left[1+0.004\cdot\left(245.6^\circ\text{C}-20^\circ\text{C}\right)\right]\cdot\sqrt{2}\cdot 16.67\text{kA} = 75.3\text{mV}$$

is reached. A comparable contact temperature thus results, simplified and neglecting the resistance increase due to internal heating in the contact (Eq. 4.54) of

$$\vartheta_K = \sqrt{\frac{(0.0753\text{V})^2}{4\cdot 2.2\cdot 10^{-8}\text{V}^2/\text{K}^2}+(245.6^\circ\text{C}+273\,\text{K})^2}-273\,\text{K} = 304.4^\circ\text{C}.$$

Since the thermal load on the compression connection and the contacts in the event of a short circuit is higher than in continuous operation, it is recommended to test the behavior of the compression connection under load with a short-term current. Test criteria can be sensibly oriented based on the highest temperature in the event of a short circuit of 245.6 °C, by setting the test current I_K accordingly for a duration of 1 s as follows:

$$I_K =$$

$$\sqrt{\frac{36\,\text{MS/m}\,910\,\text{J/(kg K)}2700\,\text{kg/m}^3\left(0.000265\text{m}^2\right)^2}{1.01\;0.004/\text{K}\;1.0\,\text{s}}\ln\frac{1{+}0.004/\text{K}\left(245.6^\circ\text{C}{-}20^\circ\text{C}\right)}{1{+}0.004/\text{K}\left(30^\circ\text{C}{-}20^\circ\text{C}\right)}}$$

$$I_K = 30471\,\text{A}$$

The test of the current-carrying fittings must therefore be carried out with $I_k = 30.5$ kA for $t_k = 1$ s. This also exceeds the specified peak short-circuit current of 41.7 kA ($i_p \geq$ 43.1 kA) and a separate peak short-circuit current test is not necessary. To avoid excessive mechanical stress during the short-term current test due to a too high peak current value, a symmetrical current ($\varkappa = 1$) should be used in the test with $I_k = 30.5$ kA. Further boundary conditions of the test and the basics for evaluating the results can be carried out according to IEC 61284:1997.

6.7.2.6 Pretreatment and Assembly

Before assembly, film layers, especially on the outer layer of the aluminum conductor cables, must be removed. For this purpose, they should be treated abrasively with a wire brush to break up the oxide layer on the aluminum. In addition, it is also recommended to pretreat the soft aluminum wires of the inner layers. To do this, they need to be exposed, also break up the oxide layers abrasively with a wire brush, and then compact the individual wires of the conductor again. The quickly forming thin oxide layers can be broken up by the elastoplastic deformation during pressing (see Sect. 6.3.2). Additional cleaning of the crimp sleeve is usually not possible, as it is already filled with grease for corrosion protection. During compression, it is particularly important to ensure that the compression inserts always close completely. If this is not achieved, the compression sleeve and conductor are only incompletely deformed and lower compression and contact forces as well as smaller contact areas are produced. The friction between the compression dies and the sleeve can be reduced, for example, by wetting with a viscous liquid. This also reduces the required compression force and facilitates the closing of the compression inserts.

6.7.3 Connection 3: Movable Contact Piece to Support Element in a High-Voltage Switching Device

A high-voltage switching device (circuit breaker) has a movable contact piece, which essentially takes the form of a cylindrical conductor tube with an outer diameter of 50 mm. At the end of the movable contact piece, the arcing contact and the rated current contact system are integrated, which technically realize the actual switching process through movement. The arcing contact and the rated current contact system are not considered in this example. In the example, the cylindrical body (outer diameter 50 mm) of the movable contact piece is to be connected with the stationary support element, which is part of the main current path. The support element is also cylindrical with an inner diameter of 55 mm. Furthermore, the following requirements are known:

- Carrying of the rated and short-circuit current (Assumption: $I_n = 2000$ A and $I_K = 50$ kA for 1 s as well as peak current $i_p = 125$ kA)
- Sliding path when moving the contact piece is 100 mm
- Lifespan: 2000 cycles,

- Low movable weight (drive power)
- Ambient medium: Assembly in normal air, operation in insulating gas sulfur hexafluoride SF_6

6.7.3.1 Type of Connection

To select the type of connection, the procedure according to Fig. 6.2 is applied again:

- Assembly **with tool** in the production process of the switching device
- Relative movement of the contact partners occurs during operation (switching)
- A complete separation of the contact partners does not occur (arcing contact system and the rated current contact system are onsidered separately)
  → **Sliding connection**

According to Fig. 6.3, no auxiliary connection is necessary for the connection of two solid conductors with a plug connection.

6.7.3.2 Materials

In principle, aluminum and copper materials are possible as the base material to carry a high continuous current. For reasons of economy and to allow a low mass of the movable contact with sufficient heat resistance up to a temperature of 115 °C, an aluminum alloy is advantageous (Sect. 3.1), we choose AlMgSi1 in the example. A contact element is needed to establish the contact between the plug and socket. To avoid oxidation, this is silver-plated galvanically. Therefore, to ensure a long-term stable contact, both the plug and socket must also be silver-plated (Sect. 3.2and 5.2). If the connection is used in an oxygen-containing environment, a diffusion barrier between silver and the aluminum base material must be applied to prevent the diffusion of oxygen through the coating to the aluminum (Sect. 5.1.2). Typical for this are (2…4) μm nickel or copper, which is applied galvanically. This intermediate layer is usually used anyway in the galvanic silver plating of aluminum to improve adhesion.

The thickness of the silver top layer can be defined based on the sliding path (Sect. 5.1.4). The following boundary conditions are assumed:

- Contact force per contact point of the contact element: 5 N
- Temperature of the switch during the plugging process: 55 °C
- Contact hardness: 300 N/mm^2(soft silver)
- Point contact with contact radius 3 mm
- Specific friction wear coefficient unlubricated $1.5 \cdot 10^{-8}$ $mm^3/(N\ mm)$; lubricated $1.5 \cdot 10^{-9} \dots 6.3 \cdot 10^{-11} mm^3/(N\ mm)$

Mathematically,this results in a minimum layer thickness of 55 μm for the unlubricated galvanic silver coating for a total friction path of 400 m (2000 cycles x 100 mm friction path x 2). Such thick silver coatings are expensive, require a long galvanic process, and are therefore almost never used in practice. The friction wear can be greatly reduced by using grease. With a lubricated surface of the plug and socket—depending on the assumed test conditions—a layer thickness of 2.9 μm (high friction speed) over 4.9 μm (medium friction speed) up to 17 μm (very slow friction process) is necessary. Since circuit breakers have a high plug and pull speed, a design with at least 6 μm of galvanic silver layer and the use of a suitable contact lubricant is sensible.

6.7.3.3 Design

Geometry of the Connection

A contact lamella is used as the contact element. The properties of the contact element are specified by the manufacturer (Sect. 6.7.3.2).

To contact the movable contact piece with a 50 mm outer diameter, a mean socket diameter of approximately 63 mm is determined for installation in the carrier element based on the manufacturer's specifications. This is achieved by incorporating a 4 mm deep recess into the carrier element with a 55 mm inner diameter, into which the contact element is then inserted. (Figure 6.24). This results in a number of 39 period units for the contact spacing of the specific contact element.a grid dimension

Friction Wear

According to the manufacturer's specifications, up to 10,000 plug-in cycles and a sliding path of 6600 m are possible with the contact element. In the previous design step Sect. 6.7.3.2, it was already calculated that the maximum sliding path is approximately 400 m. The number of required plug cycles is 2000, which is less than the achievable cycle number. Therefore, reserves are available that allow longer operation without maintenance.

6.7.3.4 Continuous Operation

The carrier element and movable contact piece in the insulating gas SF_6 have a sufficient conductor cross-section to carry the continuous current of 2000 A without exceeding the

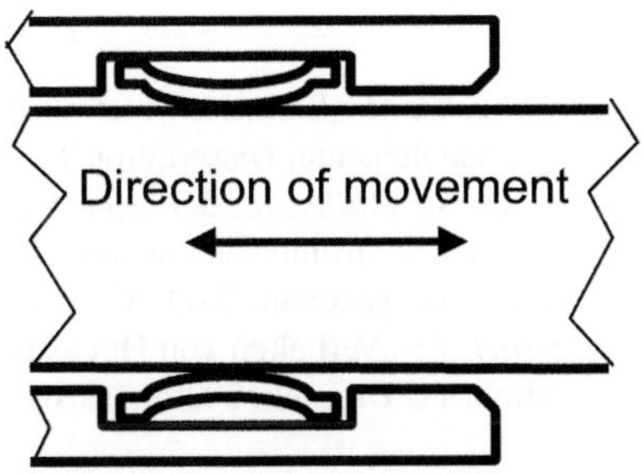

Fig. 6.24 Example Connection 3: movable contact piece to support element in a high-voltage switching device

limit temperature of 115 °C. In heating tests with comparable conductor arrangements, temperatures of approximately 85 °C were determined.

It is now checked whether the voltage drop across the contact element in continuous operation is <18 mV. A period unit has a through resistance of 0.105 mΩ, which results in a total through resistance of 2.7 μΩ. The voltage drop at rated current 2000 A is therefore approximately 5.4 mV. The criterion is thus met. Since the unaffected conductors at 85 °C are significantly cooler than the limit temperature of 115 °C, the additional power loss through the contact system can be absorbed without exceeding the limit temperature.

6.7.3.5 Short-Term Load

Based on the rated values of a period unit (valid for galvanically silver-plated copper):

- 1-s short-circuit current 2200 A
- 3-s short-circuit current 1700 A
- peak short-circuit current 5600 A

considering a derating factor for aluminum base material (Sect. 4.5.3) of 0.8 and 39 period units, a rated short-circuit current of

- 1-s short-circuit current: $I_{K\,1s} = 39 \cdot 0.8 \cdot 2200\ \text{A} = 68.6\ \text{kA}$
- 3-s short-circuit current: $I_{K\,3s} = 39 \cdot 0.8 \cdot 1700\ \text{A} = 53.0\ \text{kA}$
- Shock short-circuit current $i_p = 39 \cdot 0.8 \cdot 5600\ \text{A} = 174.7\ \text{kA}$.

The required short-circuit carrying capacity ($I_K = 50$ kA for 1 s and shock current $i_p =$ 125 kA) is thus achieved.

6.7.3.6 Pretreatment and Assembly

To achieve the required plug cycles, lubrication is necessary (see Sect. 6.3.3). The lubricant recommended by the manufacturer should be used. Before applying the lubricant as a thin film on the conductor tubes, these should be degreased with ethanol or similar. The application should be carried out by trained personnel.

References

1. *Gatzsche, M.*: Elektrisch-thermisches Betriebs- und Langzeitverhalten hochstromtragfä-higer Kontaktelemente. Dissertation TU Dresden, Shaker Verlag, 2016. – ISBN 978-3-84404-894-0
2. *Israel, T.; Müller, R.; Schlegel, S.; Schmidt, H.; Ludwig, M.:* Untersuchung von Steckverbindungen bei Strombelastung im Milli- bis Mikrosekunden-Bereich. Elektrische und Optische Verbindungstechnik 2021, VDE-Verlag. 8. Symposium Connectors, Lemgo, 23.-24.3.2021.
3. *Israel, T.:* „Verhalten von Hochstrom-Steckverbindungen bei kurzer Strombelastung". Dissertation, TU Dresden, 2020. – ISBN 978-3-86780-653-4.

4. *Israel, T.;* Schlegel, S.; Großmann, S.; Kufner, T.; Freudiger, G.: Recommendations for Testing the Short Circuit Behaviour of Power Plug-In Connectors. 30th International Conference on Electrical Contacts, Rohrschach, CH, 2020.
5. *Israel, T.; Oberst, M.; Hildmann, C.;* Schlegel, S.: Selbstschmierende Beschichtungen für Hochstrom-Steckverbindungen. 26. Albert-Keil-Kontaktseminar, Karlsruhe, 30.3-1.4.2022.
6. *Oberst, M.; Israel, T.; Hildmann, C.; Schlegel, S.; Egetenmeyer, A.-K.; Kappl, H.; Willing, H.:* Elektrische und tribologische Eigenschaften von Silber-Dispersionsschichten für die Anwendung in Steckverbindungen. Elektrische und Optische Verbindungstechnik 2021, VDE-Verlag. Gehalten auf dem 8. Symposium Connectors, Lemgo, 23.-24.3.2021.
7. *Greiner, F. et al.:* Hochbeständige, physikalisch abgeschiedene Kontaktoberfläche auf Silberbasis für die Anwendung in Buchsenkontakten. Elektrische und Optische Verbindungstechnik 2021, VDE-Verlag. Gehalten auf dem 8. Symposium Connectors, Lemgo, 23.-24.3.2021.
8. *Yuan, H., Song, J.*: Fretting behavior of tin plated contacts under various lubrications. Elektrische und Optische Verbindungstechnik 2021, VDE-Verlag. Gehalten auf dem 8. Symposium Connectors, Lemgo, 23.-24.3.2021.
9. EFB-Forschungsbericht zum BMWI/IGF-Projekt 16952 BR: Elektrisches Eigenschaftsprofil umformtechnischer Fügeverbindungen. Europäische Forschungsgesellschaft für Blechverarbeitung e.V. (EFB-Nr. 389); Auflage: 1, 2014. – ISBN 978-3-86776-432-2
10. EFB-Forschungsbericht zum BMWI/IGF-Projekt 18617 BR: Optimierung umformtechnischer Fügeverfahren zur Kontaktierung elektrischer Leiter. Europäische Forschungsgesellschaft für Blechverarbeitung e.V. (EFB-Nr. 40/214); Auflage: 1, 2019. – ISBN 978-3-86776-559-6
11. *Ramonat, A.*: Untersuchungen zum elektrischen Kontakt- und Langzeitverhalten von Fügetechnologien mit zylindrischen Leitern aus Aluminium. Dissertation, TU Dresden, 2018. – ISBN: 978-3-86780-604-6
12. *Bäuml, K.*: Fügetechnologien für stromführende Verbindungen aus normal- und supralei-tenden Werkstoffen für den Einsatz bei tiefen Temperaturen. Dissertation, TU Dresden, 2021.
13. *Rühl, M.*: Prozessmodellierung von Reaktiv-Multischicht-Systemen (RMS). Dissertation, TU Dresden, 2015. – https://nbn-resolving.org/urn:nbn:de:bsz:14-qucosa-198818
14. *Reisgen, U.; Olschok, S.; Jakobs, S.; Holtum, N.*: Influence of the degree of dilution with laser beam vacuum-welded Cu-Al mixed joints on the electrical properties. Procedia CIRP, Volume 74, 2018, Pages 23-26. – ISSN 2212-8271
15. *Stritt, P.; Hagenloch, C.; Kizler, C.; et al.*: Laser spot welding of copper-aluminum joints using a pulsed dual wavelength laser at 532 and 1064 nm. Physics Procedia, Volume 56, 2014, Pages 759-767. – ISSN 1875-3892
16. Patentschrift DE 10 2013 009 234 B4 2019.10.02 – Verfahren zur Herstellung elektrisch leitender Verbindungen zwischen Fügepartnern sowie eine Verwendung eines Polymers oder Polymergemischs.
17. *Langer, M.; Jansen, I.; Beyer, E.*: Untersuchungen zum Kleben metallischer Leiter zur Übertragung elektrischer Energie am Beispiel von Lithium-Ionen-Zellen. Doktorandenseminar Klebtechnik, DVS Media, DVS-Berichte 313, 2016. – ISBN 978-3-945023-54-9
18. *Kämpfer S., Kopatsch G., ABB AG (Hrsg.)*: Schaltanlagen-Handbuch, 12. Aufl., Cornelsen, 2011. – ISBN 978-3-06450-726-5
19. VDE 0151 (Ausg. 1986-06): Werkstoffe und Mindestmaße von Erdern bezüglich der Korrosion.
20. *Kloos, K.-H., Thomala, W.*: Schraubenverbindungen – Grundlagen, Berechnung, Eigenschaften, Handhabung. 5. Aufl., Berlin, Heidelberg. 2007. – ISBN 978-3-540-21282-9
21. VDI-Richtlinie 2230: Systematische Berechnung hochbeanspruchter Schraubenverbindungen – Zylindrische Einschraubenverbindungen, 2003.

22. DIN 43673-1 (Ausg. 1982-02): Stromschienen-Bohrungen und -Verschraubungen; Stromschienen mit Rechteck-Querschnitt.
23. DIN 43673-1 (Ausg. 1971-06): Anschlüsse für elektrische Betriebsmittel Flachanschlüsse ≥ 40 A Hauptmaße und Zuordnung.
24. *Hildmann, C.*: Zum elektrischen Kontakt- und Langzeitverhalten von Pressverbindungen mit konventionellen und Hochtemperatur-Leiterseilen mit geringem Durchhang. Dissertation, TU Dresden 2016. – ISBN 978-3-00-056211-2.
25. DIN 7190-1 (Ausg. 2017-02) Pressverbände –Teil 1: Berechnungsgrundlagen und Gestaltungsregeln für zylindrische Pressverbände.
26. *Timsit, R. S.:* Formation of Tubular Crimp Connections: Elementary Considerations. Tagungsbd. 54th IEEE Holm Conference on Electrical Contacts. Orlando, 27.10.-29.10.2008, S. 161-167.
27. *Salewski, A.*: Grundlagenuntersuchungen zur Herstellung und zu dem Betriebsverhalten elektrisch leitender Umfangpreßverbindungen. Dissertation, Technische Hochschule Leipzig, 1985.
28. *Israel, T. et al.*: Zum elektrisch-thermisch-mechanischen Verhalten von silberbeschichte-ten Hochstrom-Steckverbindern. Das I^2t-Kriterium bei der Kurzschlussprüfung Teil II. Elektrische und Optische Verbindungstechnik, 7. Symposium Connectors, Lemgo, 19.-20.3.2019
29. *Novak, B.*: Electrodynamic Forces in Circular Contact Systems. 28th International Conference on Electric Contacts, Edinburgh, UK, 2016.
30. *Böhme, H.*: Mittelspannungstechnik-Schaltanlagen berechnen und entwerfen.2. stark bearbeitete Auflage Berlin: Verlag Technik 2005. – ISBN 3-341-01495-0
31. *Bergmann, R.*: Zum Langzeitverhalten des Widerstands elektrischer Stromschienenverbindungen. Dissertation, TU Dresden, Fortschritt-Berichte VDI Nr. 195. VDI-Verlag, 1996.- ISBN 3-18-319521-6
32. *Braunovic, M.*: Aluminum Connections: Legacies of the Past. 39th IEEE Holm Conference on Electrical Contacts, Chicago, 17.-19.10.1994.
33. *Großmann, S.; Löbl, H.; Schlegel, S.*: Erfahrungen zum Langzeitverhalten von Aluminium-Verbindungen in Starkstromanlagen. 21. Albert Keil Kontaktseminar, Karlsruhe, 2011.
34. *Öberg, Å.; Bohlin, Å; Olsson, K.-E.*: The influence of contact surface preparation on the performance of copper and aluminium connectors. 16th International Conference on Electrical Contacts, Loughborough, UK, September 7-11, 1992, pp. 407-412.
35. *Schlegel, S.*: Langzeitverhalten von Schraubenverbindungen mit Stromschienen aus Reinkupfer in der Elektroenergietechnik unter besonderer Berücksichtigung der Temperatur. Dissertation, TU Dresden, 2011.
36. *Gatzsche, M.; Lücke, N.; Großmann, S.; Ledermann, T.; Freudiger, G.*: Impact of the Temperature-Induced Reduction of Joint Force on the Long-Term Behavior of Contact Elements with Material-Allocated Electrical and Mechanical Function. 27th International Conference on Electrical Contacts, 22.-26.6.2014, Dresden, Germany.
37. *Oberst, M.*: Einfluss von Sauerstoff auf die Alterung stromführender Bimetall-Verbindungen. Dissertation, TU Dresden, 2021. – ISBN 978-3-86780-655-8
38. DIN EN 60865-1 VDE 0103:2012-09: Kurzschlussströme – Berechnung der Wirkung – Teil 1: Begriffe und Berechnungsverfahren.
39. *Gatzsche, M.; Lücke, N.; Großmann, S.; Ledermann, T.; Blumenroth, F.; Freudiger, G.*: Untersuchungen zu Stromaufteilung in Hochstrom-Steckverbindungen mit mehreren parallelen Kontaktelementen. 22. Albert Keil Kontaktseminar, Karlsruhe 2013.
40. *Ostrowski J; Buffoni M.; Gatzsche M.; Samul N.*: A Statistical Approach to Electro-thermal Simulations with Contacts. 65th IEEE Holm Conference on Electrical Contacts, 2019.

41. IEC 62271-1:2017+AMD1:2021: High-voltage switchgear and controlgear – Part 1: Common specifications for alternating current switchgear and controlgear.
42. IEC 61439-1:2020: Low Voltage switchgear and controlgear assemblies – Part 1: General rules.
43. *Pomsel, M.; Israel, T.; Hildmann, C.; Schlegel, S.; Kufner, T.*: Electromagnetic Modelling of Current Density Distribution in Power Plug-In Connectors. 31st IEEE International on Conference on Electrical Contacts, Sapporo, 13.-16.6.2022
44. *Lücke, N.*: Zum Langzeitverhalten elektrischer Steckverbindungen mit Kupfer-Beryllium-Kontaktlamellen. Dissertation, TU Dresden, TUDpress 2004. – ISBN 978-3944331751
45. *Oberst, M.; Israel, T.; Hildmann, C.; Schlegel, S.; Egetenmeyer, A.-K.; Kappl, H.; Willing, H.*: Elektrische und tribologische Eigenschaften von Silber-Dispersionsschichten für die Anwendung in Steckverbindungen. 8. Symposium Connectors, VDE-Verlag, Lemgo, 23.-24.3.2021.
46. *Israel, T.; Oberst, M.; Hildmann, C.; Schlegel, S.*: Selbstschmierende Beschichtungen für Hochstrom-Steckverbindungen. 26. Albert-Keil-Kontaktseminar, Karlsruhe, 30.3.-1.4.2022.
47. *Israel, T.; Hildmann, C.; Schlegel, S.; Schmidt, H.; Ludwig, M.*: Methodik zur Bewertung der Kontaktgüte von Ladesteckverbindungen unter Labor- und Realbedingungen. 26. Albert-Keil-Kontaktseminar, Karlsruhe, 30.3.-1.4.2022.
48. *Yuan, H.; Song, J.*: The effect of lubrication and thermal storage on the fretting behavior of electrical contacts. Tribologie und Schmierungstechnik Band 67. S. 14-20. DOI 10.30419/TuS-2020-0024. Narr Francke Attempto Verlag GmbH + Co. KG2020
49. *Löbl, H.*: Zur Dauerstrombelastbarkeit und Lebensdauer der Geräte der Elektroenergieübertragung. Dresden, Technische Universität Dresden, Sektion Elektrotechnik, Dissertation B, TU Dresden, 1985.
50. Icore International Ltd., „Datenblatt Kontakt-Band 8“ und „Datenblatt Kontakt-Band 12“. 2009.

Test Procedures 7

The function of a current-carrying connection in electrical energy technology is demonstrated in standardized procedures under uniform requirements. For almost all test procedures, no proof of lifetime is provided, even if numerous test procedures are partially referred to as aging tests. The purpose of testing the function of current-carrying connections is therefore rather to meet minimum requirements for current-carrying connections with the same application area. Accordingly, the electrical function of the respective connection is demonstrated application-specifically. The basis for this are reproducible load and uniform evaluation criteria, with which current-carrying connections for an application area but especially from different manufacturers can be comparably tested. The loads of the respective application in operation are the basis for testing the connection. In addition, accelerated aging in the test compared to the rated operation is often aimed at, with the goal of not activating additional aging mechanisms or unacceptably forcing or suppressing them [2]. Internationally, depending on the type of connection and the corresponding application area, numerous procedures for testing the electrical function of current-carrying connections are standardized. The following chapter will therefore first consider the load variables in the test procedures and their basic mode of operation with regard to the activated aging mechanisms. In Sect. 7.2, tests according to valid standards are exemplified and in Sect. 7.3 framework conditions for development-accompanying tests are presented.

S. Schlegel et al., *Contact and Long-Term Behavior of Current-Carrying Connections in Electrical Power Engineering*, https://doi.org/10.1007/978-3-662-69644-6_7

7.1 Influencing Factors in Test Procedures

The long-term behavior and the life time characteristic of a current-carrying connection depend in particular on the dominant aging mechanisms and their interactions (Sect. 5.2). The life time characteristics are specific for the different connection technologies and material combinations, as the respective effect of the aging mechanisms can differ significantly for these. An approach to developing test procedures tailored to the respective connection is based on assumptions for the sequential and combined action of the dominant aging mechanisms. For example, such an approach for current-carrying screwed connections with busbars made of aluminum was developed by *Braunovic* [1]. Empirical studies are often found in the literature, with which the most critical load for a type of connection is to be determined. In these considerations [2], the underlying physical relationships play a subordinate role. The aim of the following consideration is to first explain and discuss the physical effect of the primary load variables in the test procedures and their relation to the aging mechanisms. The results of tests from the literature are taken into account. The same procedure is followed with the evaluation variables.

7.1.1 Load Variables

Electrical Loads

The testing of the electrical function through standardized test procedures is carried out almost exclusively in current-carrying tests. The current is the cause of the heating of conductors and connections due to the generated current heat losses (Eq. (7.1)).

$$P_V(\vartheta) = kI^2 R_V(\vartheta, F_K, f) \tag{7.1}$$

To test the operating behavior, alternating currents with a defined frequency are required according to the predominant distribution of three-phase systems in electrical energy technology. A load with direct current leads to the same heating of the conductors and the connection, as long as the current displacement in the conductor carrying alternating current can be neglected. An influence of the electromigration possible in tests with direct currents on the long-term behavior of current-carrying connections in electrical energy technology has not yet been proven (see Sect. 5.1.5). For example, the experiments from [2] with direct current load show no changed resistance curves of the tested connections compared to those that were loaded with alternating current. In the definition of the test criteria for connections of electrical power engineering, electromigration was accordingly considered secondary.

The amperage used to test operational behavior is based on the required thermal stress of the conductor or connection under the thermal boundary conditions permissible for each test method. Most tests require thermal stresses that exceed the rated temperature. This means that the current in the test is to a certain extent higher than the rated current. The main effect of the current in a test is therefore the joule heating of the current-carrying connection. It is permissible to set the required temperature for testing operational behavior with an increased current in a shorter time, as this does not result in additional aging. The level of current can also influence the heating of the contact area (see Sect. 4.4.2). The measure of the heating of the contact interface is the contact voltage U_K (Eq. (4.55)). Since the contact voltage in current-carrying connections with surface contacts is usually in the single-digit mV range during rated operation, significant overtemperatures in the contact area are not to be expected even with overcurrents up to twice the rated current (e.g., Fig. 4.42). In a water-cooled contact system, it was shown that the contact interface is a maximum of 15 K warmer than the connected conductor when the connecting conductors are heated symmetrically to 105 °C and loaded with up to three times the rated current [24]. Accordingly, for most current-carrying connections, the temperature in the contact interface is close to the surface temperature of the current-carrying connection, provided the contacts are of sufficient quality (see Sect. 4.1.2).

In addition to proving the long-term stability of the connection under load with its rated current, the behavior in the event of a fault must be tested. Fault currents, for example in the event of a short circuit, can be many times higher than the rated current. An electrical conductor heats up almost adiabatically during a short-circuit current load. This means that the generated loss power is mainly stored in the material [3].

$$\vartheta_K = \frac{1}{\alpha_{20}}\left[1 + \alpha_{20}\left(\vartheta_a - 20^\circ C\right)\right]\left(e^{\frac{k\ I_K^2\ \ \alpha_{20} t_K}{c_p\ \kappa_{20}\delta\ A_L^2}} - 1\right) + \vartheta_a \tag{7.2}$$

The maximum temperature ϑ_K of the conductor can be calculated during adiabatic heating (Eq. (7.2)). The temperature coefficient α_{20}, the density ρ, the specific electrical resistance ρ_{20}, the specific heat capacity c_p of the conductor material, the cross-section of the conductor A_L, the current displacement factor k and the duration and the short-term current must be known. The term I^2t is a measure for the heat generated during a short circuit. If the effective value of the short-circuit current I and the short-circuit duration t are changed so that I^2t is kept constant, the heating of a homogeneously current loaded conductor (Eq. (7.2)) is the same, as long as the amount of heat dissipated by the conductor can be neglected. This is the so-called I^2t criterion. The I^2t criterion often does not apply to the a-spots (Sect. 6.5.3; Fig. 6.15). These can heat up significantly more than the bulk of the connection or the conductor during the same short-term load with a fault current due to their low heat capacity and the time-dependent heat conduction. The I^2t criterion is therefore not universally valid for the contacts. Numerous tests according to standards allow the short-term current to be reduced if the duration is increased. With the correct ratio of time and current and consequently the same energy input, the heating of a homogeneously current loaded conductor and the temperature on the surface of connections are the same. For example, short-term currents of 3 s or even 5 s are allowed, starting from a rated short-circuit duration of 1 s [10].

Example

Short-circuit testing of cable lugs for a 300-mm²cable conductor (Class 2) made of aluminium

Short-circuit tests are to be carried out as part of the electrical test according to DIN EN IEC 61238-1-3:2020-11. For this, a current is required that heats the reference conductor from a maximum initial temperature of 35 °C to (250 … 270) °C. The following are permissible:

$$I_K \leq 25\ \text{kA}\ \ \text{mit}\ t_k = \left(1 \begin{array}{c} +0{,}5 \\ -0{,}1 \end{array}\right) \text{s}$$

$$25\ \text{kA} < I_K \leq 40\ \text{kA}\ \ \text{mit}\ t_k \leq 5\ \text{s}$$

With the material characteristic data for aluminium and an initial temperature $\vartheta_a =$ 30 °C, the temperature of the conductor with $I_K = 25$ kA and the longest permissible short-circuit duration of 1.5 s is

$$\vartheta_K = \frac{1+0{,}004/\text{K}\left(30^\circ\text{C}-20^\circ\text{C}\right)}{0{,}004/\text{K}} \left(e^{\frac{1{,}01\ (25000\ \text{A})^2\ 0{,}004/\text{K}\ 1{,}5\ s}{35\ \text{MS/m}\ 910\ \text{J/(kg K)}2700\ \text{kg/m}^3\ \left(0{,}0003\text{m}^2\right)^2}} - 1 \right) + 30^\circ\text{C}$$
$$= 194{,}1^\circ\text{C}$$

Accordingly, a current >25 kA must be set to achieve the required thermal load. Eq. (7.2) can be rearranged for the current I_K:

$$I_K = \sqrt{\frac{c_p\ \kappa_{20}\delta\ A_L^2}{k\ \alpha_{20} t_K} \ln \frac{1+\alpha_{20}\left(\vartheta_K - 20^\circ\text{C}\right)}{1+\alpha_{20}\left(\vartheta_a - 20^\circ\text{C}\right)}}$$

If a short-circuit duration of one second is chosen, it must be tested with

$$I_K = \sqrt{\frac{35\ \text{MS/m}\ 910\ \text{J/(kg K)}2700\ \text{kg/m}^3\ \left(0{,}0003\ \text{m}^2\right)^2}{1{,}01\ 0{,}004/\text{K}\ 1{,}0\ \text{s}} \ln \frac{1+0{,}004/\text{K}\ \left(250^\circ\text{C}-20^\circ\text{C}\right)}{1+0{,}004/\text{K}\ \left(30^\circ\text{C}-20^\circ\text{C}\right)}}$$

$$I_K = 34272\ \text{A}$$

Table 7.1 Rules for setting the test short-circuit current of high-current contact systems

	1. Condition	2. Condition
Test **Connection can** $I_{\mathrm{K\,r}}$ **carry for** $t_{\mathrm{K\,r}}$ **if:**	$I_{\mathrm{K\,test}} \geq I_{\mathrm{K\,r}}$ für $t_{\mathrm{K\,test}} \geq t_{\mathrm{K\,r}}$ or $I_{\mathrm{K\,test}} \geq I_{\mathrm{K\,r}}\sqrt{\frac{t_{\mathrm{K\,r}}}{t_{\mathrm{K\,test}}}}$ für $t_{\mathrm{K\,test}} \leq t_{\mathrm{K\,r}}$	$i_{\mathrm{p\,test}} \geq i_{\mathrm{p\,r}}$

or with a short-circuit duration of 1.5 seconds

$$I_{\mathrm{K}} = \sqrt{\frac{\dfrac{35\ \mathrm{MS/m}\ \ 910\ \mathrm{J/(kg\ K)}2700\ \mathrm{kg/m^3}\ \left(0{,}0003\ \mathrm{m^2}\right)^2}{1{,}010{,}004/\mathrm{K}\ \ 1{,}5\ \mathrm{s}}}{\ln\dfrac{1+0{,}004/\mathrm{K}\ \left(250^\circ\mathrm{C}-20^\circ\mathrm{C}\right)}{1+0{,}004/\mathrm{K}\ \left(30^\circ\mathrm{C}-20^\circ\mathrm{C}\right)}}}$$

$$I_{\mathrm{K}} = 27983\ \mathrm{A}.$$

◀

The stress on the a-spots is significantly lower at smaller amplitudes of transient currents, as lower maximum contact temperatures are reached . This is particularly important to consider in connections with contact elements. However, this statement generally applies to all current-carrying connections. If fault currents occur in practical operation that are larger than those tested in the standard, the resistance of a connection can increase due to significantly higher thermal stress, even if the test according to the standard was passed. High operational safety can therefore only be achieved if testing is carried out according to the conditions in Table 7.1. Proof of the short-term current carrying capacity has not been provided in full if the test was carried out with $I_{\mathrm{K\,r}} > I_{\mathrm{K\,test}}$ and $t_{\mathrm{k\,r}} < t_{\mathrm{K\,test}}$ as well as $i_{\mathrm{p\,r}} > i_{\mathrm{p\,test}}$ [15].

Thermal Stresses

Temperature is a significant influencing factor when testing current-carrying connections, as it determines the type and speed of the physical aging processes. This applies equally to all aging mechanisms (Sect. 5.1). Thermal stress can be generated by current or external heating , for example in a heating cabinet. For the aging of current-carrying connections, it is secondary in most cases how the thermal stress is generated. Studies with various connections have found comparable results in current-flowing tests and those at the same temperature in the heating cabinet [16–18], so long-term stable connections can be demonstrated equally with both methods. In current-flowing tests, the change in connection resistance can be higher if the temperature at the connection is not kept constant. This is the case, for example, when the resistance of the test specimens increases over time in current-controlled tests [19]. Practical test methods for long-term behavior predominantly do not specify a specific thermal load at the connection, but a temperature range that is to be set on the reference conductor. In general, the temperature at which a connection should be stressed in a test cannot be chosen arbitrarily high, for example to reduce the number of stress cycles. If the physical aging processes change, this can influence the result of the test. The conditions of the respective application are no longer mapped and possibly incorrect conclusions are drawn from the result of the test.

In tests, either permanent or cyclic thermal loads are required. A cyclic thermal load is defined by a temperature-time profile for the heating and cooling phase, which is repeated N times. The number of cycles N is specific to the test procedure. In standardized test methods, cyclic thermal loading is predominantly used due to its higher correlation with practical load profiles. Compared to a permanent thermal load, the duration in which the connections are subjected to very high thermal loads is less if the same test time is assumed. However, cyclic thermal loading causes a higher number of thermal expansions of the contact partners and possibly the joining elements. If the contact force in a stationary current-carrying connection is sufficiently high, the forces resulting from the thermal expansions can be taken up. If the contact force is too low, displacements of the contact partners can occur. Only when very low contact forces are present can it be expected that a-spots can break open due to thermal expansions. Whether correspondingly higher resistance changes in a connection are achieved by a cyclic thermal load than with permanent thermal load depends largely on whether the current-carrying connection can absorb the mechanical stresses or whether relative movements of the contact members are to be forced. Connections with high connection forces, which secure the integrity of the microcontacts, must therefore first be tested for their resistance to force degradation.

The force degradation, chemical reactions, and diffusion are activated to a lesser extent with cyclic thermal loads, as the speed of both aging mechanisms is primarily temperature-dependent. If the same test time is assumed for both types of load, stronger aging due to oxidation, force degradation, and diffusion is to be expected with continuous load (Fig. 7.1).

For a large number of force-fitted current-carrying connections in power engineering, comparative tests have shown that the thermal continuous load is more critical. For example, screw connectors with multi-wire conductors made of pure aluminum (Al 99.5) for cable accessories were examined, in which a permanently high temperature led to faster aging than the changing thermal load during the same duration of the test (Fig. 7.2) [21]. If only the time $t_{>80°C}$ is considered, during which the temperature at the connection was higher than 80 °C, the aging process is approximately the same (Fig. 7.3). The same behavior was also observed with screw connections with a busbar made of aluminum and one made of copper (Fig. 7.4) [5–7].

For the specific technical implementation of a current-carrying connection, it is therefore specific whether a thermal continuous load or alternating load leads to stronger aging. Depending on which aging mechanisms are dominant, this assessment or proof must be made on a case-by-case evaluation.

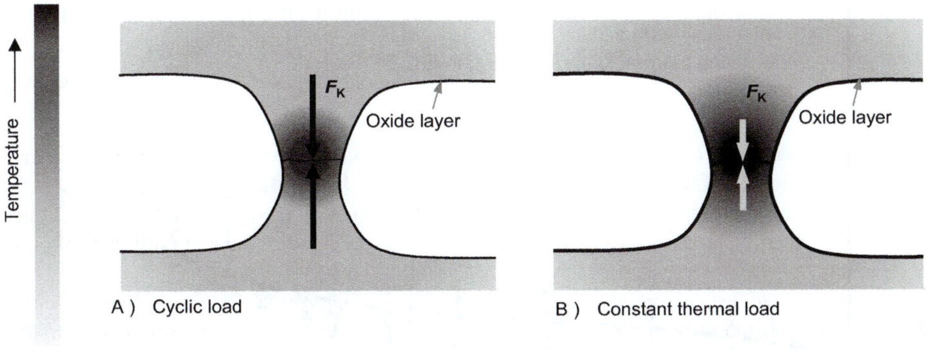

Fig. 7.1 Comparison of the effect of **a**) cyclic and **b**) constant thermal load on the integrity of a force-fitted a-spot after the same test time (schematic)

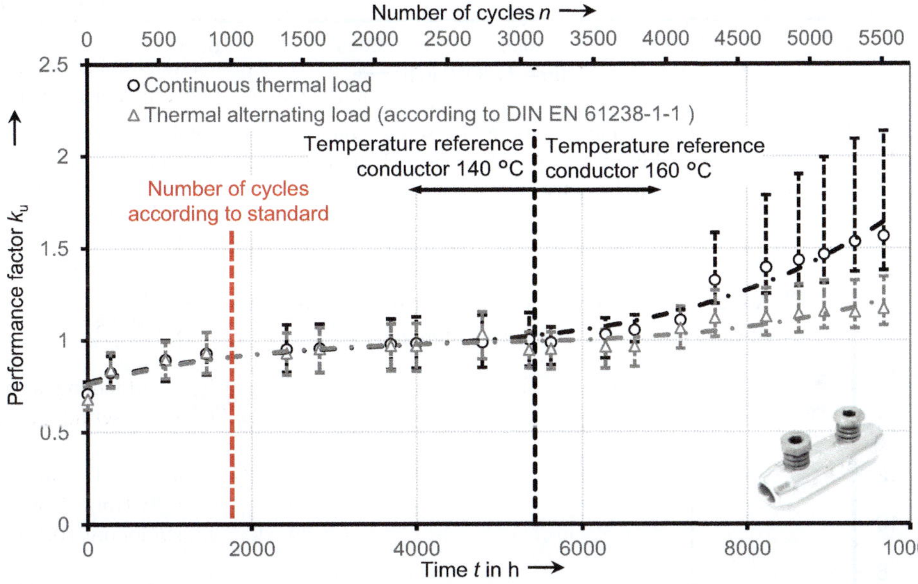

Fig. 7.2 Course of the performance factor k_u of screw connectors for cable accessories with stranded conductors made of Al99.5 under continuous and alternating load (test conditions according to [9, 10])

Mechanical Stresses

The requirements for current-carrying connections in practical operation can involve mechanical functions in addition to electrical ones. For example, this can be relevant for cable accessories if thermal expansions of the cables in practical operation cannot be compensated, resulting in tensile or compressive forces [11]. Another example are

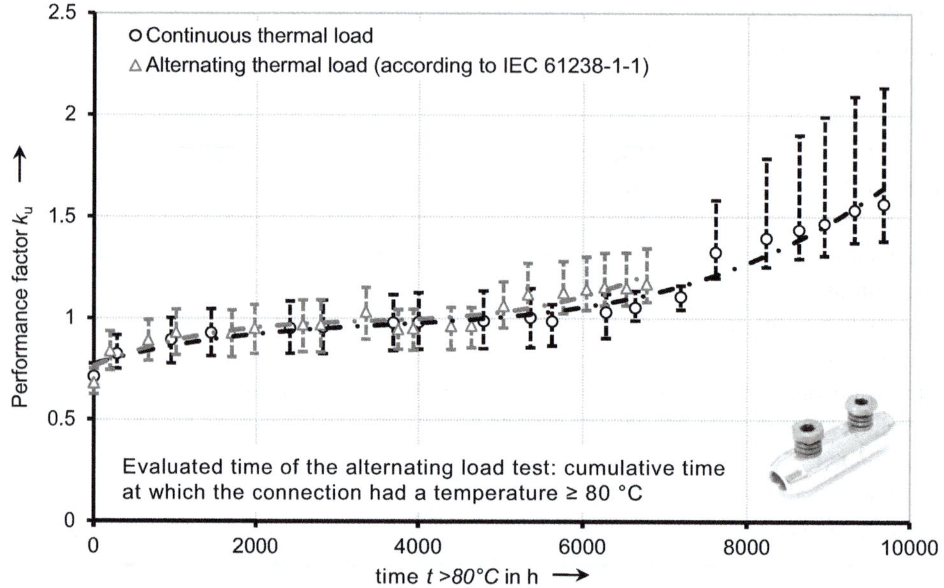

Fig. 7.3 Course of the performance factor k_u of screw connectors for cable accessories with stranded conductors made of Al99.5 under continuous and alternating load depending on the warm time (test conditions according to [9, 10])

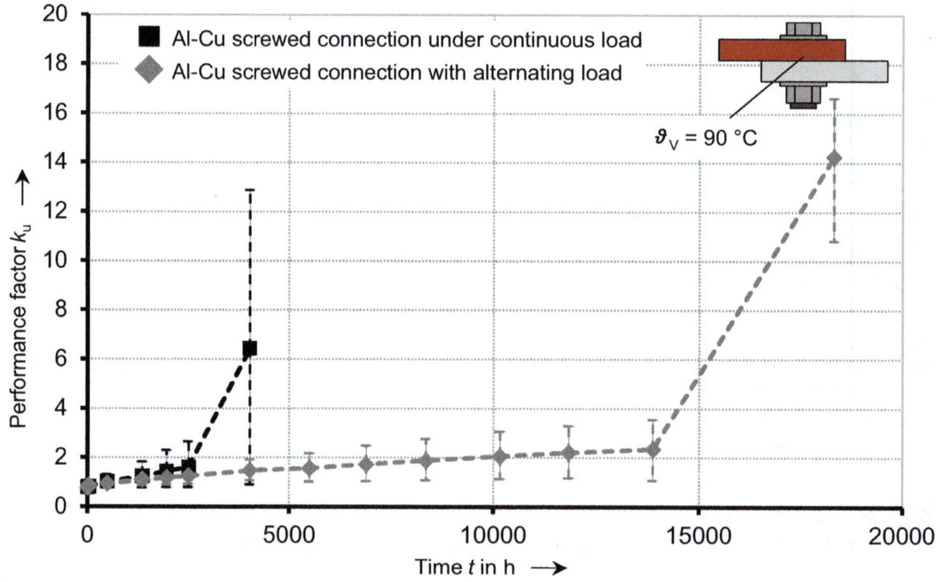

Fig. 7.4 Course of the performance factor k_u of screwed connections with one busbar made of aluminum and one made of copper under continuous and alternating electrical-thermal load with the same maximum connection temperature of 90 °C

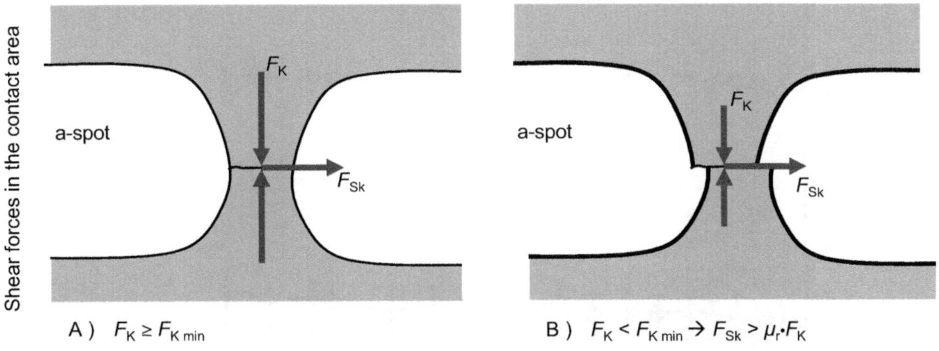

Fig. 7.5 Effect of mechanical stresses on the integrity of the a-spots (schematic)

tensile-proof current-carrying fittings of overhead line conductors, which also have to take-up significant mechanical stresses. These are operational tensile forces that can be significantly increased by the effect of wind or the weight of ice on the transmission line [12]. The mechanical stresses can equally affect the contact area of the current-carrying connections and hence influence the electrical contact and long-term behavior [21]. Based on the funtional principle of force and form-fitted connections, tensile or compressive forces in the contact area of the current-carrying connection an be taken up as long as the contact force is sufficiently high (Fig. 7.5—A). The integrity of the microcontacts is still ensured, as there are no displacements of the contact members. In current-carrying connections with too low contact forces, a displacement of the contact partners can initially occur in the micrometer range (Fig. 7.5—B). If the contact partners move against each other, microcontacts can break open and consequently the true metallic contact surface can decrease or microcontacts can reform on oxidized surfaces. A significant degradation of the connection is the result. The fundamental influence of a mechanical load on the electrical long-term behavior was determined in comparative tests (Fig. 7.6). Already in the formation phase , a higher increase in the connection resistance and the specific transverse resistance was determined when a mechanical load acts on the connection in addition to the thermal one [4, 13].

In the existing test procedures of electrical power engineering, the function of carrying current in an electrical-thermal test is tested separately from the other requirements (for example mechanical strength, corrosion resistance, etc.). Currently, there is no test procedure that mandatorily considers combinations with the other (operational) loads when proving the electrical function. For example, the electrical test according to IEC 61284:1997 can optionally be carried out with tensile load on the test specimens. For current-carrying connections that are operationally mechanically loaded, it is advisable to consider mechanical loads in the electrical test.

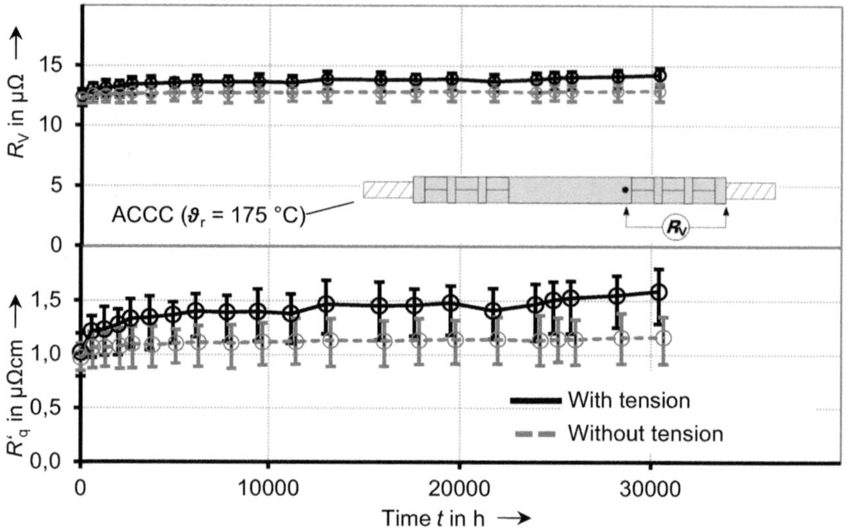

Fig. 7.6 Connection resistance and specific transverse resistance of compression connections for an ACCC overhead line conductor under permanent electric-thermal load with and without mechanical tensile force (average tensile stress in the range of the permissible every-day stress (limit of the tensile stress for a overhead line conductor in practical operation, to avoid vibration damage) [11]

Classification of the Test in Relation to Long-Term Behavior

The findings from investigations show that a reliable current-carrying connection must have a low value of the connection resistance after assembly and thus good contact behavior as a prerequisite for stable long-term behavior (see Fig. 5.3). The connection resistance only increases slowly within the lifetime. The aging is divided into characteristical phases (Fig. 5.2). If the duration of the test only covers the formation phase , no clear statements can be made about long-term stability. At least the phase of relative stability should be reached, which is not predominantly required in the test procedures of current standards. The standardized test procedures are therefore to be regarded as functional and qualification tests, as the test duration is too short to reliably capture the long-term behavior (Fig. 7.7) [8]. Late failures after a few 1000 hours of load duration are usually also not recorded. Therefore, specific aging tests are recommended, the duration of which, depending on the type of connection examined, safely reaches the phase of relative stability considering the field of application.

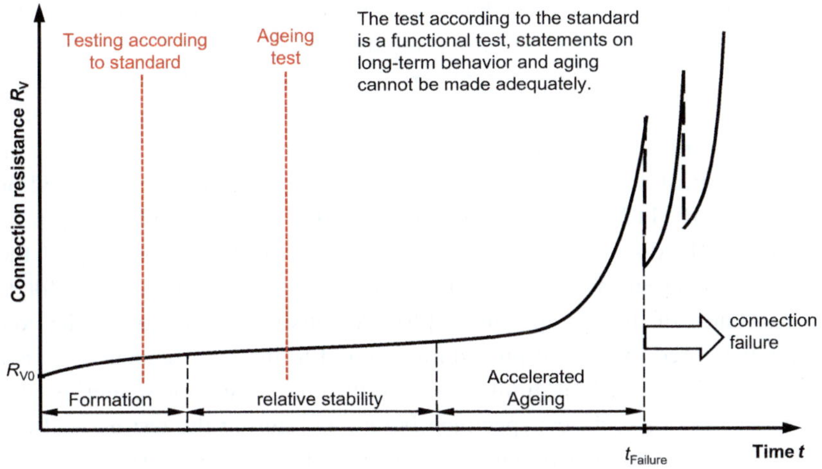

Fig. 7.7 Testing according to standard and testing of long-term behavior

7.1.2 Evaluation Parameters

In testing procedures, the connection resistance, the performance factor, and the connection temperature or, in the case of plug-in connectors, the contact voltage are used as evaluation parameters (Sect. 4.1.3). In general, the connection resistance, the contact resistance and the contact voltage are the most sensitive parameters that can be used to evaluate the behavior of current-carrying connections in the test. Accordingly, the absolute value, the change in the mean value, and the range of the connection resistance or the performance factor are considered. Unacceptably high increases are identified by suitable limit values of the mentioned parameters. The limit values to be complied with always apply to the type of connection to be tested under the specific loads of the test procedure and are based on the experiences in the test procedure or the operational behavior of the connection as well as the thermal requirements of the application. The limit values defined in the test procedures are usually not derived from the lifetime curves of the respective connections.

A fundamental minimum requirement when evaluating the results of tests is that the temperature of the connection during and after the test must not be higher than that of the connected conductor with the smallest cross-section. Based on this, a limit performance factor can be estimated for nearly rotationally symmetric systems via the ratio of the diameter of conductor and connection (Fig. 4.75) [10]. Exceeding the limit performance factor inevitably leads to temperatures at the connection that are higher than those of the conductor. For connections with rectangular conductors, the limit performance factor can be calculated via the surface ratio (Table 4.21). In addition, the change in temperature or the temperature rise are also used as evaluation criteria in test procedures.

Even if standardized test procedures address the same application area, the evaluation of the respective results can differ significantly. By way of example, the specifications of standards for testing current-carrying fittings of overhead line are compared (Table 7.2). In the various procedures, the defined measurement locations and the acceptance criteria are specific. In some cases, maximum values for the performance factor are defined, which must not be exceeded after the test sequence has been carried out, such as in the GOST R 51177-98 or the (withdrawn) DIN VDE 0212-52:1991 [28, 29]. The ANSI C119.4:2011 dispenses, apart from the evaluation of the scattering of the resistances of the test group at each measurement point, with further criteria for the connection resistance of the test specimens [30]. A statistical evaluation of the connection resistances measured in the test procedure, tailored to the load sequence, is required in the DIN EN 61284:1998. In it, limit values for the change and the scattering of the connection resistances of each test specimen before and during the test as well as general maximum values for the performance factor are defined (Table 7.2). In the test procedures that take into account the temperature of the fittings in the evaluation criteria, the maximum temperature of the connection is always set to that of the reference conductor. Only the ANSI C119.4:2011 defines additional limit values for the permissible change in the difference between connection and reference conductor temperature.

The range of evaluation criteria for this selected application field illustrates that a specific stress sequence of a test procedure always requires adapted limit values of the evaluation parameters. For example, a higher electrical-thermal load in the test compared to the rated operation can justify higher limit values for the joint resistance or the performance factor or their change. Direct comparison of results from different procedures is usually not possible without further ado. Fundamentally, the life time curve should be used for setting the evaluation criteria in test procedures, so that connection-specific aspects can be taken into account in particular. Accordingly, it may be necessary to initially determine individual evaluation criteria in long-term tests, which must be carried out significantly longer than the tests in the standardized test procedures. Furthermore, practical experience values must be taken into account.

7.2 Standardized Test Procedures in Power Engineering according to Application Areas

First, an overview of the general structure of standardized test procedures will be given. Furthermore, it will be classified how the corresponding test procedures are to be applied and evaluated for the possibly adapted application. Important standards of power engineering are presented based on a categorization (Fig. 7.8). No claim is made to the completeness of the procedures presented for the respective connection technologies.

The structure of a test procedure according to standard can be divided into the *general prerequisites for the test*, the *test setup*, the *test execution* and the *evaluation of the test*

Table 7.2 Comparison of evaluation parameters of test procedures for current-carrying fittings of overhead line conductors [28 –32]

	ANSI C119.4:2011	DIN EN 61284:1998		GOST R 51177-98	TGL 16554/02 TGL 16554/03	VDE 0212-52:1991
		Class A	Class B			
Criterion for the (mean) resistance in the initial state				$k_u \leq 0.8/k_u \leq 1.0$ (Pressv./Screwv.)		
Criterion for the span of resistance in the initial state		±30 %				
Criterion for the limit value of resistance/performance factor		after each 0.1 *N* LW $k_u \leq 0.75$		$k_u \leq 1.0/k_u \leq 1.2$ (Pressv./Screwv.)	$k_u \leq 1$ (Cl. 1) $k_u \leq 3$ (Cl. 2) $k_u \leq 10$ (Cl. 3)	
Criterion for the change of resistance/performance factor per test specimen		On average in the second half of the test less than or equal to 1.5 times the initial value of resistance				Range in which performance factor may change (see Table 7.4 and 7.5) for defined measurement points
Criterion for the span of resistance/performance factor per test specimen	Range per measurement point ±5 %	For the last 11 measuring times less than 15% of the respective average resistance of the test group				
Criterion for the change/limit value of the average resistance/performance factor of the test group						

(continued)

Table 7.2 (continued)

	ANSI C119.4:2011	DIN EN 61284:1998		GOST R 51177-98	TGL 16554/02 TGL 16554/03	VDE 0212-52:1991
		Class A	Class B			
Criterion for the change/ limit value of the span of the resistance of the test group						
Criterion for the temperature of the fitting	lower than the temperature of the reference conductor	after each 0.1 *N* LW lower than the temperature of the reference conductor			lower than the temperature of the reference conductor	
Criterion for the change in temperature of the fitting	Change in the difference between conductor and connection temp.$\leq$ 10 K in defined intervals					
Criterion Short-Circuit-test			Change R_V <50%			

1 General requirements for the examination
- General requirements for connection technology and conductors in the application area
- Application-specific classification for testing purposes
- Basic framework conditions of the audit

2 Experimental setup
- Requirements for the test setup and test technology
- Requirements for the temperature and resistance measuring locations and the measurement technology
- Ambient conditions during the test

3 Test execution
- Load profile(s) - type, level and duration of the electrical-thermal and, if applicable, mechanical load variables for the continuous and, if applicable, short-circuit current test
- Temperature and resistance measurement procedure

4 Evaluating the test results
- Compliance with the permissible limit values for connection and contact resistance, temperature and -voltage after the electrical test(s)
- Adherence to the permissible limit values for the temporal change of connection, Contact resistance and temperature after the electrical test

Fig. 7.8 Classification of the specifications in standardized test procedures of power engineering

results (Fig. 7.8). In the *general prerequisites*, the scope of the corresponding standard, general requirements for the equipment and basic conditions of the test are defined. In these sections, basic requirements for the design of the respective connection are already presented in part. For example, IEC 61284:1997 mentions general requirements for the construction of and the materials for fittings [31].

Specifications describing the arrangement of the test specimens or the circuit, required test and measurement technology, as well as the measuring points, are described in the *test setup* part. These provisions aim to achieve comparable testing of the electrical function as a primary requirement. To ensure this, precise specifications for the distances of the connections within the circuit or to neighboring objects, for the environmental conditions and the locations of measuring points are necessary. These specifications reduce the leeway with which the thermal boundary conditions in the test can be influenced and are intended to ensure comparable results in the same tests at different test laboratories. Furthermore, it should be noted that the specifications for the test setup aim to minimize the thermal influence on the test specimens.

The load sequences as the core of each test procedure and the required measurements are defined in the *test execution*. The principal effect of the load sizes in test procedures and the corresponding evaluation criteria oriented towards them were discussed in Sect. 7.1.1 and 7.1.2.

Methods for testing current-carrying connections are independently defined in power engineering for individual equipment or systems, as well as for different voltage levels. It is not taken into account that in different application areas with essentially comparable electrical-thermal loads, the same connection technologies with comparable dimensions are sometimes used. For example, the test method for electrical function from IEC 61284:1997—Overhead lines—Requirements and tests for fittings is applied to both current-carrying screwed connections with flat terminals, screwed, clamp and compression connections. The provisions of this standard for demonstrating electrical function also apply to other fittings that use other connection technologies (e.g., form-fitting current-carrying connections with spiral rods). The standard series IEC 61238-1, which is valid for the connection technology of conductors of cables in low and medium voltage technology, is equally applied to screw and compression connectors as well as to screwed connections with flat terminals [9, 26, 27]. In operation, the thermal stress of the connections in the applications is partly comparable, e.g., with flat terminals, and partly significantly different, e.g., due to the better cooling of the fittings of overhead lines. Accordingly, the mentioned standards define the respective requirements for the connection technology specifically for the application and take into account the peculiarities in the operation of the corresponding systems. In addition to the electrical-thermal properties, this involves different requirements for the mechanical function, which are to be considered in or in addition to the respective electrical test. Application-specific peculiarities in power engineering also include the required service life of the equipment in the range of decades with partly reduced electrical loads of the systems due to their generally redundant operation in practice (N-1 safety[1]). The proof of the electrical function is nevertheless provided in the procedures by testing at increased loads compared to the rated operation in a relatively short time (see Sect. 7.1.1). The tests therefore do not reflect the actual long-term behavior.

In general, the specifications in the test procedures of power engineering must be considered in the context of the application of connection technology in the facilities of electrical energy supply and cannot be transferred one-to-one to changed operating conditions of other or new applications. In applications that, in contrast, have changed operating conditions and deviating stresses (Table 7.3), corresponding functional tests are required. For example, for the connection technology in classic low-voltage applications, a lower utilization and load in operation is to be assumed. These previous requirements are to be tested and possibly supplemented if the load in low-voltage networks increases due to new applications, such as charging devices for electric vehicles. Compared to applications for high power transmission, the previous requirements in these tests are lower (e.g., DIN EN 60352-2:2014 33). For example, fewer thermal cycles are required (Table 7.4). On the other hand, certain climatic tests or vibration tests may be relevant for connections in low-voltage applications. Test procedures adapted to the respective operating conditions are therefore appropriate (see Sect. 7.3). When testing plug-in

[1] Redundant operation of electrical energy technology systems with a maximum relative utilization of less than 70%, in which the transmission function can be taken over by the remaining systems in the event of a system failure.

Table 7.3 Standardized temperature rise and endurance tests for screwed and compression connections (selection)

Application	High-voltage overhead lines	Low and medium voltage cables	Connection material—Screwed and screwless terminals	Crimp connections (solder-free)	High-voltage switchgear and controlgear
Standard (-series)	DIN EN 61284:1998 [31]	DIN EN 61238-1-1: 2020 [9] DIN EN 61238-1-2: 2020 [26] DIN EN 61238-1-3: 2020 [27]	DIN EN 60999-1:2000 [41] DIN EN 60999-2:2000 [42]	DIN EN 60352-2:2014 [33]	DIN EN 62271:2018 [34]
Rated voltage	> 45 kV	≤1000 V AC/DC (Part 1)	≤1000 V AC	No restriction	>1000 V
		(1 … 36) kV AC/DC Part 3	≤1500 V DC		
Conductor cross-section	No restriction	(2.5 … 1200) mm^2 (Cu) (16 … 1200) mm^2 (Al)	Part 1: (0.2 … 35) mm^2 Part 2: (35 … 300) mm^2	(0.05 … 10) mm^2	No restriction
Temperature rise test at rated current	Not provided	Not provided	Not for the components, possibly required within the scope of the "relevant product standards"	Not provided	Contacts: Cu: ≤75 °C Ag/Ni: ≤115 °C Sn: ≤90 °C ($\vartheta_0 = 40$ °C)
Endurance Test (Electrical-Thermal)	$1000^1/500^2/100^3$ cycles with I 70 K^1/100 K^2/130 K^3 subsequently I_K with <180 °C (short-term current) ($\vartheta_0 = (15 … 30)$ °C)	1000 cycles at reference conductor temperature (120 … 140) °C short-circuit test after 200 cycles; reference conductor temp.: (250 … 270) °C ($\vartheta_0 = (15 … 30)$ °C)	196 cycles of approx. 1 h, 30 min under load, 30 min cooling, storage in the heating cabinet: $\vartheta_0 = 20$ °C → 40 °C during the load	The provisions of DIN EN IEC 60512-9-5:2021 apply 20/100/500 cycles e.g. 125 °C at the connection	none
Mechanical Tests (without el. load)	specific for each type of fitting (see Table 2.1 in DIN EN 61284:1998)	Tensile test with: 60 N/mm^2 (Cu-conductor) 40 N/mm^2 (Al-conductor) max. 20 kN	Minimum tensile strength: Part 1:(10 … 190) N Part 2: (236 … 578) N	Tensile strength according to DIN EN IEC 60512-16-4:2009	no requirements for screwed and compression connections

Table 7.4 Standardized temperature rise and endurance tests for connectors in atmospheric air (selection)

Application	High-voltage switchgear and controlgear	Low-voltage switchgear	High-voltage contacts in automobiles	Charging connectors for electric vehicles,	Connectors for industrial applications	Connectors for electrical and electronic equipment	Other connectors
Standards (series)	DIN EN 62271:2018 [34]	DIN EN 61439-1:2021 [37]	Delivery specifications for automotive suppliers LV 214, LV 215 [45–47]	DIN EN 61851-23:2014 [39] DIN EN 62196-1:2014 [40]	DIN EN 60309-1:1999 [44]	DIN EN 60512:2019 [43]	DIN EN 61984:2009 [38]
Rated current	no restriction	no restriction	LV 215: (10 … 405) A	$\leq$250 A AC $\leq$400 A DC	$\leq$800 A	no restriction	$\leq$125 A, up to 500 A as a guideline
Rated voltage	>1000 V	$\leq$1000 V AC $\leq$1500 V DC	(60 … 1500) V DC (30 … 1000) V AC	$\leq$690 V AC $\leq$ 1500 V DC	$\leq$ 1000 V AC or DC	no restriction	50 V to 1000 V
Temperature-rise test at rated current (ambient temperature ϑ_U)	Contacts: Cu: $\leq$ 75 °C Ag/Ni: $\leq$ 115 °C Sn: $\leq$ 90 °C ($\vartheta_0 = 40$ °C)	Heating Cu-conductor 105 K, connections 70 K ($\vartheta_U = 35$ °C)	Maximum temperature in contact 180 °C ($\vartheta_0 = (65 \ldots 140)$ °C)	Temperaturerise max. 50 K at the terminals (max. $\vartheta_0 = 25$ °C)	Temperaturerise max. 50 K at the terminals (max. $\vartheta_0 = 25$ °C)	Part-5-1: temperature rise max. 30 K for the specified current, one hour persistence	Limit temperature (70 … 125) °C according to manufacturer's specification

(continued)

Table 7.4 (continued)

Application	High-voltage switchgear and controlgear	Low-voltage switchgear	High-voltage contacts in automobiles	Charging connectors for electric vehicles,	Connectors for industrial applications	Connectors for electrical and electronic equipment	Other connectors
Endurance test (electrical-thermal)	none	none	Temperature range (−40 … 180) °C 60 cycles, each 4 h current loaded	none	none	-9-2: el. load at max. $\vartheta_0{}^a$ -9-5: cyclic current load[b]	none
Endurance test (mechanical) without el. load	circute breaker: 2000 cycles; GIL: 10,000 sliding cycles	200 cycles or according to manufacturer's specification	Ag, Sn: 50 mating cycles; Au: 100 cycles	10,000 mating cycles	1000 (32 A) to 125 (>200 A)	Mating cycles according to manufacturer's specification	Mating cycles according to manufacturer's specification

[a]Test duration according to manufacturer's specification, preferred values 250 h, 500 h, 1000 h or 2000 h
[b]Number of cycles according to manufacturer's specification, preferred values 20, 100 or 500 cycles; 45 min on +15 min off

and sliding connections, the resistance to a high number of mating or sliding cycles and moderate limit temperatures must be demonstrated (Table 7.5). The proof of long-term electrical behavior is not required here, but should definitely be provided by development-accompanying testing. In contrast, for applications in electromobility, significantly

Table 7.5 Selected use cases of electrical screw, screwed, compression and plug-in connections—load currents, types used and testing standards

Application	Electric power generation and transmission/distribution		Electrical systems of consumers		Special cases
	High and medium voltage	Low voltage	Building and industry	Electromobility	e.g. particle accelerators [12], test facilities
Example	Cable and overhead line routes, circuit breakers, grounding switches and disconnect switches, expansion compensation, plug-in connection, device connections		Terminal connection, crimp connections, CEE connectors	plug-in charging connection, components of the wiring harness	Expansion compensation, Assembly plug connection grounding system, creation of isolation distances
Continuous load current	AC or DC; some (10 … 1000) A, according to daily load profile (EEV) or usage categories (industry)		in a vehicle mostly AC and DC systems, some (10 … 100) A; volatile driving profile with current peaks		specific to the system
Short-term load current	AC: close to generator or far fromgenerator AC short circuit DC: strongly dependent on the type of current sources; standardized short-circuit durations $t_K = (0.5 \ldots 3)$ s		No generally valid description known, when using fuses at AC $t_K < 5$ ms [13]		mostly impulsive, specific to the system
Exemplary selected testing standards	High-voltage switchgear and controlgear DIN EN 62271 -1/-100/-204 [34–36]	Low-voltage switchgear DIN EN 61439-1 [37], other plug-in connections DIN EN 61984 [38]	Delivery specifications for automotive suppliers LV 214, LV 215 [45–47], plug-in charging connections DIN EN 61851-23 [39], DIN EN 62196-1:2015-6 [31, Sect. 12.3] Crimp connections DIN EN 60352-2:2014 [40] clamped connections DIN EN 60999-1:2000 [41]		not standardized

higher limit temperatures and more mating or sliding cycles are required. In industrial and other low-voltage applications of plug connections, the standardized requirements for connection technology are again specific (Table 7.5).

7.3 Development Accompanying for Special Applications

In a test procedure according a standard, the basic proof is provided that the minimum requirements for the function of a component for a specific application are provided. Depending on the application, different test standards and test methods have been established (Sect. 7.2). If requirements in the application change, the test methods should also be adapted in or against the standards. The revision of standardized test methods takes place in cycles lasting several years to decades, as this is done by working groups with participants from different interest groups and must be coordinated or reconciled with national and international committees. This has repeatedly led to connections available on the market and tested for a specific application failing in practice. The reason for this is that the test conditions no longer accurately reflected the possibly changed requirements or the application no longer matched the one considered as the basis for the test procedure. An example are connectors in cable joints, which were used and worked in the electrical power supply network for decades, but led to problems in wind energy plants. Also, current-carrying connections, which are used in electromobility at significantly higher temperatures than in classic electrical power supply, lead to new challenges. This often raises questions about the expected lifetime, which many test methods cannot adequately answer. It is therefore often necessary, especially during development, to describe specific test methods that take into account the essential stress factors in the application. Development-accompanying tests can be carried out at different points in the development process and at different levels. In the early stages of component development, this can be accompanied by model tests and initial tests at the component level. Even these tests should be application-specific. In further stages, system tests are carried out that take into account device or asset-specific requirements. In many application fields, these are supplemented by field tests under real conditions.

In development-accompanying tests at the component level, depending on the type of connection, the influence of the relevant aging and degradation mechanisms must be evaluated and weighted (Fig. 7.9). This requires a deep understanding of the contact behavior of the relevant type of connection, taking into account the electrical and mechanical stress factors as well as the environmental conditions in operation. If the design values of the loads are used as a guide, it can be ensured that the test runs in the area of effective aging mechanisms. It must also be decided whether loads must be considered in separate or superimposed in common tests, as there are often interactions between these and thus the test result is significantly influenced. On this basis, specific test sequences can be derived. The procedure for force-fit current-carrying connections is exemplified below.

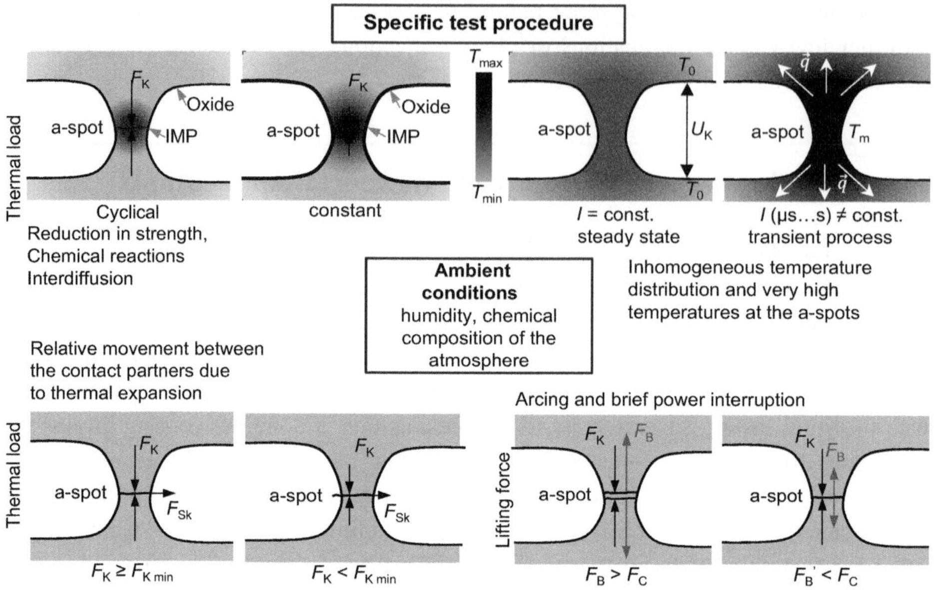

Fig. 7.9 Specific test methods—Influence of stress factors on the electrical contact

Based on the considerations in Sect. 7.1, force-fitted connections should be subjected to the rated temperature for a defined period of time in the first phase of the test in order to test the resistance of the connection to force reduction in particular. Since the physical processes involved in force reduction, as well as oxidation and interdiffusion, depend exponentially on temperature according to the Arrhenius law (Sect. 5.1.1), a consistently high thermal stress should be chosen in this phase (Fig. 7.9—top left). After the preloading of the first phase forced the force reduction, the resistance of the connection to possible deterioration mechanisms is tested in the subsequent phase of the test procedure. In force-fited connections with low contact forces, these are, for example, displacements of the contact partners, caused by different thermal expansion or mechanical additional loads (Fig. 7.9—bottom left). Accordingly, cyclic thermal loads are applied in this phase and mechanical additional loads are superimposed, which, for example, are based on those of practical operation [4]. The latter can be derived for fittings for overhead overhead line conductors, for example, from the operational tensile stress or from wind and ice loads [11]. For screwed connections with busbars, additional mechanical loads are, for example, transverse forces F_Q or torques M_T in operation (Sect. 4.1.2). In a final phase of the test procedure, pre-aged connections are subjected to application-specific short-circuit currents. When loaded with short-circuit currents, the contacts are thermally and mechanically stressed (Fig. 7.9—right). Application-related environmental conditions such as increased humidity or a special chemical composition of the atmosphere can be taken into account in the specific test sequence or carried out as individual tests.

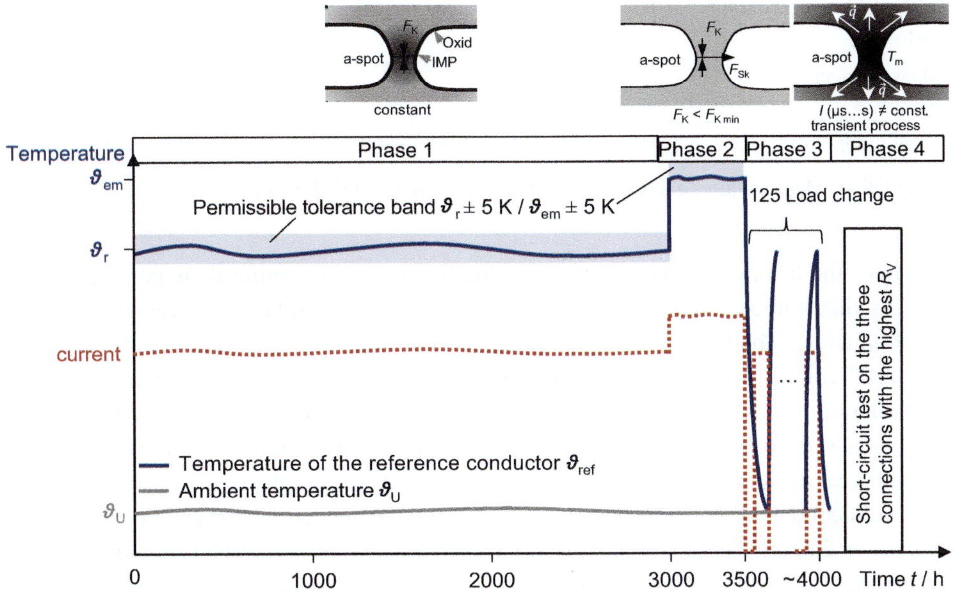

Fig. 7.10 Specific test procedure for current-carrying fittings for high-temperature overhead line conductors (1st phase: continuous load at the rated temperature of the conductor over a period of 3000 h, 2nd phase: continuous load over 500 h at the highest permissible operating temperature of the conductor, 3rd phase: alternating load cycles and 4th phase: short-circuit test) [22]

A specific test, which is strongly oriented towards testing force-fited connections, was developed, for example, for current-carrying fittings with High-Temperature-Low-Sag (HTLS) overhead lines [22].

The test sequence is divided into four phases (Fig. 7.10) [22]). In the 1st and 2nd phase of the test sequence, at the permanently maximum permissible temperature of the overhead line conductor in normal and emergency operation, aging is forced by power reduction and oxidation (Fig. 7.9). Subsequently, in the 3rd phase, thermomechanical stresses are to be generated in the connections by electrical-thermal alternating load. Since the contact force is already reduced by the previous thermal continuous load, a stronger influence of the cycles is to be expected at this point than in the initial state. In all three phases, a static mechanical tensile stress is superimposed, which is oriented towards that in practical application. After this pre-aging, a short-circuit test is finally carried out on the three connections with the highest connection resistances and the effect of the high thermal and mechanical load is tested. The test procedure thus represents the essential electrical-thermal and mechanical loads that can occur during operation. The test sequence is chosen so that the occurring thermal load in real N-1 operation is accumulated by the high proportion of continuous load and at the same time conceptually follows the sequence of relevant aging and damage mechanisms. In the further

Table 7.6 Specifications for the timing of resistance measurements in the test procedure according to [22]

Measurement No. n	1	2	3	4	5	6	7	8	9	10	11	12	13
Time t in h	0	24	50	100	250	500	1000	1500	2000	2500	3000	3500	4000*

* Guideline; Measure after each 125 load cycles as well as after short-circuit tests

course of the life cycle curve, particular attention is paid to acting damage mechanisms, such as thermomechanical stresses and possible resulting relative movements between the conductors. The behavior of the connection in the test is evaluated with the cyclically measured connection resistance (Table 7.6) and the continuously measured connection temperature. Since the measured connection resistance can be significantly erroneous under certain boundary conditions in connections with multi-wire conductors [23], thestatistical evaluation of the temperatures of the test specimens was focused.

As a second example, a test sequence for screwed connections for use in electric vehicles is considered. These connections are force-fitted, so that the basic considerations for testing such connections can be applied equally. Accordingly, the test sequence begins with 2000 h of continuous thermal load at the rated temperature of the conductor, in which primarily the force reduction, diffusion and oxidation are to take place intensively, so that an intensive pre-aging of the connections is achieved (Fig. 7.11). This is followed by rapid temperature shocks. In this process, the connections are cyclically heated to rated temperature ϑ_{max} at a very high rate of change and then cooled to minimum temperature $\vartheta_{0\ min}$. This examines the effect of different thermal expansion coefficients between the joining elements and the conductors, which can have a major influence especially on pre-aged screwed connections, especially when small clamping lengths/thin conductors are used [25]. The further loads are a practical vibration test and a climatic test before finally testing the resistance to fault currents (Fig. 7.11).

Furthermore, a generalized test procedure was developed as part of a publicly funded project. This harmonizes and extends the requirements of existing test procedures, so that the function of the current-carrying connection under current requirements of power engineering in operation can be tested and initial statements on long-term behavior can be made [10]. Accordingly, the processes of the test procedure were generally described. This includes general criteria for the test setup, the load, and for evaluating the results. Specific properties, such as thermal expansion or additional mechanical loads in practical operation can be equally considered and should be replicated in the test. This uniform test procedure for current-carrying connections in power engineering could be supplemented e.g. in standards that do not currently provide specifications for endurance testing taking into account specific requirements. Accordingly, uniform test procedures and evaluation criteria could be created in these standards, leading to significantly more objective and comparable results. Advantages arise from this for both the manufacturers and the operators of the systems.

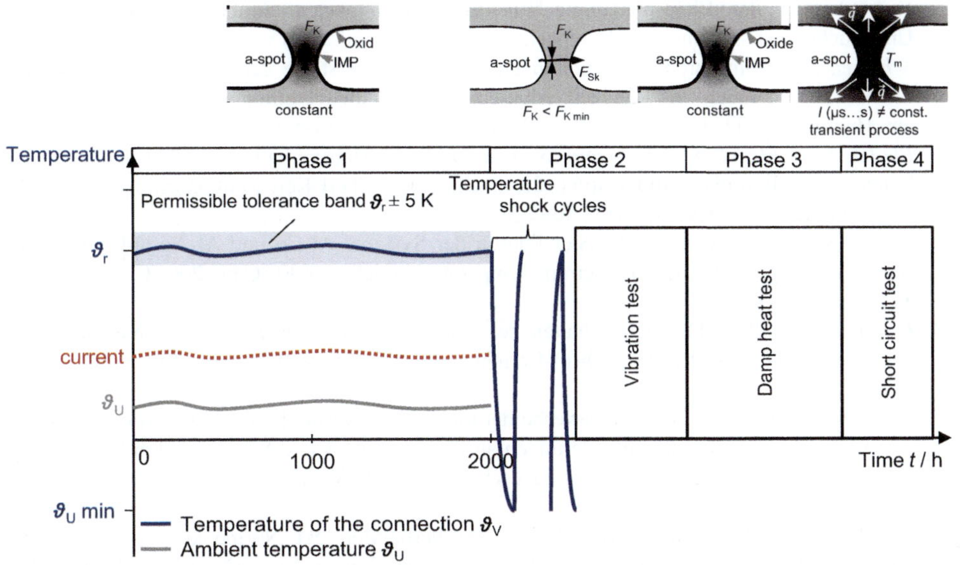

Fig. 7.11 Exemplary specific test procedure of a screwed connection with busbars in the vehicle (1st phase continuous load at maximum permissible temperature of the conductor over a period of 2000 h, 2nd phase simulation of aging through the temperatures and vibrations occurring in practice, 3rd phase simulation of the climatic conditions 4th phase short-circuit test)

These examples show that with knowledge of the functioning and the resulting aging mechanisms of the relevant connections as well as the operating conditions, specific test procedures can be specifically developed. If the loads are limited to the rated values, statements about the long-term stability of the connections can be made with the test procedure. The considerations in ths chapter are strongly related to Chap. 5 and derives applicable basics for testing current-carrying connections from the physical processes of aging for practical use.

References

1. *Braunovic, M.*: Aluminium connections: legacies of the past. Konferenzbd. 40th IEEE Holm Conference on Electrical Contracts, Chicago, 17.–19.10.1994, pp. 1–31.
2. *Öberg, Å.; Olsson, K.-E.; Bohlin, A.*: Testing of power connectors-influence of testing parameters. 36th IEEE Holm Conference on Electrical Contacts and the 15th International Conference on Electrical Contacts, Montreal, 20.–24.08.1990.
3. *Böhme, H.*: Mittelspannungstechnik-Schaltanlagen berechnen und entwerfen. 2. Stark bearbeitete Auflage Berlin: Verlag Technik 2005. – ISBN 3-341-01495-0.
4. *Hildmann, C.; Schlegel, S.; Lücke, N.; Großmann, S.; Gorschakov, S.; Brandenburg, R.; Plonus, F.; Schoenemann, T.*: Vergleich genormter elektrischer Alterungsprüfungen für

Verbindungen der Elektroenergietechnik mit Erkenntnissen aktueller wissenschaftlicher Untersuchungen. 5. Connector Symposium, Lemgo, 18.–19.3.2015.

5. *Schlegel, S.*: Stromführende Verbindungen und Leiterwerkstoffe der Elektroenergietechnik – Theorie zum Kontakt- und Langzeitverhalten von Schraubenverbindungen mit Flächenkontakten. Habilitation, TU Dresden 2019. – ISBN 978-3-86780-614-5.
6. *Pfeifer, S.; Schlegel, S., Großmann, S., Günter, F.*: Betriebsverhalten von Verbindungen mit Leitern aus Aluminium und Kupfer. 23. Fachtagung Albert-Keil-Kontaktseminar, Karlsruhe 7.–9.10.2015. – ISBN 978-3-8007-4092-5.
7. *Pfeifer, S.*: Einfluss intermetallischer Phasen der Systeme Al-Cu und Al-Ag auf den Widerstand stromtragender Verbindungen im Temperaturbereich von 90 °C bis 200 °C. Dissertation, TU Dresdeb 2015.
8. *Bergmann, R.*: Zum Langzeitverhalten des Widerstands elektrischer Stromschienenverbindungen. Diss. TU Dresden, Fortschritt-Berichte VDI Nr. 195. VDI-Verlag 1996. – ISBN 3-18-319521-6.
9. DIN EN 61238-1-1:2020-11: Pressverbinder und Schraubverbinder für Starkstromkabel – Teil 1-1: Prüfverfahren für und Anforderungen an Pressverbinder und Schraubverbinder für Starkstromkabel für Nennspannungen bis zu 1 kV (U_m = 1,2 kV), geprüft an nicht isolierten Leitern (IEC 61238-1-1:2018).
10. Bericht zum BMWi-Projekt (FKZ 01FS13038): Verbundprojekt: Normenkonzept für Verbindungen der Elektroenergietechnik zur Bewertung der Lebensdauer (NOVEL), Teilvorhaben Analyse und Bewerten des Langzeitverhaltens elektrischer Verbindungen. 2016.
11. Cigré TB 758. Test regimes for HV and EHV cable connectors. https://e-cigre.org/publication/758-test-regimes-for-hv-and-ehv-cable-connectors. 06.10.2020.
12. *Kießling, F.*: Freileitungen – Planung, Berechnung, Ausführung. Edition Number 5, Springer, Berlin, Heidelberg, 2001. – ISBN 978-3-642-56738-4.
13. *Hildmann, C.*: Zum elektrischen Kontakt- und Langzeitverhalten von Pressverbindungen mit konventionellen und Hochtemperatur-Leiterseilen mit geringem Durchhang. Dissertation, TU Dresden 2016. – ISBN 978-3-00-056211-2.
14. *T. Israel, M. Gatzsche, S. Schlegel, S. Großmann, T. Kufner and G. Freudiger*: The impact of short circuits on contact elements in high power applications. 63rd IEEE Holm Conference on Electrical Contacts, Denver, CO, 2017, pp. 40–49, doi: https://doi.org/10.1109/HOLM.2017.8088061.
15. *Gatzsche, M., Israel, T., Lücke, N., Großmann, S., Kufner, T., Freudiger, G.*: Grenzen der Gültigkeit des I^2t-Kriteriums beim Prüfen der Kurzschlussfestigkeit von Kontaktsystemen. 6. Connector Symposium, Lemgo, 15.–16.3.2017.
16. *Blumenroth, F.*: Zum Langzeitverhalten von Steckverbindungen mit Schraubenfedern in Anlagen der Elektroenergietechnik. Dissertation, TU Dresden, Shaker Verlag, 2010. – ISBN 978-3-8322-9506-6.
17. *Lücke, N.*: Zum Langzeitverhalten elektrischer Steckverbindungen mit Kupfer-Beryllium-Kontaktlamellen. Auflage 1, TUDpress 2004. – ISBN 978-3944331751.
18. *Schneider, R.*: Langzeitverhalten geschraubter Stromschienenverbindungen in der Elektroenergietechnik. Dissertation, TU Dresden 2012.
19. Ramonat, A.: Untersuchungen zum elektrischen Kontakt- und Langzeitverhalten von Fügetechnologien mit zylindrischen Leitern aus Aluminium. Dissertation, TU Dresden, 2018. – ISBN 978-3-86780-604-6.
20. *Israel, T., Schlegel, S.; Großmann, S., Kufner, T.; Freudiger, G.*: Zum elektrisch-thermisch-mechanischen Verhalten von silberbeschichteten Hochstrom-Steckverbindern. Das I^2t-Kriterium bei der Kurzschlussprüfung Teil II. 7. Connector Symposium, Lemgo, 19.–20.3.2019.

21. *Hildmann, C.; Israel, T.; Schlegel, S.; Großmann S.*: Recent Findings in the Long-term and Short Circuit Behavior of Power Connections and their Relevance for Testing. 64th IEEE Holm Conference on Electrical Contacts, Albuquerque, NM, 2018, pp. 270–277.
22. Bericht zum BMWi-Projekt (FKZ 03TNG009): Entwurf für ein Prüfverfahren zum Nachweis der elektrischen Funktion von stromführenden Verbindungen für HTLS-Leiterseile.
23. *Hildmann, C.; Schlegel, S.; Großmann S. Schmidt, C.*: Measuring Joint Resistances of Power Connections with Stranded Conductors – Fundamentals, Principles and Common Reasons for Measurement Errors. 65th IEEE Holm Conference on Electrical Contacts, Milwaukee, MI, 2019 pp. 57–63.
24. *Gatzsche, M.*: Elektrisch-thermisches Betriebs- und Langzeitverhalten hochstromtragfähiger Kontaktelemente. Dissertation TU Dresden, Shaker Verlag, 2016. – ISBN 978-3-84404-894-0.
25. *Poudeu F.*: Langzeitverhalten stromführender Schraubenverbindungen mit Stromschienen kleiner Dicke bei wechselnden Umgebungsbedingungen, Dissertation, TU Dresden 2022 (in press).
26. DIN EN 61238-1-2: 2020-11. Pressverbinder und Schraubverbinder für Starkstromkabel – Teil 1-2: Prüfverfahren für und Anforderungen an isolationsdurchdringende Verbinder für Starkstromkabel für Nennspannungen bis zu 1 kV (U_m = 1,2 kV), geprüft an isolierten Leitern (IEC 61238-1-2:2018)
27. DIN EN 61238-1-3: 2020-11. Pressverbinder und Schraubverbinder für Starkstromkabel – Teil 1-3: Prüfverfahren für und Anforderungen an Pressverbinder und Schraubverbinder für Starkstromkabel für Nennspannungen über 1 kV (U_m = 1,2 kV) bis zu 36 kV (U_m = 42 kV), geprüft an nicht isolierten Leitern (IEC 61238-1-3:2018)
28. GOST R 51177-98. Lineare Armaturen – Allgemeine Spezifikationen.
29. DIN VDE 0212-52:1991. Armaturen für Freileitungen und Schaltanlagen. Elektrisches Kontaktverhalten von betriebsmäßig stromführenden Armaturen
30. ANSI C 119.4:2011. American National Standard for Electric Connectors — Connectors for Use Between Aluminum-to-Aluminum and Aluminum-to-Copper Conductors Designed for Normal Operation at or Below 93°C and Copper-to-Copper Conductors Designed for Normal Operation at or Below 100°C.
31. DIN EN 61284:1998-05. Freileitungen – Anforderungen und Prüfungen für Armaturen (IEC 61284:1997)
32. TGL 16554/02 /03. Armaturen für stromführende Verbindungen. Technische Forderungen.
33. DIN EN 60352-2:2014-04. Lötfreie Verbindungen – Teil 2: Crimpverbindungen – Allgemeine Anforderungen, Prüfverfahren und Anwendungshinweise (IEC 60352-2:2006 + A1:2013)
34. DIN EN 62271-1:2018-05. Hochspannungs-Schaltgeräte und -Schaltanlagen – Teil 1: Gemeinsame Bestimmungen für Wechselstrom-Schaltgeräte und -Schaltanlagen (IEC 62271-1:2017)
35. DIN EN 62271-100:2018-14. Hochspannungs-Schaltgeräte und -Schaltanlagen – Teil 100: Wechselstrom-Leistungsschalter (IEC 62271-100:2008 + A1:2012 + A2:2017 + A2:2017/COR1:2018)
36. DIN EN 62271-204:2012-05. Hochspannungs-Schaltgeräte und -Schaltanlagen – Teil 204: Starre gasisolierte Übertragungsleitungen für Bemessungsspannungen über 52 kV (IEC 62271-204:2011)
37. DIN EN 61439-1:2021-10. Niederspannungs-Schaltgerätekombinationen – Teil 1: Allgemeine Festlegungen (IEC 61439-1:2020)
38. DIN EN 61984:2009-11. Steckverbinder – Sicherheitsanforderungen und Prüfungen (IEC 61984:2008)
39. DIN EN 61851-23:2014-11. Konduktive Ladesysteme für Elektrofahrzeuge – Teil 23: Gleichstromladestationen für Elektrofahrzeuge (IEC 61851-23:2014)

40. DIN EN 62196-1:2015-6. Stecker, Steckdosen, Fahrzeugkupplungen und Fahrzeugstecker – Konduktives Laden von Elektrofahrzeugen – Teil 1: Allgemeine Anforderungen (IEC 62196-1:2014, modifiziert)
41. DIN EN 60999-1:2000-12. Verbindungsmaterial – Elektrische Kupferleiter; Sicherheitsanforderungen für Schraubklemmstellen und schraubenlose Klemmstellen – Teil 1: Allgemeine Anforderungen und besondere Anforderungen für Klemmstellen für Leiter von 0,2 mm² bis einschließlich 35 mm² (IEC 60999-1:1999)
42. DIN EN 60999-2:2004-04. Verbindungsmaterial – Elektrische Kupferleiter – Sicherheitsanforderungen für Schraubklemmstellen und schraubenlose Klemmstellen – Teil 2: Besondere Anforderungen für Klemmstellen für Leiter über 35 mm² bis einschließlich 300 mm² (IEC 60999-2:2003)
43. DIN EN IEC 60512-16-4:2009-03. Steckverbinder für elektronische Einrichtungen – Mess- und Prüfverfahren – Teil 16-4: Mechanische Prüfungen an Kontakten und Anschlüssen – Prüfung 16d: Zugfestigkeit von Crimpverbindungen (IEC 60512-16-4:2008)
44. DIN EN 60309-1:2013-02. Stecker, Steckdosen und Kupplungen für industrielle Anwendungen – Teil 1: Allgemeine Anforderungen (IEC 60309-1:1999 + A1:2005, modifiziert + A2:2012)
45. LV 214:2010-03. Motor Vehicle Connectors – Test Specification.
46. LV 215-1:2013-02. Electrical/Electronic Requirements for HV Contacts – Performance Specification.
47. LV 215-2:2013-03. Motor Vehicle High-Voltage Contacts – Test Specification.

Conclusion 8

The contents of the book are the result of the authors' many years of work at the Chair for High Voltage and High Current Engineering at the Dresden University of Technology. We see ourselves as the spokespersons of a research group that has been active in this field for more than 40 years. The high quality of the research is reflected in the numerous qualification works, which in turn represent a unique treasure of knowledge. The essential findings of this unique feature in research at the chair are summarized in this book. It is intended to provide students, engineers, experts, and interested parties with a basis for utilizing and preserving the knowledge about the contact and long-term behavior of current-carrying connections.

As a community of authors, we have greatly enjoyed compiling the contents of the book over the last three years and fondly remember the numerous critical discussions, many hours of work, and the joint collection and shaping of insights.

We hope to have brought all readers closer to an interdisciplinary field of study that is so interesting to us and look forward to feedback and suggestions that we may be able to incorporate into a future revised edition. Furthermore, we hope that the research and further development of current-carrying connections will continue, as good electrical contacts are the basis for a stable electrical energy system, which in turn is the lifeblood of modern society, in which electrical energy will become even more important.

Our special thanks go to all those who have supported us with questions, content, and corrections. We would particularly like to thank PD Dr.-Ing. habil. Helmut Löbl and Prof. Dr.-Ing. Steffen Großmann.

Dresden and Zurich, May 2022

S. Schlegel et al., *Contact and Long-Term Behavior of Current-Carrying Connections in Electrical Power Engineering*, https://doi.org/10.1007/978-3-662-69644-6_8

GPSR Compliance

The European Union's (EU) General Product Safety Regulation (GPSR) is a set of rules that requires consumer products to be safe and our obligations to ensure this.

If you have any concerns about our products, you can contact us on ProductSafety@springernature.com

In case Publisher is established outside the EU, the EU authorized representative is:

Springer Nature Customer Service Center GmbH
Europaplatz 3
69115 Heidelberg, Germany

Batch number: 10775940

Printed by Printforce, the Netherlands